FUNDAMENTAL ASPECTS OF ELECTROMETALLURGY

FUNDAMENTAL ASPECTS OF ELECTROMETALLURGY

Konstantin I. Popov
University of Belgrade
Belgrade, Yugoslavia

Stojan S. Djokić
Westaim Corporation
Fort Saskatchewan, Alberta, Canada

Branimir N. Grgur
University of Belgrade
Belgrade, Yugoslavia

Springer Science+Business Media, LLC

ISBN 978-1-4757-8720-7 ISBN 978-0-306-47564-1 (eBook)
DOI 10.1007/978-0-306-47564-1

Originally published by Kluwer Academic / Plenum Publishers, New York in 2002.
Softcover reprint of the hardcover 1st edition 2002

http://www.wkap.com

10 9 8 7 6 5 4 3 2 1

A C.I.P. record for this book is available from the Library of Congress

CONTRIBUTORS

Konstantin I. Popov
Faculty of Technology and Metallurgy, University of Belgrade, Belgrade, Yugoslavia

Stojan S. Djokić
The Westaim Corporation, Fort Saskatchewan, Alberta, T8L 3W4, Canada

Branimir N. Grgur
Faculty of Technology and Metallurgy, University of Belgrade, Belgrade, Yugoslavia

FOREWORD

Electrometallurgy is a broad field but it is not a new one. It was the great Faraday in the 1830s who discovered laws covering the electrodeposition of metals and its relation to the current passed and equivalent weight of the metal undergoing deposition. Since that time, applications and developments of his discoveries have spread to many areas of technology. Electrowinning is the most well known, partly because it embraces the process by which aluminum is extracted from its ores. In electrorefining, the impure metal is made into anode and the pure metal dissolved therefrom is deposited on a cathode. Electroplating is exemplified by its use in the manufacture of car bumpers. Finally, in electroreforming, objects may be metallized, often with a very thin layer of the coating desired.

The numerous technologies vary greatly in the degree to which they are intellectualized. Until the work of Popov et al., electrometallurgy has been regarded as largely empirical, an activity in which there was much art and little science. This will all change with the publication of this book. Several aspects of the background of its senior author, Konstantin Popov, make him uniquely suited to the job of intellectualizing electrometallurgy. First, he had as his mentor the great Alexander Despić, surely the leading electrochemist in Eastern Europe since the death of Frumkin. Second, he has had ample experience with the leading electrochemical engineer in America, Ralph White. And third, he has an admirable track record of a series of publications aimed at showing the remarkable variety of forms which may be made to arise in electrodeposition.

Dr. Popov's contributions are characterized by a comprehensive mathematical treatment of the phenomena he has discovered. Co-author Stojan Djokić, too, has had much relevant experience in applying the ideas so extensively developed in Belgrade at Canadian companies.

The contents of the present volume illustrate how certain fields of science become settled in certain countries. Industrial organic chemistry grew up in Germany, and developed predominantly there until World War II. In electrometallurgy, the intellectual development has been largely in Serbia, with the most fundamental studies in nucleation being carried out in the neighboring country of Bulgaria (Kaischev, Busevski). This is not to imply that there have been few American contributors (one recalls Brenner, Kardos, Lowenheim), but that the contributions made here in this field have come from commercial laboratories and have been near the commercialization stage. University or research institute work on electrometallurgical topics in the United States has been nonexistent since the 1950s.

There are many figures in this splendid book of Popov et al. which impress me. The first is the strong, broad contents of its arrangement. There is a fine first chapter on the principles of application to electrochemical kinetics—the equations being written in a form modified for use in electrometallurgical situations (e.g., deposition on the tips of growing crystals of minimal radius of curvature and on corners and edges). Here, it is encouraging to find authors applying the electrochemical version of Kelvin's equation relating vapor pressure as a function of the radius of drops to the phenomena during the electrogrowth of dendrites.

I personally find the treatments of the effects of current varying regimes (e.g., pulse, reverse pulse, square wave, sinusoidal, etc.) the most exciting for I have long thought that instead of the use of chemical additives to the solution, the type of surface finally produced—even the crystal shape—could be achieved by electrical variations only. This book contains much toward the realization of this approach.

In the second half of the book, one finds the mathematical treatments of practical situations in electrowinning, electrorefining, electroplating, and electroreforming. What is the difference all this will make? It should enable to engineer to set up regimes to achieve what he wants with a minimum of prefatory experiments.

This book has no competitor. There are certainly books on electroplating, but they are largely recipes for what to do which eschew the important question of why.

Getting the intellectual side over to the practical engineer, of course, requires great lucidity, for he will not puzzle over material delivered over his head. I think the required clarity has been attained herewith, particularly in the early chapters where the concepts of exchange currents and overpotential are being added to the weary thermodynamics which covers most of what engineers are likely to know about electrochemistry.

A great strength is in the photographs of electrodeposited crystals in all their variety. Such photographs can be found in the usual journals, but I have not previously seen such a collection accompanied by textual rationalization.

Lastly, I was impressed by the application of the theory to areas which normally receive little more than a definition. I would cite electropolishing, where theory is seldom presented; electromachining; and electroless plating.

This book is a feast simply to read, but I believe its main importance is that it gives for the first time an educational tool. It will surely lead to translations and its use will feed back upon the economics of electrometallurgical processes—with a reduction in cost many orders of magnitude greater than the total in purchases of the books.

John O'M. Bockris
Texas A&M University

PREFACE

In their preface to the book *Fundamental Aspects of Electrocrystallization*, Bockris and Razumney in 1966 wrote: "Electroplating, the electrochemical extraction of metals from ores and mixtures, and electrodeposition from nonaqueous solutions, together make up a large area in technology. It is relevant to distinguish two classes of technologies: those in which, the fundamentals and theory were understood first and the applications followed, and those in which the applied side began as a kind of art and the theory limped behind the art, sometimes dragged back by it and occasionally given a little push. Of course, the atomic energy and electronic industry are ideal example of the first two. The electrodeposition and electroextraction industry is a fair example of the second. It is clear enough that the rate of development of the first type of industry was much greater than that of the second, as the possibilities could be estimated or at least the direction in which to push defined, whereas the second type relied in the past on what might be called "inspired groping." It is hoped that this small monograph may provide a basis for the training of research workers who can then perform the conversion of what remains an art in technological electrocrystallization into a technology with a largely rational basis." In the meantime, although the electrochemistry of metals has significantly advanced, the message of Bockris and Razumney, especially in the light of education professionals, is practically the same. The bridge between high level theory and highly developed practice is still necessary, probably even more than previously. The purpose of this book is to enable this. It is written in order to explain the principles of electrometallurgical technologies, not to give their technical details. It should be easy to follow by a reader with a graduate or an undergraduate degree in either engineering or science. It assumes the knowledge of the basic facts required on the level of *Modern Electrochemistry* by Bockris and Reddy.

The book comprises twelve chapters, which can tentatively be divided into three parts: Chapters 1, 2, 6, and 7 are written by Grgur and Popov, Chapters 3–5 and 8 by Popov and Grgur and Chapters 9–12 by Djokić.

In the first part (Chapters 1 and 2) the significance of electrometallurgy in science and technology is discussed. The fundamentals of electrochemistry necessary for an understanding of electrometallurgical processes are also given. In the second part (Chapters 3–6) the mechanisms of metal deposition are discussed at a high scientific level although efforts were made to simplify them in an approachable way. For example the text related to the morphology of electrodeposited metals is separated into three levels: the mathematical approach, the physical model and, finally, a realistic system. The current distribution in electrochemical cells is described in a new way. Using this approach, classical complicated mathematical methods are not only avoided, but also calculations of typical examples for electrochemical cell are performed using simple relations of electrochemical kinetics (described in Chapter 2). The reader is also introduced to the effects of additives and periodically changing rate at a level which allows an understanding of metal deposition under these conditions. In addition, in the third part (Chapters 6–12), theoretical aspects of the electrowinning and electrorefining of metals, as well as electro- and electroless deposition of thin films used in electronics, automotive and aerospace applications are discussed. Electrodeposition of thin films of metals, alloys or composite coatings under direct or periodically changing current conditions from aqueous or non-aqueous solutions, as well as molten salts is described. Attempts have been made to up-to-date the current knowledge with kinetics and mechanisms of electroless deposition and to critically evaluate the similarities and differences of deposition processes with and without an external current source.

In the next stage of technological development, we hope, this book will initiate efforts for the advancement of theoretical fundamentals of electrometallurgy at a level, which will permit the planning of technological processes without or with a minimum of experimental data. This is in a way similar to challenge of Bockris and Razumney, a half-century ago, which led to the appearance of this book.

ACKNOWLEDGMENTS

Chapters 1–8 of this book are based on a few classical studies and on research in the field of metal electrodeposition performed in the Department of Physical Chemistry and Electrochemistry at the Faculty of Technology and Metallurgy, the University of Belgrade, Yugoslavia. K. I. Popov and B. N. Grgur would like to acknowledge Professor A. R. Despić, who initiated this research, and colleagues Professors M. D. Maksimović, M. G. Pavlović, N. V. Krstajić, S. K. Zečević, Dr. R. M Stevanović, and S. M Pešić, M.Sc., as well as numerous other colleagues and students who participate in it.

Chapters 9 to 12 represent a discussion of various aspects of electrodeposition of alloys, composite coatings, electroforming, electrochemical oxidation of metals, electroless deposition and electrodeposition from molten salts, found in the open literature. S. Djokić acknowledges the cooperation of numerous colleagues and coworkers from Sherritt Inc., The Westaim Corporation, and Dr. B. E. Conway at the University of Ottawa. For the help in the literature search and manuscript preparation, S. Djokić is thankful to Ms. K. Djokić and Mrs. C. Lepard.

Finally, we wish to express our thanks to Mr. Ken Howell for helpful advice and to P. M. Živković, M. Sc. for the preparation of the camera-ready manuscript.

CONTENTS

Chapter 1

WHAT IS ELECTROMETALLURGY

Electrometallurgy deals with technical aspect of metal electrodeposition. The electrometallurgical processes can be categorized into four main groups:

1. Electrowinning,
2. Electrorefining,
3. Electroplating,
4. Electroforming.

The schematic representation of electrometallurgical processes is shown in Fig. 1.1.

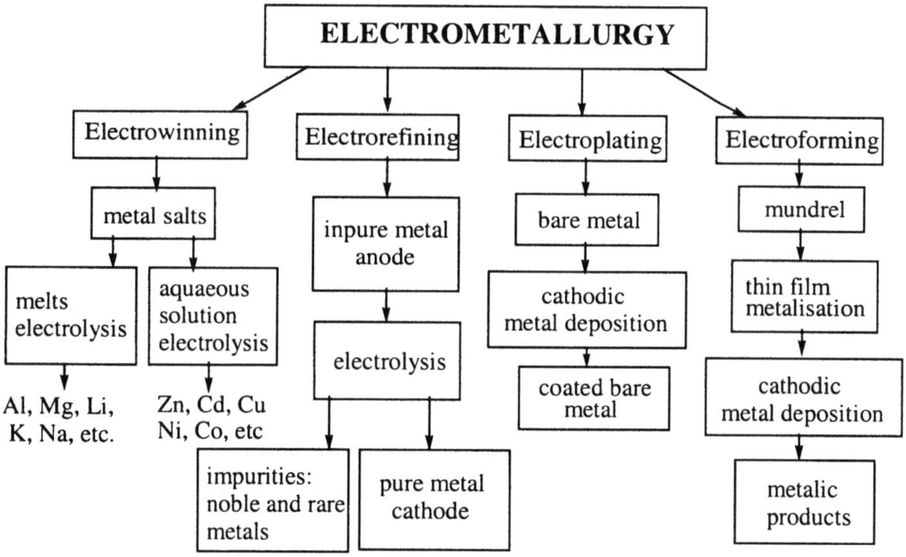

Figure 1.1 Schematics representation of electrometallurgical processes

Electrowinning is the extraction of metals by electrodeposition from aqueous solution or melts of their salts. On a large scale electrodeposition from molten salts is used for extraction of electronegative elements which cannot be electrodeposited from aqueous solutions, such as aluminum and magnesium, as well as very pure copper, zinc and cadmium by electrodeposition from an aqueous solutions of the metal salts[1].

Electrorefining is the purification of metals by electrolysis. The impure metals is dissolved anodically and pure metal is deposited catodically, while the impurities being left as anode sludge or as ions in the solution. Many metals are electrorefined such as copper because of conducting application and precious metals because of theirs cost. Electrorefining is also a part of processes in recycling of metals.[1]

It should be noted that large electrolytic plants for metal production are heavy consumers of electric energy[1]. In the metal electrorefining and electrowinning the main requirements are to produce pure and compact deposits. This is done at lower current densities. From qualitatively the same, but less concentrated solutions at higher current densities metal deposits in form of powder are obtained. Powder electrodeposition can also be treated as kind of electrowinning or electrorefining, which produces the metal deposits in forms suitable for sintering and various different applications.

Electroplating can be defined as a treatment that modifies the surface of a metal or occasionally a nonmetal, without changing its bulk properties, in order to improve the appearance of a surface, to increase the corrosion and abrasion resistivity, etc. The improving the appearance was the aim of electroplating earlier, now it is mainly the change of surface properties from those of substrate material to those of electroplated metal. Obviously, the coating can successfully change the surface properties of substrate only if it is compact and nonporous, as well as good adherent[1-3]. Metal objects we meet in everyday life are often electroplated, but it seams that the most important application of electroplating technology is the manufacture of electronic components (circuit breakers and contacts). Electroplating can be performed from molten salts and aqueous and non-aqueous solutions, depending on the nature of electrodeposited metal, but most frequently from aqueous solutions[1-3].

Electroforming is the manufacture of articles by electrodeposition. If deposit is good from electroplating point of view

> except adhesion, and can be removed from the cathode as an entity in itself, it has been electroformed. Electroforming is a branch of electroplating technology, but involve some additional steps, as for example the production, preparation and extraction of the master[2,4].

Electroless metal deposition and anodic oxidation of metals can also be include in the field of electrometallurgy.

Empirically is known what type of deposit can be obtained under specific conditions, however how and why this can be achieved still remains a mystery in some cases.

The aim of this book is to give answers to some of open questions.

1.1 FURTHER READINGS

1. Pletcher, Derek, *Industrial Electrochemistry*. New York: Chapman and Hall, 1984.
2. Lowenheim, Frederick, *Electroplating (Fundamentals of Surface Finishing)*. New York: McGrow-Hill book company, 1978.
3. Lowenheim, Frederick, *Modern Electroplating*. New York: John Wiley & Sons, 1974.
4. Spiro, Peter, *Electroforming,* Teddington: Robert Draper, 1971.

Chapter 2

DEFINITIONS, PRINCIPLES AND CONCEPTS

2.1 BASIC FACTS

2.1.1 Electrodes and electrochemical reactions, cell and circuit

In a metallic conductor free conduction electrons transport the charge whereas in an electrolytic conductor it is ions. In order to include an electrolytic conductor in an electrochemical circuit it is necessary to make electrical contacts to and from the electrolyte by metallic conductors. A metallic conductor immersed in an electrolyte solution is an electrode, and two electrodes connected electrolytically represent an electrochemical cell[1,2].

The simplest electrochemical circuit is shown in Fig. 2.1.

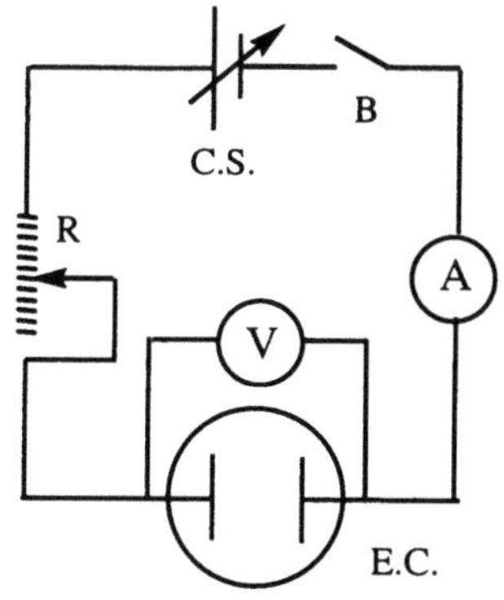

Figure 2.1. An electrochemical circuit. C.S.-current source, R-ohmic resistance, V-voltmeter, A-ampermeter, E.C.-electrochemical cell, B-circuit breaker.

The electrochemical circuit consists of a current source, metallic connecting wires, an electrochemical cell, ohmic resistance, current and voltage measuring instruments and a circuit breaker. In technical practice more complicated circuits are used, but in principle all of them are the same as the one shown in Fig. 2.1.

Obviously, a steady current flow in the circuit from Fig. 2.1 can only be maintained if there is a change of charge carrier at the metal-electrolyte interface by a chemical transformation involving the transfer of electrons across the interface, i.e., by an electrochemical reaction. It constitutes the bridge between the current of electrons in the metallic part of the electrochemical circuit and the current of ions in the electrolytic part of the circuit[1,2].

2.1.2 The electrochemical double layer and possible electrochemical reactions

If the metal ions in the solution are the same as in the electrode metal lattice, or if the same substance is present in the electrolyte in two oxidation states, an electron transfer reaction can occur at the metal-electrolyte interface and lead to the development of a potential difference. Such an interface behaves like an electrical circuit consisting of a resistor and a capacitor in parallel. The electron transfer takes place until a dynamic equilibrium is reached. In the case of metal electrodes, depending on the system, this process begins with either the deposition of ions from solution onto the metal electrode or with the dissolution of the metal electrode. In equilibrium the electrode is more positive than the solution in the first case and more negative in the second one. A number of electrochemical reactions are possible at such an interface, as for example

1. The reduction of metal cations to the metal and vice versa

$$Cu^{2+} + 2e^- \rightleftarrows Cu$$

2. The reduction of hydrogen cations to gaseous hydrogen and vice verse

$$2H^+ + 2e^- \rightleftarrows H_2$$

3. The decrease of the oxidation state of the cations and vice versa

$$Fe^{3+} + e^- \rightleftarrows Fe^{2+}$$

4. The reduction of anions to metal and vice versa

$[Cd(CN)_4]^{2-} + 2e^- \rightleftarrows Cd + 4CN^-$

5. The reduction of molecules to anions and vice versa

$2H_2O + O_2 + 4e^- \rightarrow 4OH^-$ (in alkaline media)

6. The oxidation of molecules to cations and vice versa

$2H_2O \rightarrow O_2 + 4H^+ + 4e^-$ (in acid media)

2.2 SELF DRIVING CELLS

2.2.1 The Nernst equation and energy producing cells

For the electrochemical reaction

$$O + ne^- \rightleftarrows R \tag{2.1}$$

where O is the oxidized state which accepts n electrons and R is the reduction state or the donor of electrons, in equilibrium, the Nernst equation[3] is written in the form

$$E_{O/R} = E^{\theta}_{O/R} + \frac{RT}{nF}\ln\frac{a_O}{a_R} \tag{2.2}$$

where $E_{O/R}$ is the equilibrium electrode potential, $E^{\theta}_{O/R}$ is the standard electrode potential and a_O and a_R are the activities of the electron acceptor and donor, respectively[1,2].

In Table 2.1. the standard potentials for some electrode reactions are given.

Table 2.1. Standard potentials of metal deposition and dissolution reactions.

Electrode reaction	Standard potentials, V	Electrode reaction	Standard potentials, V
$Li^+ + e^- = Li$	-3.045	$Co^{2+} + 2e^- = Co$	-0.277
$K^+ + e^- = K$	-2.925	$Ni^{2+} + 2e^- = Ni$	-0.250
$Na^+ + e^- = Na$	-2.714	$Sn^{2+} + 2e^- = Sn$	-0.163
$Mg^{2+} + 2e^- = Mg$	-2.363	$Pb^{2+} + 2e^- = Pb$	-0.125
$Al^{3+} + 3e^- = Al$	-1.662	$Cu^{2+} + 2e^- = Cu$	0.337
$Zn^{2+} + 2e^- = Zn$	-0.763	$Cu^+ + e^- = Cu$	0.521

Electrode reaction	Standard potentials, V	Electrode reaction	Standard potentials, V
$Cr^{3+} + 3e^- = Cr$	-0.744	$Ag^+ + e^- = Ag$	0.799
$Cr^{3+} + 3e^- = Cr$	-0.744	$Ag^+ + e^- = Ag$	0.799
$Fe^{2+} + 2e^- = Fe$	-0.440	$Au^{3+} + 3e^- = Au$	1.498
$Cd^{2+} + 2e^- = Cd$	-0.403	$Au^+ + e^- = Au$	1.691

The signs of two electrodes connected to a cell can be determined by using the values of standard electrode potentials. Only, if they are close to each other the Nernst equation should be used to determine the polarity of them.

The equilibrium potential difference for the cell

$$Zn \mid ZnSO_4, H_2O, H_2SO_4 \parallel CuSO_4, H_2O, H_2SO_4 \mid Cu$$

which is illustrated in Fig. 2.2, can be evaluated as follows.

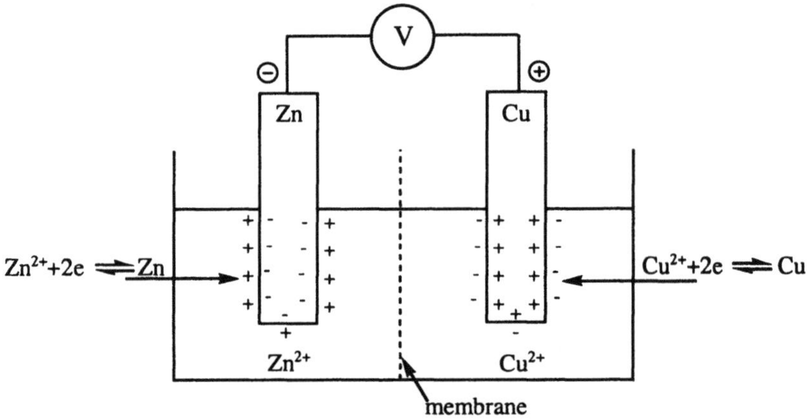

Figure 2.2. Schematic representation of a cell $Zn \mid ZnSO_4, H_2O, H_2SO_4 \parallel CuSO_4, H_2O, H_2SO_4 \mid Cu$ in equilibrium. V is a high impedance voltmeter. (According to[3]).

Obviously, the equilibrium concentration of electrons in the more negative electrode will be larger than in the more positive one.

According to Table 2.1, the possible reactions on the electrodes are:

$$Zn \rightleftarrows Zn^{2+} + 2e^- \qquad E^\theta = -0.763 \text{ V} \tag{2.3}$$

and

$$Cu^{2+} + 2e^- \rightleftarrows Cu \qquad E^\theta = 0.337 \text{ V} \tag{2.4}$$

The equilibrium potential difference E is given by

$$E = \left(E^{\theta}_{Cu^{2+}/Cu} + \frac{RT}{2F} \ln a_{Cu^{2+}/Cu} \right) - \left(E^{\theta}_{Zn^{2+}/Zn} + \frac{RT}{2F} \ln a_{Zn^{2+}/Zn} \right) \tag{2.5}$$

If the electrodes are connected as in Fig. 2.3, the reaction on the electrodes are:

$$Zn \rightarrow Zn^{2+} + 2e^{-} \tag{2.6}$$

and

$$Cu^{2+} + 2e^{-} \rightarrow Cu \tag{2.7}$$

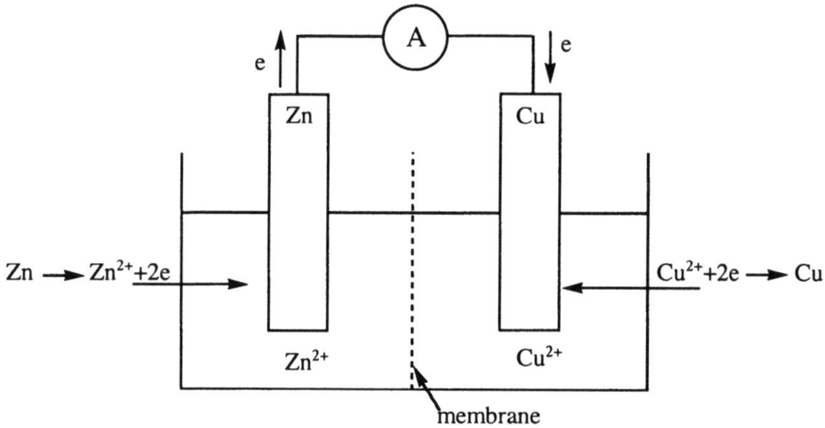

Figure 2.3. The cell Zn | $ZnSO_4$, H_2O, H_2SO_4 || $CuSO_4$, H_2O, H_2SO_4 | Cu as a energy producer, A is the low impedance ampermeter. (According to[3]).

The overall reaction in is then:

$$Cu^{2+} + Zn \rightarrow Cu + Zn^{2+} \tag{2.8}$$

Obviously, oxidation will take place on the more negative electrode, making it an electron sink, and reduction will occur on the more positive electrode, making it an electron source.

Such electrochemical transformations at the two interfaces provides a steam of electrons available for external use, which is the essence of energy producing cells (electrochemical power sources).

Reaction, given by Eq. 2.8 will go from left to right until the potentials of the electrodes become equal, which corresponds to the zero cell voltage and equilibrium activities $a_{e,Zn^{2+}}$ and $a_{e,Cu^{2+}}$ of zinc and copper ions, respectively. It follows then from Eq. 2.5 that if $E = 0$

$$E^{\theta}_{Cu^{2+}/Cu} - E^{\theta}_{Zn^{2+}/Zn} = \frac{RT}{2F} \ln K^{\theta} \tag{2.9}$$

where

$$K^{\theta} = \frac{a_{e,Zn^{2+}}}{a_{e,Cu^{2+}}} \tag{2.10}$$

represents the equilibrium constant for reaction 2.8. The value of $K^{\theta} \sim 10^{37}$ means that the Cu^{2+} ions can be completely removed from the electrolytic solution by reaction 2.8.

2.2.2 Cementation

If a piece of zinc is immersed in a copper sulfate solution the reaction 2.6 occur at one local area of the electronic conductor and reaction 2.7 at another local area. Reaction 2.6 and 2.7 occur spontaneously, and the overall reaction is given by Eq. 2.8. The system is self-driven and produces power, but the corresponding electrochemical energy is unavailable because the two interfaces are short-circuited[3]. This reaction is used in purification of zinc sulphate solution from more positive metallic ion impurities in zinc electrowinning process.

2.3 ELECTROLYSIS

2.3.1 Decomposition voltage

The self-driving cell from Fig. 2.3 can be rearranged to a driven cell by connecting the power supply in the circuit as shown in Fig. 2.4[3].

The reactions on the electrodes will be

$$Zn^{2+} + 2e^- \rightarrow Zn \tag{2.11}$$

and

$$Cu \rightarrow Cu^{2+} + 2e^- \tag{2.12}$$

and the electron steam will flow in the opposite direction to that in the same cell working as a self driven one only if the power supply voltage is larger than the equilibrium potential difference of the same cell, working as a self driven one.

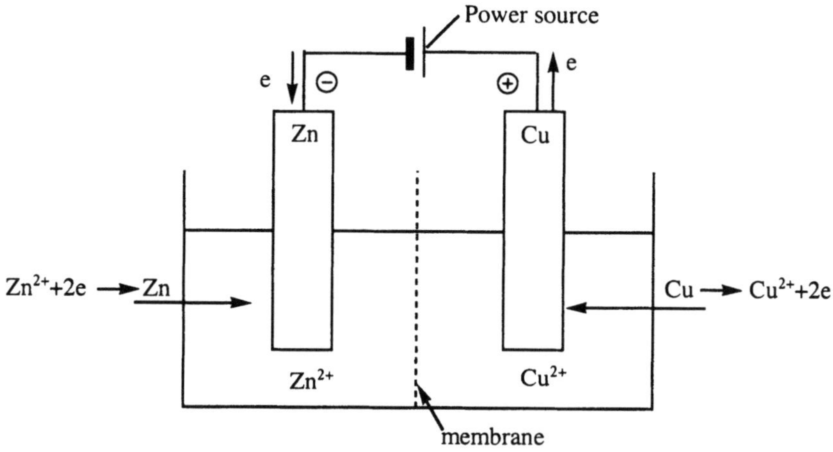

Figure 2.4. Schematic representation of the cell Zn | $ZnSO_4$, H_2O, H_2SO_4 || $CuSO_4$, H_2O, H_2SO_4 | Cu as a driven cell. (According to[3]).

If two or more anodic and cathodic reactions are possible in some driven cell, the reactions with the lowest equilibrium potential difference will take place on the electrodes first. *This means that the reaction with the most positive equilibrium potential will take place first on the cathode (the electrode connected with the negative terminal of power supply, at which reduction occurs), and the reaction with the most negative equilibrium potential on the anode (the electrode connected with the positive terminal of the power supply, at which oxidation occurs).* It is important to remember that the terms anode and cathode are connected with the nature of the reaction (oxidation or reduction) at the electrode and not with their polarity. Thus, in a self driven cell, the anode is the negative terminal and the cathode is the positive terminal of the cell, a situation which is precisely the opposite of that which exists in an externally driven cell[3].

2.3.2 A cell with an insoluble anode

In the driven cell

(-) Cu | $CuSO_4$, H_2SO_4, H_2O | Au (+)

the following reactions on the anode (Au) are possible:

$$2\,SO_4^{2-} \rightarrow S_2O_8^{2-} + 2e^- \qquad E^\theta = 2.010\ \text{V}$$

$$Au \rightarrow Au^{3+} + 3e^- \qquad E^\theta = 1.498\ \text{V}$$

$$2H_2O \rightarrow O_2 + 4H^+ + 4e^- \qquad E^\theta = 1.229\ \text{V}$$

and on the cathode (Cu)

$$O_2 + 4H^+ + 4e^- \rightarrow 2H_2O \qquad E^\theta = 1.229\ \text{V}$$

$$Cu^{2+} + 2e^- \rightarrow Cu \qquad E^\theta = 0.337\ \text{V}$$

$$2H^+ + 2e^- \rightarrow H_2 \qquad E^\theta = 0.000\ \text{V}$$

$$SO_4^{2-} + 2H^+ + 2e^- \rightarrow SO_3^{2-} + H_2O \qquad E^\theta = -0.221\ \text{V}$$

According to the rule derived in the conclusion from section 2.3.1, on the anode the oxygen evolution reaction will occur, and on the cathode copper ions will be reduced to the metal phase, because the most positive catodic reaction can be neglected due to the low oxygen concentration in the electrolyte solution. Hence the reaction:

$$2H_2O \rightarrow O_2 + 4H^+ + 4e^- \qquad E^\theta = 1.229\ \text{V} \tag{2.13}$$

will take place at the anode and the reaction

$$Cu^{2+} + 2e^- \rightarrow Cu \qquad E^\theta = 0.337\ \text{V} \tag{2.7}$$

at the cathode.

Obviously, the minimum external cell voltage for electrolysis to occur in this case is 0.893 V.

The overall reaction in the cell is

$$2Cu^{2+} + 2H_2O \rightarrow 2Cu + O_2 + 4H^+ \tag{2.14}$$

It follows from the Eq. 2.14 that in a cell with an insoluble anode the concentration of depositing ions decreases and the hydrogen ion concentration increases during electrolysis.

The mechanism of the extraction of metals from ionic solutions and the essence of the electrowinning process are well explained by Eq. 2.14.

2.3.3 A cell with a soluble anode

In the cell:

(-) Au | $CuSO_4$, H_2SO_4, H_2O | Cu (+)

the following reactions are possible on the anode (Cu):

$$2SO_4^{2-} \rightarrow S_2O_8^{2-} + 2e^- \qquad E^\theta = 2.010\ V$$

$$2H_2O \rightarrow O_2 + 4H^+ + 4e^- \qquad E^\theta = 1.229\ V$$

$$Cu \rightarrow Cu^{2+} + 2e^- \qquad E^\theta = 0.337\ V$$

and on the cathode (Au):

$$O_2 + 4H^+ + 4e^- \rightarrow 2H_2O \qquad E^\theta = 1.229\ V$$

$$Cu^{2+} + 2e^- \rightarrow Cu \qquad E^\theta = 0.337\ V$$

$$2H^+ + 2e^- \rightarrow H_2 \qquad E^\theta = 0.000\ V$$

$$SO_4^{2-} + 2H^+ + 2e^- \rightarrow SO_3^{2-} + H_2O \qquad E^\theta = -0.221\ V$$

Hence, the reaction

$$Cu^{2+} + 2e^- \rightarrow Cu \qquad E^\theta = 0.337\ V \qquad (2.7)$$

occurs on the anode, and

$$Cu \rightarrow Cu^{2+} + 2e^- \qquad E^\theta = 0.337\ V \qquad (2.12)$$

at the cathode, and so the composition of the electrolyte solution remains constant if the anode is made of pure copper and oxygen is removed from the solution. The lowest cell voltage at which electrolysis can start in this cell is zero.

It is obvious that electrorefining processes are based on electrolysis in cells with soluble anodes.

2.3.4 Current efficiency

Metal deposition can be accompanied by any other cathodic reaction, most frequently hydrogen evolution.

This leads to the situation in which metal is deposited but metal deposition uses only a part, I_{Me}, of the total current, I, through the cell. The current efficiency

$$\eta_I = \frac{I_{Me}}{I} \tag{2.15}$$

indicates which part of the total current is used for the deposition of metals. It is a very important parameter of an electrodeposition process[3].

2.3.5 Faraday's law

Faraday's law relates the quantity of electricity passed through the cell and the quantity of chemical substances which react on the electrodes. It states that the mass of metal, m, electrodeposited on the cathode is given by

$$m = \frac{ItM}{nF}\eta_I \tag{2.16}$$

where I is the total current, t is the deposition time, M is the molar mass of the deposited metal, η_I is the current efficiency and nF is the number of Faradays per mole of consumed ions. It follows from Eq. 2.16 that η_I can be easily determined by measuring the electrodeposited mass of metals and supplied quantity of electricity[1,2].

2.3.6 The current density – overpotential relationships

2.3.6.1 Basic equations

The most complete discussion of the current density – overpotential relationship was given by Bockris[4]. The approach suitable for use in metal electrodeposition will be given here.

The general form of current density – overpotential relationship in electrodeposition of metals for the reaction

$$\mathrm{O} + ne^- \rightleftarrows \mathrm{R} \tag{2.1}$$

taking cathodic current density and overpotential as positive, is given by

$$j = j_0\left[\left(\frac{a_j^{\mathrm{O}}}{a_{\mathrm{e}}^{\mathrm{O}}}\right)f_{\mathrm{c}} - \left(\frac{a_j^{\mathrm{R}}}{a_{\mathrm{e}}^{\mathrm{R}}}\right)f_{\mathrm{a}}\right] \tag{2.17}$$

where j_0 is the exchange current density and a_j is the activity of the oxidized (O) or reduced (R) state at a current density j and a_{e} is the activity in the equilibrium state.[5]

On the other hand

$$f_{\mathrm{c}} = \exp\left(\frac{\alpha_{\mathrm{c}}F\eta}{RT}\right) = \exp\left(\frac{2.3}{b_{\mathrm{c}}}\eta\right) \tag{2.18}$$

and

$$f_{\mathrm{a}} = \exp\left(\frac{-\alpha_a F\eta}{RT}\right) = \exp\left(\frac{-2.3}{b_{\mathrm{a}}}\eta\right) \tag{2.19}$$

where α_{c} and α_{a} are the cathodic and anodic transfer coefficient, b_{c} and b_{a} corresponding Tafel slopes and η is the overpotential and

$$b_{\mathrm{c}} = \exp\frac{2.3RT}{\alpha_{\mathrm{c}}F} \tag{2.20}$$

and

$$b_{\mathrm{a}} = \exp\frac{2.3RT}{\alpha_{\mathrm{a}}F} \tag{2.21}$$

The ratio of the activities for the cathodic reaction may be written as:

$$\frac{a_j^{\mathrm{O}}}{a_{\mathrm{e}}^{\mathrm{O}}} = 1 - \frac{j}{j_{\mathrm{L}}} \tag{2.22}$$

where j_{L} is the limiting diffusion current density and for the reverse anodic reaction as:[5]

$$\frac{a_j^{\mathrm{R}}}{a_{\mathrm{e}}^{\mathrm{R}}} = \exp\frac{2\gamma V}{RTr_{\mathrm{el}}} \tag{2.23}$$

taking into account the Kelvin term which becomes appreciable at low values of electrode radii[6]. In Eq. 2.23 γ is the surface energy, V is the molar volume of the electrodeposited metal and r_{el} is the radius of the electrode. Eq. 2.23 is valid for two electron reactions[5], other possibilities are discussed in Ref. 7.

For a spherical electrode Eq. 2.17 can be written as:

$$j = j_0 \left[\left(1 - \frac{j}{j_{L,sp}} \right) f_c - f_a \exp\left(\frac{2\gamma V}{RTr_{el}} \right) \right] \tag{2.24}$$

or

$$j = \frac{j_0 \left[f_c - f_a \exp\left(\frac{2\gamma V}{RTr_{el}} \right) \right]}{1 + \frac{j_0 f_c}{j_{L,sp}}} \tag{2.25}$$

where

$$j_{L,sp} = \frac{nFDC_0}{r_{el}} \tag{2.26}$$

where C_0 is the bulk concentration and D diffusivity coefficient of a depositing ions.

A somewhat modified Eq 2.25 is necessary for an understanding of electrodeposition on the tip of dendrites inside the diffusion layer of a macroelectrode, especially in the case of electrodeposition at a periodically changing rate[7-9].

For sufficiently large r_{el} to make surface energy term negligible Eq. 2.25 can be rewritten in the form:

$$j = \frac{j_0 (f_c - f_a)}{1 + \frac{j_0 f_c}{j_{L,sp}}} \tag{2.27}$$

For a sheet electrodes and sufficiently large spherical electrodes Eq. 2.25 becomes:

$$j = \frac{j_0 (f_c - f_a)}{1 + \frac{j_0 f_c}{j_L}} \tag{2.28}$$

where

$$j_L = \frac{nFDC_0}{\delta} \quad (2.29)$$

and δ is the diffusion layer thickness[3].

2.3.6.2 Some approximations

Equation 2.27 is valid generally but it is more convenient to use some approximative relations derived from them, so for flat surface if

$$\frac{j_0 f_c}{j_L} << 1 \text{ and } f_c > f_a \quad (2.30)$$

Eq. 2.28 can be rewritten in the form:

$$j = j_0 (f_c - f_a) \quad (2.31)$$

which becomes

$$j = j_0 \frac{nF}{RT} \eta \quad (2.32)$$

at very low overpotentials by expanding the exponential terms in Eq. 2.31 and retaining the two terms of expansion of each exponential terms, if

$$\alpha_c = \beta n \quad (2.33)$$

and

$$\alpha_a = (1 - \beta)\, n \quad (2.34)$$

where β isa the symmetry factor.

When

$$f_c >> f_a \text{ and } \frac{j_0 f_c}{j_L} << 1 \quad (2.35)$$

the relation:

$$j = j_0 f_c \quad (2.36)$$

or

$$\eta = \frac{b_c}{2.3} \ln \frac{j}{j_0} \tag{2.37}$$

is valid.

If $f_c >> f_a$, Eq. 2.28 becomes:

$$j = \frac{j_0 f_c}{1 + \frac{j_0 f_c}{j_L}} \tag{2.38}$$

or

$$\eta = \frac{b_c}{2.3} \ln \frac{j}{j_0} + \frac{b_c}{2.3} \ln \frac{1}{1 - \frac{j}{j_L}} \tag{2.39}$$

For

$$f_c > f_a \text{ and } \frac{j_0 f_c}{j_L} >> 1 \tag{2.40}$$

Eq. 2.28 can be rewritten in the form:

$$j = j_L \left(1 - \frac{f_a}{f_c}\right) = j_L \left[1 - \exp\left(-\frac{nF\eta}{RT}\right)\right] \tag{2.41}$$

which is valid if Eqs. 2.33 and 2.34 are valid and finally, if

$$f_c >> f_a \text{ and } \frac{j_0 f_c}{j_L} >> 1 \tag{2.42}$$

Eq. 2.39 becomes

$$j = j_L \tag{2.43}$$

The range of validity of Eqs. 2.31 and 2.32, 2.36 and 2.37 and 2.38 and 2.39 and 2.43 can be easily determined from η – log(j) and $\eta - j$ plots. The equations 2.36 and 2.37 are valid from the beginning to the end of Tafel linearity (Tafel line). At lower overpotentials, Eqs. 2.31, and 2.32 are valid and at higher ones, Eq. 2.38, 2.39 and 2.43. In the Fig. 2.5 the simulated polarization curve for cathodic metal electrodeposition together with the Tafel plot and the range of validity of mentioned equations are shown. Usually, in electrochemistry the marked regions are called: 1 and 2- activation controlled region, 3- mixed activation-diffusion controlled region, 4- pure diffusion controlled region.

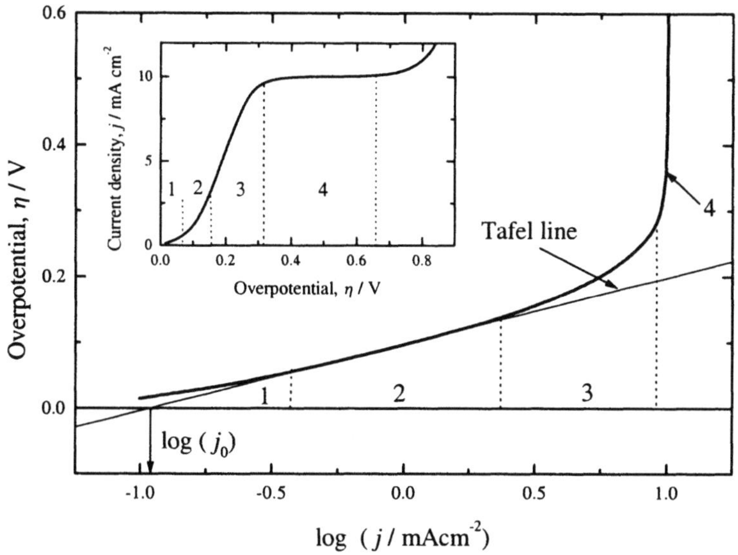

Figure 2.5. Simulated Tafel plot for the metal deposition (j_0=0.11 mA cm^{-2}; b_c=118 mV dec^{-1}) and range of validity of the equation: 1) Eqs. 2.31 and 2.32; 2) Eqs. 2.36 and 2.37; 3). Eqs. 2.38 and 2.39; 4) Eq. 2.43. Insert: polarization curve.

2.3.7 The cell voltage

It is to be noted that the difference of cathodic and anodic overpotentials is the sum of their absolute values. Hence, if anode overpotential is considered in the anode Tafel region the absolute values of overpotentrials and current density should be used, and equation

$$\eta = \frac{b_a}{2.3} \ln \frac{j}{j_0} \tag{2.44}$$

is valid.

There are not mass transport limitations in anodic dissolution of metals and there is not equation analoguous to Eq. 2.39.

It is now possible to define the cell potential in electrolysis. The cell voltage U of a driven electrochemical cell is given by:

$$U = E + \eta_a + \eta_c + I \sum_i^n R_i \qquad (2.45)$$

where E is the equilibrium potential difference between the anode and the cathode, η_a and η_c are the absolute values of the anodic and cathodic over-potentials respectively, I is the current, and $\sum_i^n R_i$ is the sum of the Ohmic resistance of the electrolyte, electrodes, contacts and connecting wires.

2.3.8 Specific energy consumption

The electrical work W required to deposit a quantity m of metal on the cathode is given by:

$$W = U\,I\,t \qquad (2.46)$$

On the other hand, the quantity of deposited metal and the required quantity of electricity for deposition are related by the equation:

$$m = \frac{ItM}{nF}\eta_I \qquad (2.16)$$

By combining Eq. 2.16 with Eq. 2.46, the specific energy consumption, w, is then given by:

$$w = \frac{W}{m} = \frac{nFU}{M\eta_I} \qquad (2.47)$$

The specific energy consumption, w, is the most important energetic parameter in metal electrowinning and electrorefining technologies.

2.4 SOME ASPECTS OF ELECTROCRYSTALLIZATION

Figure 2.6 shows the originally published in-situ STM images of copper clusters which were formed during electrodeposition on Au(111) substrate[10].

The top image shows the gold substrate at a potential positive of the Nernst potential for bulk copper deposition. This area of the surface is characterised by two atomically smooth terraces separated by a monoatomic high step edge. Upon stepping the electrode potential to a value negative of the Nernst potential, distinctive copper clusters form at the step edges. By contrast, the terraces in Fig 2.6b remain free from copper clusters. The clusters are seen to grow from Fig. 2.6b to the subsequent image 2.6c. These STM images are a visual verification of the important role which defects can play in the initial stages of electrocrystallization. Clearly, they are in good accordance with the textbook model of Kossel and Stranski[3,10]. Nucleation occurs preferentially at the step edges, where an ad-atom is more high coordinated with the surface than an ad-atom on the atomically flat surfaces. The energetics of copper nucleation on flat gold terraces are clearly less favorable, occurring only at longer times or higher overpotentials[10].

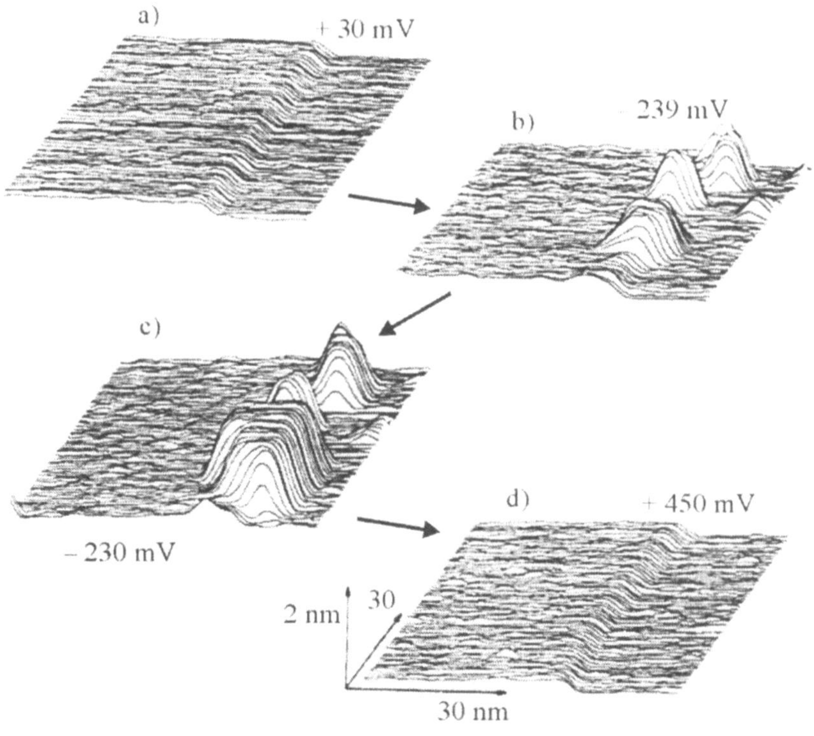

Figure 2.6. In-situ STM images taken a) before; b) and c) during and after bulk copper deposition on Au(111) electrode in 0.1 mol dm^{-3} $HClO_4$ in 5×10^{-5} mol dm^{-3} $Cu(ClO_4)_2$.[10] (Reprinted with permission from Elsevier Science).

Hence, the monoatomic high step edges, the microsteps, are required for continuos metal electrocrystalization. Possible sources of microsteps on a surface are shown in Figs. 2.7 and 2.8, i.e. the low-index planes, two dimensional nuclei, emergent screw dislocations and indestructible reentrant groves.[7,11]

It is obvious from Figs. 2.7c and 2.7a that after the formation of a low index plane, two-dimensional nucleation is necessary for the growth to be continued.

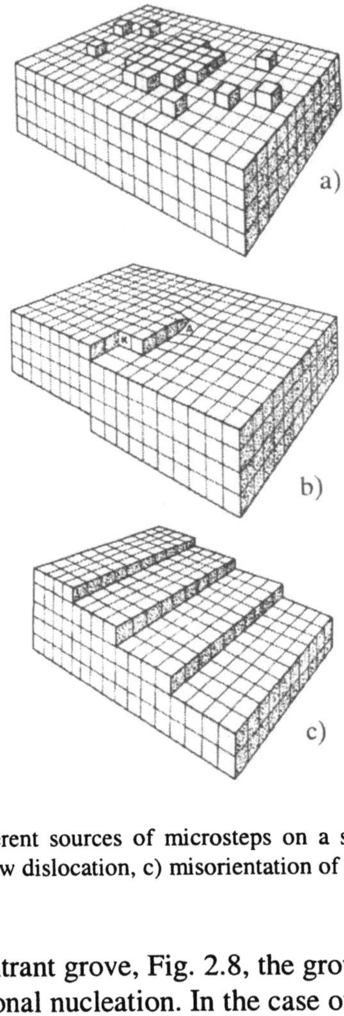

Figure 2.7. Models of different sources of microsteps on a surface: a) a two-dimensional nucleus, b) an emergent screw dislocation, c) misorientation of the surface with respect to the ideal low-index plane.[11]

In the case of a reentrant grove, Fig. 2.8, the growth of new layers can be started by one-dimensional nucleation. In the case of a screw dislocation, the step provokes the growth by retaking itself with one end fixed at the point where the screw dislocation emerges.

If a crystal plane lacks steps and kinks, i.e., points of growth, or if growth at these sites is sufficiently inhibited so that a large concentration of ad-ions builds up compared to the equilibrium concentration, the probability increases that new growing centers will form as two-dimensional nuclei. A very convincing illustration of such a situation was made by Budevski et al.[12] after a remarkable achievement of preparing metal surfaces free of any dislocations. At constant current, polarizations much larger than those upon ordinary planes of such metals were obtained[12]. Moreover, they were distinguished by periodic oscillations. These phenomena are ascribed to fluctuations in the formation of two-dimensional nuclei. Under potentiostatic conditions, a current can be observed on a dislocation-free surface only at overpotentials exceeding 8 to 12 mV, whereas the cell is electrically cut off at lower overpotentials. When a short voltage pulse in excess of this value is applied to a cell in the cut off condition, supersaturation by ad-ions is achieved and a nucleus of a new lattice net is formed. The propagation of the produced step is accompanied by a certain current flow. When the new layer has spread out over the whole surface the current again drops to zero, since the steady-state potential is insufficient to form a new nucleus[13].

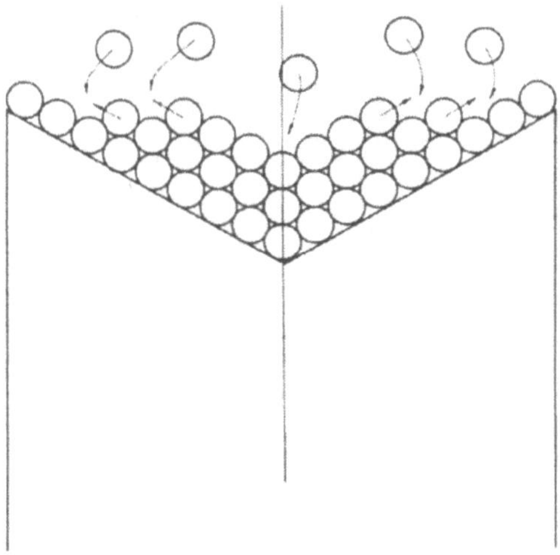

Figure. 2.8. Schematic representation of the formation of an "indestructible reentrant grove".[7]

The current-time curves for a series of successive voltage pulses vary in form, but the integral of the current over time has the same value in all cases. The amount of electricity given by this integral corresponds exactly to the amount required for the completion of a monoatomic layer over the cubic plane[11,13].

The growth rates in the cases of one-dimensional and two-dimensional nucleation as rate determining steps can be compared to each other by considering the growth of two dimensional flat cadmium dendrites from Fig 2.9.

The tip of the twined cadmium dendrite precursor from Fig. 2.9a represents the physical equivalent of the scheme of the growth site from Fig. 2.8. As shown in Fig. 2.8 an layer of atoms advance in the direction determined by twining laws, an edge is constantly renewed, in which the new layers can be started by one-dimensional nucleation. Further growth and branching of precursor like that from Fig. 2.9a produces the dendrites shown in Figs. 2.9b and 2.9c. The deposition on the lateral flat dendrite surfaces takes place by repeated two-dimensional nucleation, as in earlier described deposition on dislocation free surface. This makes the deposition rate in the direction of tip motion many times larger, which results in dendrite shape like that from Fig. 2.9c.

Let it be supposed now, that an advancing microstep suddenly stops advancing. The movement may cease, e.g., owing to the adsorption of impurities from the solution at the step. On a solid surface with its hierarchy of sites, there will be a hierarchy of free energies of adsorption and it may occur that impurities seek adsorption at steps in preference to adsorption on flat planes.

Figure 2.9. The cadmium deposits obtained by deposition from 0.1 mol dm^{-3} $CdSO_4$ in 0.5 mol dm^{-3} H_2SO_4 onto a cadmium electrode. Deposition overpotential 50 mV. Deposition times: a) 2 min, b) 2 min, c) 10 min.[14] (Reprinted with permission from Elsevier Science).

So think of a microstep which, for a reason such that given above, has stopped advancing somewhere within the boundaries of the crystal (Fig. 2.10). Now imagine that a layer B of atoms is growing on top of the layer A. The step B will keep advancing until it comes to the point where the advance of step A was blocked. The layer B will then act as though it has reached the edge of the crystal. If the same process is repeated with another layer C on top of layer B, and then another layer on top of layer C, and so on, then there is a pile-up of layer upon layer. Microsteps bunch into macrosteps and sometimes the pile-up reaches such proportions that it can be seen under a microscope as a macrostep.[3]

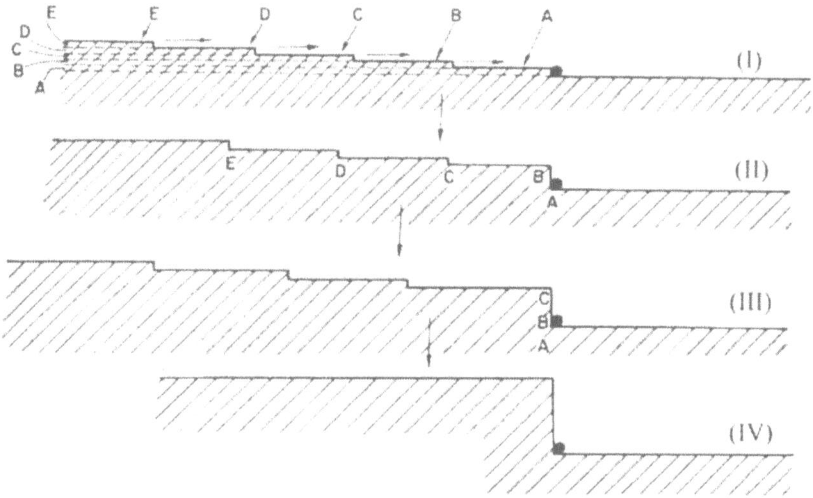

Figure 2.10. Representation of the four successive stages in the formation of a macrostep by microstep bunching[3].

The macrosteps can be clearly seen on the lateral surfaces of flat cadmium dendrites as shown in Fig. 2.11.

Figure 2.11. The detail from Fig. 2.10c.[14] (Reprinted with permission from Elsevier Science).

The flat dendrite lateral surfaces behaves as monocrystals ones and the questions arises: How valid is the picture of deposition, developed above, when the electrodes are polycrystalline metals? The answer is simple. One can consider the surface exposed to the solution by each grain as a single-crystal macrosubstrates. That is, one would have charge transfer followed by surface diffusion, transfer to steps, then to kinks. etc., and one would also have rotating addition, however steps, resulting from screw dislocation, growth spirals, faceting, etc. In addition, however, at the grain boundaries where the single-crystal microsubstrates meet and the periodic atomic arrangement of each grain is interrupted, the deposition and growth processes will be abnormal. But the actual area of an electrode surface occupied by the grain boundaries is so negligible that the abnormal processes occurring there can be largely ignored. In conclusion, therefore, the basic picture of deposition and growth developed for single crystals is valid as a basis for understanding the electrogrowth of polycrystals[3].

2.5 CONCLUSIONS

In this chapter some basic definitions, principles and concepts necessary for understanding of the following chapters have been treated. Assuming that riders have some electrochemical knowledge, for all necessary questions we strongly recommends further reading of excellent books of J. O'M Bockris and A. K.N. Reddy, Modern Electrochemistry.

2.6 FURTHER READINGS

1. Bockris, O'M John; Reddy, K.N. Amulya, *Modern Electrochemistry,* Vol 1, New York: Plenum Press, 1970.
2. Bockris, O'M John; Reddy, K.N. Amulya, *Modern Electrochemistry,* Vol 1, second edition, New York: Plenum Press, 1998.
3. Bockris, O'M John; Reddy, K.N. Amulya, *Modern Electrochemistry,* Vol 2, New York: Plenum Press, 1970.
4. Bockris, O'M John; "Electrode kinetics". *In Modern Aspects of electrochemistry,* Vol 1, John O'M Bockris, Brian E. Conway, eds. London, Butterworths, 1954.
5. Diggle J.W., Despić A.R., Bockris J.O'M. The mechanism of the dendritic electrocrystallization of zinc. J. Electrochem. Soc. 1969; 116:1503-14
6. Barton J.L., Bockris J.O'M. The electrolytic growth of dendrites from ionic solutions. Proc. Roy. Soc. London 1962; A268:485-505
7. Despić, Aleksandar; Popov, Konstantin I., "Transport Controlled Deposition and Dissolution of Metals." *In Modern Aspects of Electrochemistry,* Vol. 7, Brian E. Conway, John O'M. Bockris, eds. New York, NY: Plenum Press, 1972.
8. Popov, Konstantin I., Pavlović Miomir, "Electrodeposition of Metal Powder with Controlled Grain Size and Morphology". *In Modern Aspects of Electrochemistry,* Vol. 24, Ralph E. White, John O'M. Bockris, Brian E. Conway, eds., New York, NY, Plenum Press, 1993.
9. Popov, K.I., Stojilković, E.R., Radmilović, V., Pavlović, M.G., Morphology of lead dendrites electrodeposited by square-wave pulsating overpotential, *Powder Technology,* 1997; 93:55-67.

10. Nickols R.J., Kolb D.M., Belum R.J. STM observations of the initial stages of copper deposition on gold single crystal electrodes. J. Electroanal. Chem. 1991; 313:109-19; Nickols, J. Richard,. "Imagining Metal electrocrystallization at high resolution" In *Imagining of Surfaces and Interfaces, Series: Frontiers in Electrochemistry,* Vol IV, Philip N. Ross; Jacek Lipkovski, eds, New York: VCH Publ. Inc., 1997.
11. Bockris, O'M John, Razumney, G. A, *Fundamental Aspects of Electrocrystalization*, New York: Plenum Press, 1967.
12. Budevski, Evgeni; Georgi, Stoikov; Lorenz, Wolfgang, *Electrochemical Phase Formation and Growth,* Weinheim: VCH, 1996.
13. Budevski E., Bostanov W., Vitanov T., Stojnov Z., Kotzeva A., Kaishev R., Phys. Status. Sol. 1966; 13:577-88
14. Popov K.I., Čekerevac M.I. Dendritic electrocrystallization of cadmium from acid sulphate solution II: The effect of the geometry on dendrite precursors on the shape of dendrites. Surf. Coat. Technol. 1989; 37:435-40.

Chapter 3

SURFACE MORPHOLOGY OF METAL ELECTRODEPOSITS

Morphology is probably the most important property of electrodeposited metals. It depends mainly on the kinetic parameters of the deposition process and on the deposition overpotential or current density. The morphology of an electrodeposited metal depends also on the deposition time until the deposit has attained its final form. When the electrodeposition is from pure simple or complex salt solutions, the form of the metal electrodeposit can be:

1. compact,
2. dendritic,
3. spongy, granular (grains, boulders), etc.

There are three main cases which are divided according to the exchange current density of deposition processes.

- Deposition processes which are characterized by very large exchange current densities. Boulders are formed at lower overpotentials and dendrites at higher ones. In the limiting case, dendrites grow at practically all overpotentials in the electrodeposition of metals with low melting points (e.g. Sn, Pb),
- deposition processes which are characterized by large exchange current densities. Spongy-deposits appear at lower and dendrites at larger overpotentials,
- deposition processes which are characterized by medium and low exchange current densities. Compact deposits are obtained at lower and dendritic and spongy-dendritic deposits at larger deposition overpotentials.

On the other hand, it is known that the morphology of electrodeposits can be substantially changed if the electrodeposition is carried out in the presence of organic or inorganic additives. For example, a smooth deposit can be obtained in the presence of additives instead of rough one in the absence of additives.

The processes of metal electrodeposition can be categorized into three main groups:

1. electroplating and electroforming
2. electrorefining and electrowinning, and
3. metal powder production

each of which have different requirements with respect to the physical state of the cathodic product.

In electroplating the crystal layer is required to be fine grained, smooth, strongly adhesive, glassy, i.e. to be easily polished or bright. In refining and electrowinning relatively coarse grained, rough, but adhesive deposits are required. They have to be of high purity and firm enough to endure handling before melting and casting into shapes suitable for further processing. In metal powder production by electrodeposition a controlled product particle size is necessary and it is preferable if the product is only weakly adhesive.

The first kind of deposit can be obtained from solutions characterised by low exchange current densities at overpotentials close to the end of the Tafel linearity in the presence of different organic additives, with different effects on the cathode deposit.

The deposits in electrorefining and electrowinning are obtained from electrolytes characterised by medium exchange current density at overpotentials close to the end of Tafel linearity. Powdered electrodeposits are obtained from the same solutions, but at current densities considerably larger than the limiting diffusion current density for the electrodeposition process.

Spongy and granular electrodeposits appear during deposition by processes characterised by large exchange current densities at low overpotentials. They are not often met in electrodeposition practice. Spongy zinc can be formed during the charging cycle of same alkaline storage batteries, and can be easily removed by electrodeposition at a periodically charging rate, as can the whiskers which can be formed in some plating baths in the presence of some organic additives.

The above facts are well known, but there is no complete analysis of them so far. The main aim of this chapter is to remedy this. This analysis is performed for electrodeposition of metals from pure simple and complex salt solutions at a constant rate in the presence or in the absence of additives.

3.1 THIN COMPACT SURFACE METAL FILMS

The first stage of metal electrodeposition on an inert substrate is the formation of a thin surface film of deposited metal. It isolates the initial substrate from the solution, and the aim of this section is to show how this can be achieved using a minimum quantity of electrodeposit.

3.1.1 Crystallization overpotential

The formation of the first crystals during galvanostatic metal electrodeposition on an inert substrate is sometimes accompanied by a pronounced increase in the overpotential[1,2]. The dependence of the overpotential on time in such situations is shown in Fig. 3.1.

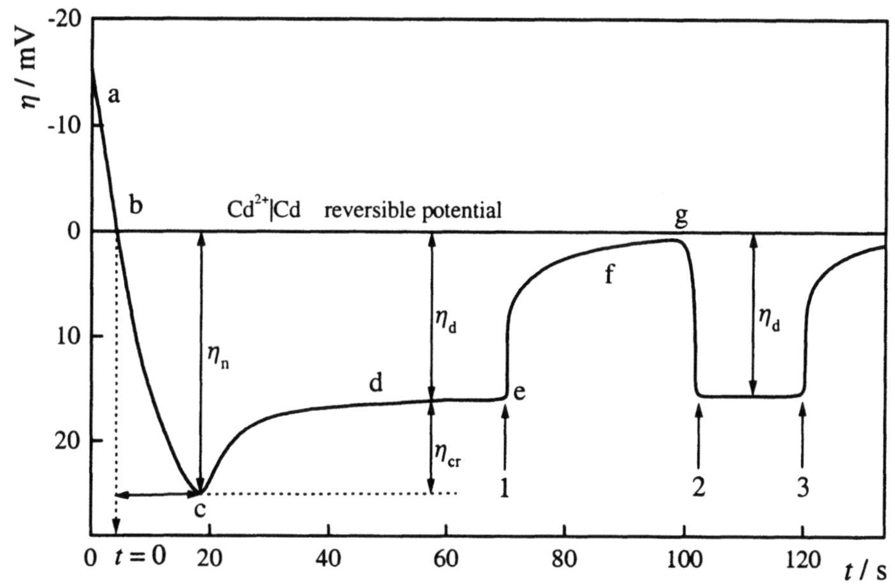

Figure 3.1. The dependence of the overpotential on time during cadmium electrodeposition on a spiral platinum cathode (electrode surface area 1.5 cm^2) from 0.5 mol dm^{-3} $CdSO_4$. The deposition current was 65 μA, η_n is the nucleation, η_{cr} is the crystallisation and η_d is the deposition overpotential. (According to [1]).

The overpotential η_t changes with deposition time from point *b* to point *c* according to the equation[1]:

$$\eta_t = \frac{RT}{nF} \ln \frac{C_{t,a}}{C_{0,a}} \tag{3.1}$$

where $C_{t,a}$ is the concentration of adatoms at time t and $C_{0,a}$ is the concentration of adatoms at $t = 0$.

On the other hand, the surface concentration of adatoms changes according to:

$$C_{t,a} = C_{0,a} + \frac{jt}{nF} \tag{3.2}$$

Substitution of Eq. 3.2 into Eq. 3.1 produces

$$\eta_t = \frac{RT}{nF}\ln\left(1+\frac{jt}{nFC_{0,a}}\right) \tag{3.3}$$

which become

$$\eta_n = \frac{RT}{nF}\ln\left(\frac{C_{cr,a}}{C_{0,a}}\right) \tag{3.3a}$$

at the moment of nucleation.

Eq. 3.3 describes the dependence of the overpotential on the deposition time from point *b* to point *c*. The overpotential changes due to the change of the surface concentration of adatoms from $C_{0,a}$ at the equilibrium potential to some critical value $C_{cr,a}$ at the critical overpotential, η_n, at which the new phase is formed. Hence, the concentration of adatoms increases above the equilibrium concentration during the cathode reaction, meaning that at potentials from point *b* to point *c* there is some supersaturation. The concentration of adatoms increases to the extent to which the boundary of the equilibrium existence of adatoms and crystals has been assumed to enable the formation of crystal nuclei. On the other hand, the polarisation curve can be expressed by the equation of the charge transfer reaction, modified with respect to the crystallisation process, if diffusion and the reaction overpotential are negligible, that is by[2]:

$$j = j_0\left\{\frac{C_a}{C_{0,a}}\exp\left(\frac{\alpha nF\eta}{RT}\right) - \exp\left[\frac{(\alpha-1)nF\eta}{RT}\right]\right\} \tag{3.4}$$

because the partial anode current density depends on the concentration of adatoms, which for $j/j_0 = 0$ becomes equal to Eq. 3.3a.

Obviously, Eq. 3.4 becomes valid at the moment of the formation of the new phase, and it can be used for the estimating the overpotential, η_n at which the nucleation takes place. In order to calculate this overpotential, the supersaturation must be known. According to Pangarov and coworkers[3-5], the work of formation of differently oriented particles can be estimated using supersaturations of 4-7. Considering the nucleation overpotential (for different supersaturations), Klapka[2] assumed 10 as the upper limit of supersaturation. The lower limit is obviously 1 and Eq. 3.4 in this case becomes identical to the equation of the charge transfer reaction.

The difference in overpotential between the curves for a given supersaturation (nucleation on an inert substrate) and for a supersaturation equal to unity (deposition on a native substrate) gives the value of the crystallization overpotential, η_{cr}.[2] It is equal to the difference of the overpotential at point *c* and at point *e* in Fig. 3.1. If the current is switched off at point *e*, the electrode potential will approach the reversible potential of the deposited metal (point *g*); after switching on the current again at point *g*, the overpotential returns to the same value as at point *e*, i.e. the deposition overpotential, η_d, meaning that a new phase is formed. On the contrary, if current is switched off before point *c*, the electrode potential will approach the initial stationary potential of the inert electrode, meaning that new phase has not been formed.[1]

Using α=0.5, t=25^{0}C, n=1 or 2 and Eq. 3.4 Klapka[2] calculated the dependencies of the nucleation overpotential on the j/j_0 ratio for $C_{cr,a}/C_{0,a}$=1, 2, 5 and 10. The calculated curves are shown in Fig. 3.2a. From these curves the dependancies of the crystallization overpotentials on the j/j_0 ratio, shown in Fig. 3.2b, can be derived.

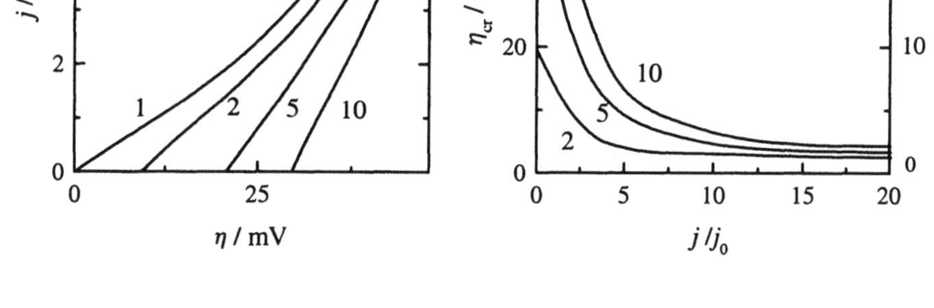

Figure 3.2. a) Dependance of the j/j_0 ratio on the overall overpotential for different $C_{cr,a}/C_{0,a}$ ratios, indicated on the curves (upper scale for n=1, lower scale for n=2). b) Dependence of the crystallisation overpotential η_{cr} (mV), for the cathodic reaction on j/j_0 for different $C_{cr,a}/C_{0,a}$ ratios, indicated on the curves (left hand scale for n=1, right-hand scale for n=2). (According to [2]).

The crystallization overpotential strongly decreases with increasing j/j_0 ratio. As a results of this, it can be measured only in the case of a metal deposition which is characterized by very high values of the exchange current density[2].

3.1.2 The nucleation exclusion zones

3.1.2.1 Basic definitions

Metal electrodeposition on inert electrodes begins with the formation of separate growth centres until a continuous or disperse deposit is produced. Once a nucleus of the depositing metal has been formed, the current flowing causes a local deformation of the electric field in the vicinity of the growing centre. As a result, an ohmic potential drop occurs along the nucleus-anode direction. Considering the high dependence of the nucleation rate on the overpotential, new nuclei would be expected to form only outside the spatial region around the initial nucleus. In that region the potential difference between the cathode and the electrolyte surpasses some critical value η_{cr}. Using simple mathematics, one obtains for the radius of the screening zone, r_{sz}, in an ohmic-controlled deposition:

$$r_{sz} = f\frac{U_\Omega}{U_\Omega - \eta_{cr}} r_N \tag{3.5}$$

where η_{cr} is the critical overpotential for nucleation to occur, U_Ω is the ohmic drop between the anode and cathode, f is a numerical factor and r_N is the radius of the nucleus. The radius of the screening zone depends on the value of both U_Ω and η_{cr}. At a constant η_{cr}, an increase in U_Ω leads to a decrease in the radius of the screening zone, the same is true if η_{cr} decreases at constant U_Ω[6].

The radius of a nucleation exclusion zone can be calculated on the basis of the following discussion, taking into account the charge transfer overpotential also. If there is a half-spherical nucleus on a flat electrode, the extent of the deviation in the shape of the equipotential surfaces which occurs around it depends on the crystallization overpotential, current density, resistivity of the solution and on the radius of the nucleus, r_N. If the distance from the flat part of the substrate surface to the equipotential surface which corresponds to the critical nucleation overpotential, η_n, is l, then this changes around defect to the extent kr_N, as is presented in Fig. 3.3.

Therefore, in this region the current lines deviate from straight lines towards the defect, thus causing an increase in the deposition rate, while in the surrounding region nucleation does not occur, *i.e.*, a nucleation exclusion zone is formed. The voltage drop between the point from which the deviation occurs and the nucleus surface. consists of the ohmic drop between these points and the charge transfer overpotential at the nucleus solution interface. The nucleation overpotential includes both the crystallization and charge transfer (deposition) overpotential:

$$\eta_n = \eta_{cr} + \eta_d \tag{3.6}$$

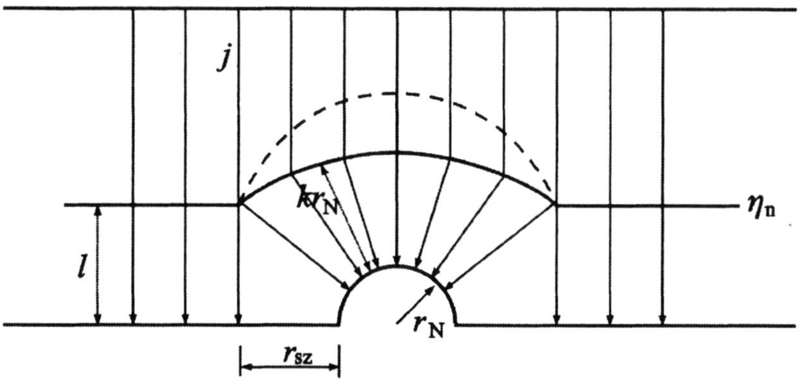

Figure 3.3. A schematic representation of the deformation of the current field around a defect or a grain grown on a foreign substrate. For an explanation of the symbols see the text[7]. (Reprinted with permission from the Serbian Chemical Society, Belgrade, Yugoslavia).

Hence, at the moment when kr_N become equal to l

$$kr_N \, j \, \rho = \eta_{cr} \tag{3.7}$$

where j is the current density along the current lines and ρ is the electrolyte resistivity. Hence, when the ohmic drop between the deviation point and nucleus surface becomes equal to the crystallization overpotential, a new nucleation becomes possible on inert substrate assuming in the both cases the same charge transfer overpotential, and the same value of the current density between the two symmetrical points on the anode and inert cathode surface and between the same point on the anode and the point at the surface of the earlier formed nucleus.

The radius of the nucleation exclusion zone, r_{sz}, corresponds to the distance between the edge of a nucleus and the first current line which not deviates (when kr_N becomes equal to l). Accordingly, nucleation will occur at distances from the edge of a nucleus equal or larger than r_{sz}, which can be calculated as:

$$r_{sz} = r_N\left(\sqrt{2k+1} - 1\right) \tag{3.8}$$

If Eq. 3.7 is taken into account, one obtains:

$$r_{sz} = r_N \left(\sqrt{\frac{2\eta_{cr}}{r_N \rho j} + 1} - 1 \right) \tag{3.9}$$

According to Eq. 3.9, a new nucleation is possible in the vicinity of a nucleus if $\eta_{cr} \to 0$ or $j \to \infty$ or $\rho \to \infty$.

The analysis of the nucleation rate around a growing grain can also be treated in a more rigorous way[8]. Regardless of this, the above model is sufficient to explain the role of nucleation exclusion zones in the first stage of electrocrystallization. This is because the nuclei formed are extremely small and the spherical diffusion control around them can be established after relatively large induction times[9]. During this induction time, the nucleation exclusion zones are due to the ohmic drop in the vicinity of a growing centre. At the same time, the nucleation process is practically terminated, because it is very fast[10]. On the other hand, the rigorous treatment of this problem is very complicated while the effect of the kinetics parameters of the deposition process in the first stage of electrocrystallization can be qualitative explained in a simple way using the described model, i.e. Klapka's concept of crystallization overpotential and the classical nucleation theory.

During the cathodic process at low j/j_0 the crystallisation overpotential is considerably high; with increasing j/j_0, however, it decreases rapidly[2]. Hence, for $j_0 \to 0$ follows that $r_{sz} \to 0$.

3.1.2.2 Physical simulation

The electrolytes used throughout the experiments were 0.1 mol dm^{-3} $AgNO_3$ in a 0.2 mol dm^{-3} HNO_3 solution and 0.1 mol dm^{-3} $AgNO_3$ in a 0.5 mol dm^{-3} $(NH_4)_2SO_4$ solution to which ammonium hydroxide had been added to dissolve the silver sulfate precipitate. The resistivity of the above solutions are almost the same[11].

It has been shown that silver deposition from a silver nitrate bath is under pure diffusion control at all overpotentials, i.e. $j_0 >> j_L$. For the ammonium complex salt bath there is a well-defined region in which the deposition process is under pure activation control (b_c=60 mV dec^{-1}; $j_0 \approx 0.25$ mA cm^{-2}; $j_L \approx 8$ mA cm^{-2}; $j_0 << j_L$)[11].

The silver grains obtained from the nitrate solution on a platinum substrate are presented in Fig. 3.4.

In Fig. 3.5 the silver deposit obtained from the same electrolyte on the substrate shown on Fig. 3.4 are presented.

In Fig. 3.6 silver deposit from ammonium complex bath on the substrate shown on Fig. 3.5 are presented. It can be seen from Fig. 3.5 that large nucleation exclusion zones are formed around the initial grains during deposition from the nitrate bath. They practically do not exist in the deposits from the ammonium complex bath. In addition, new nucleation is seen on the initial grain in Fig. 3.6.

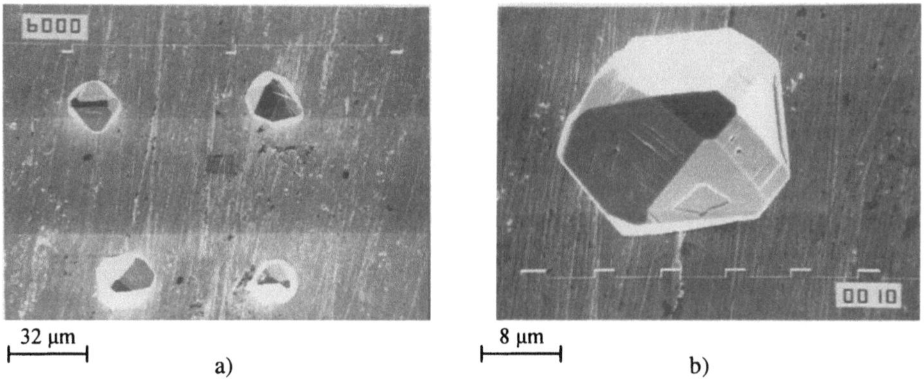

Figure 3.4. Silver deposits on a platinum plane electrode from 0.1 mol dm^{-3} $AgNO_3$ in 0.2 mol dm^{-3} HNO_3 at room temperature. Deposition overpotential 6 mV. Deposition time 45 min.[12] (Reprinted with permission from the Serbian Chemical Society, Belgrade, Yugoslavia).

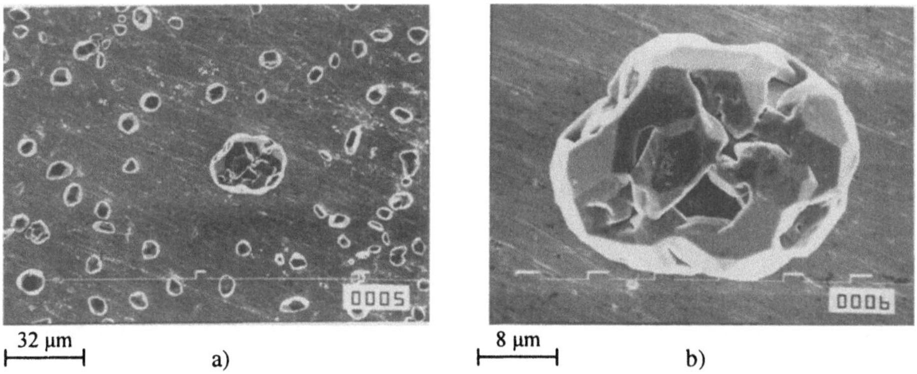

Figure 3.5. Silver deposit on the substrate shown in Fig. 3.4 from the same electrolyte. Current density 2.9 mA cm^{-2}. Deposition time 2 min.[12] (Reprinted with permission from the Serbian Chemical Society, Belgrade, Yugoslavia).

In this way the effect of exchange current density of the deposition process on the radius of the screening zone is clearly demonstrated.

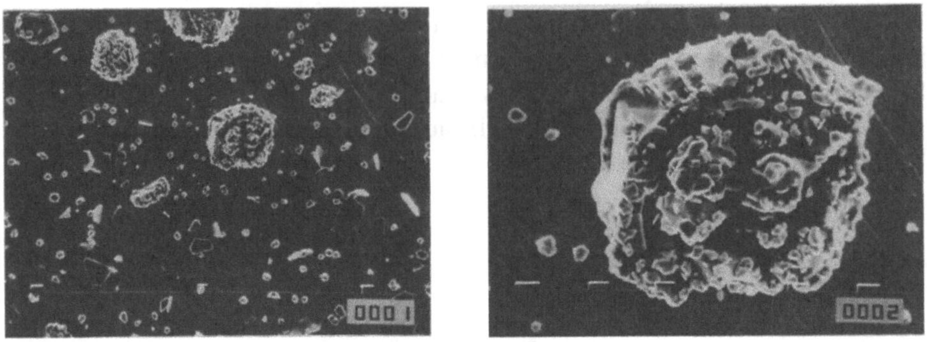

32 μm a) 8 μm b)

Figure 3.6. Silver deposit on the substrate shown in Fig. 3.5 from 0.1 mol dm^{-3} $AgNO_3$ in 0.5 mol dm^{-3} $(NH_4)_2SO_4$ to which was added ammonium hydroxide to dissolve silver precipitate. Current density 2.9 mA cm^{-2}. Deposition time 1 min.[12] (Reprinted with permission from the Serbian Chemical Society, Belgrade, Yugoslavia).

3.1.3 Nucleation rate and deposition overpotential

It has been established experimentally that the number of nuclei deposited electrolytically onto an inert electrode increases linearly with time after an induction period. After a sufficient length of time it reaches a saturated value that is independent of time. The density of the saturation value increases with the increasing applied overpotential and is strongly dependent on the concentration of the electrolyte and the state of the electrode surface[10].

Kaischew and Mutaftchiew[13] explained the phenomenon of saturation on the basis of energetic inhomogenity of the substrate surface. They assumed that the active centres have different activity, or different critical overpotential with respect to the formation of nuclei. Nuclei can be formed on those centres whose critical overpotential is lower or equal to the overpotential externally applied to the electrolytic cell. The higher the applied overpotential, the greater the number of weaker active sites taking part in the nucleation process and, hence, the greater the saturation nucleus density. The formation and growth of nuclei is necessarily followed by the formation and growth of nucleation exclusion zones. After some time, the zones overlap to cover the substrate surface exposed for nucleation, thus terminating the nucleation process[10]. This is well illustrated in Fig. 3.7. It can be seen that the deposit obtained at low current densities consist of a small number of nuclei but with increasing overpotential or current density the number of growth sites increases and the grain size of the deposit decreases.

The simultaneous action of both active centres and nucleation exclusive zones must be taken into consideration when discussing the dependence of the number of nuclei on time. In the limiting case for active centres, when screening zones are not formed, the saturation nucleus density is exactly equal to the integral number of active centres. In the limiting case for nucleation exclusive zones the saturation nucleus density is directly proportional to the nucleation rate and inversely to the zone growth rate[10]. It is obvious that the saturation nucleus density is larger in the first than in the second case, because of the deactivation of active centres by overlapping nucleation exclusive zones.

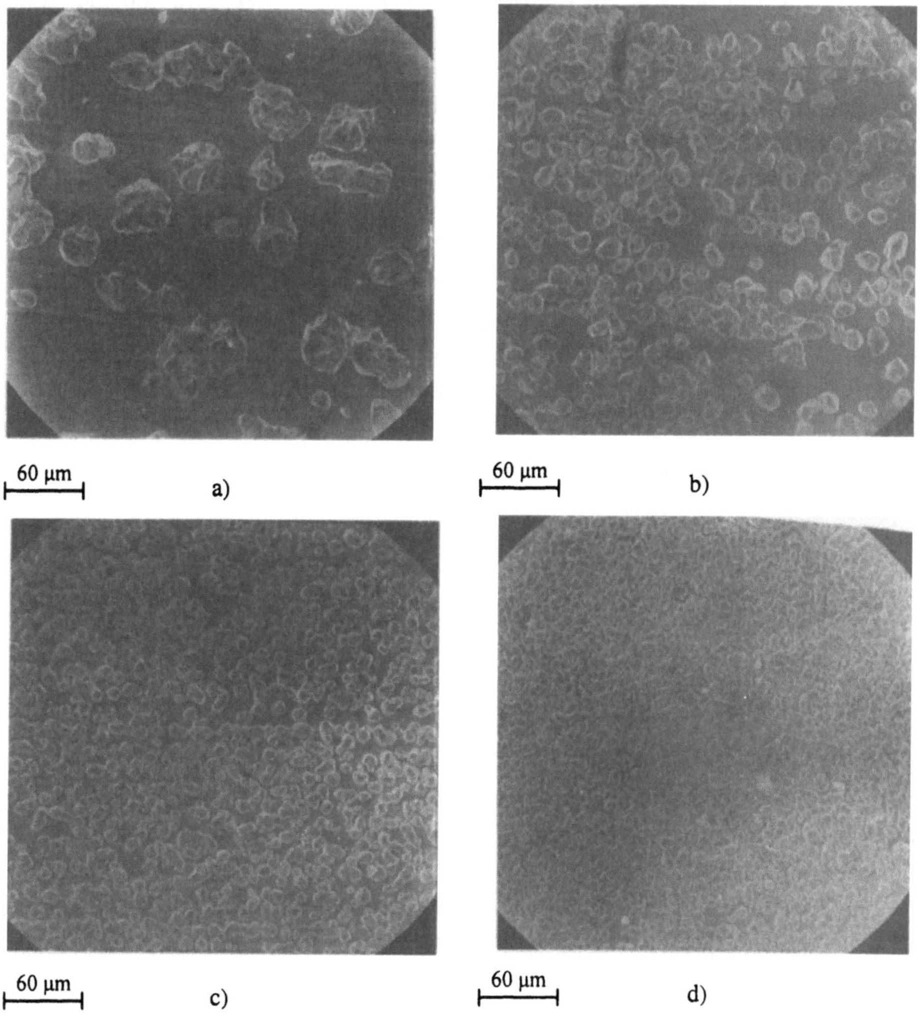

Figure 3.7. Cadmium deposits obtained from 0.1 mol dm^{-3} $CdSO_4$ in 0.5 mol dm^{-3} H_2SO_4 solution onto a copper plane electrode: a) deposition overpotential 10 mV, deposition time 24 min; b) deposition overpotential 40 mV, deposition time 4 min; c) deposition overpotential 60 mV, deposition time 2 min; d) deposition overpotential 110 mV, deposition time 80 s.[9]

The classical expression for the steady state nucleation rate, J, is given by[1,14,15]:

$$J = K_1 \exp\left(\frac{K_2}{\eta^2}\right) \tag{3.10}$$

where K_1 and K_2 are practically overpotential-independent constants. Equation 3.10 is valid for a number of systems regardless of the value of the exchange current density for the deposition process[1,15]. At one and the same deposition current density, j, decreasing j_0 leads to an increasing nucleation rate and decreasing nucleation exclusion zones radii. Hence, the limiting case for nucleation exclusion zones can be expected when $j/j_0 \to 0$, and the limiting case for active centres when $j/j_0 \to \infty$.

The saturation nucleus density, i.e., the exchange current density of the deposition process, strongly effects the morphology of metal deposits. At high exchange current densities, the radii of the screening zones are large and the saturation nucleus density is low. This permits the formation of large, well-defined crystal grains and granular growth of the deposit. At low exchange current densities, the screening zones radii are low, or equal to zero, the nucleation rate is large and a thin surface film can be easily formed. The saturation nucleus density depends also on the deposition overpotential.

The nucleation law can be written[16] as:

$$N = N_0[1 - \exp(-At)] \tag{3.11}$$

where

$$A = K_1 j_0 \exp\left(-\frac{K_2}{\eta^2}\right) \tag{3.12}$$

and N_0 is the saturation nucleus surface density (nuclei cm^{-2}), being dependent on the exchange current density of deposition process and the deposition overpotential.

The overpotential and the current density in activation-controlled deposition inside the Tafel region are related by:

$$\eta = \frac{b_c}{2.3} \ln \frac{j}{j_0} \tag{2.37}$$

Therefore, increasing b_c and decreasing j_0 leads to an increase in the deposition overpotential. According to Eq. 3.12, the value of A increases with increasing overpotential and decreases with decreasing exchange current density. It follows from all available data that the former effect is more pronounced resulting in deposits with a finer grain size with decreasing value of the exchange current density.

Nucleation does not occur simultaneously over the entire cathode surface but is a process extended in time so that crystals generated earlier may be considerably larger in size than ones generated later. This causes periodicity in the surface structure of polycrystalline electrolytic deposits, as well as coarseness of the obtained thin metal film even when formed on a ideally smooth substrate. Hence, the larger the nucleation rates, the more homogeneous is the crystal grain size distribution, which leads to a smoother deposit. Obviously, periodicity in the surface structure is a more complicated problem, as was shown by Kovarskii et al[17-19] but, for the purpose of this analysis the above conclusion is sufficient. The purpose of this work is to confirm the basic facts of the above theories and to show the effect of exchange current density on the deposition process of thin metal film formation on inert substrates.

3.1.4 Deposition from simple salt solutions

The polarisation curves for nickel, copper and cadmium deposition, corresponding Tafel plots and the results of linear polarization experiments are given in Ref. 12. The limiting diffusion currents in all cases are practically the same, but the exchange current densities (given in Table 3.1) are very different.

Table 3.1. The exchange current density for Cd, Cu and Ni deposition processes[12] (Reprinted with permission from the Serbian Chemical Society, Belgrade, Yugoslavia).

Metal	j_0 / A cm^{-2}
Cadmium	1.5×10^{-3}
Copper	3.2×10^{-4}
Nickel	1.6×10^{-9}

Electrodeposits of cadmium, copper and nickel are shown in Figs. 3.8-3.10, respectively. In the cadmium deposition, boulders were formed by the independent growth of formed nuclei inside zones of zero nucleation. As a result of the high value of j_0 the deposition overpotential is low and the crystallisation overpotential is relatively large and so the screening zone, according to Eq. 3.9, is relatively large. On the other hand, the nucleation rate is low. This results in the deposits shown in Fig. 3.8.

In the case of copper, a surface film is practically formed by a smaller quantity of electricity, as seen in Fig. 3.9, due to the lower exchange current density. The value of the deposition overpotential is larger than in the case of cadmium and the crystallisation overpotential is lower, resulting in a decrease in the zero nucleation zone radius, and hence a considerably larger nucleation rate. A further decrease in the exchange current density value, as in the case of Ni, leads to the situation shown in Fig. 3.10. A surface film is formed, but it is porous, probably due to hydrogen co-deposition.

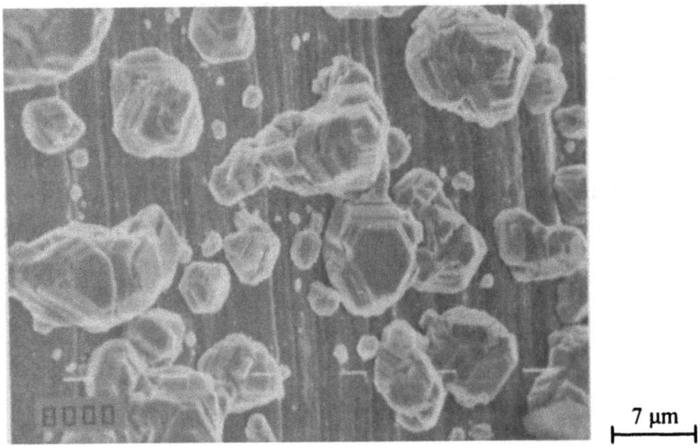

Figure 3.8. Cadmium deposit on a copper substrate obtained at a current density of 1 mA cm^{-2} from 0.07 mol dm^{-3} $CdSO_4$ in 0.5 mol dm^{-3} H_2SO_4. Deposition overpotential 15 mV. Deposition time 1200 s.[12] (Reprinted with permission from the Serbian Chemical Society, Belgrade, Yugoslavia).

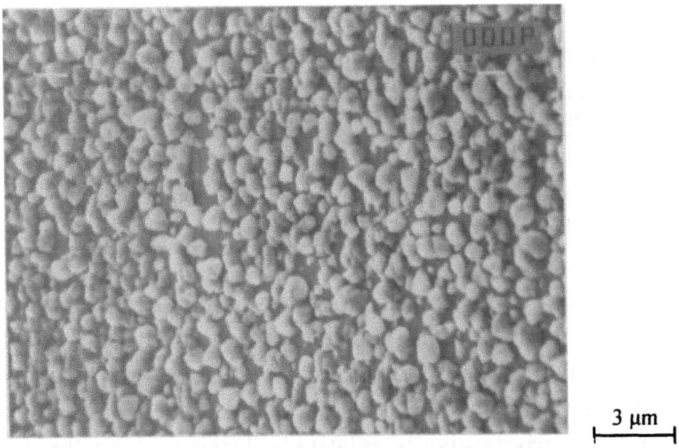

Figure 3.9. SEM microphotograph of copper deposits on a silver substrate obtained at a current density of 1 mA cm^{-2} from 0.07 mol dm^{-3} $CuSO_4$ in 0.5 mol dm^{-3} H_2SO_4. Deposition overpotential 60 mV. Deposition time 300 s.[12] (Reprinted with permission from the Serbian Chemical Society, Belgrade, Yugoslavia).

Figure 3.10. SEM microphotograph of nickel deposits on a copper substrate from obtained at a current density of 1 mA cm^{-2} from 0.07 mol dm^{-3} $NiSO_4$ in 0.5 mol dm^{-3} Na_2SO_4 + 30 g/l H_3BO_3, pH=4. Deposition overpotential 715 mV. Deposition time 120 s.[12] (Reprinted with permission from the Serbian Chemical Society, Belgrade, Yugoslavia).

Hence, a decrease in the value of the exchange current density of the deposition process enhances thin surface metal film formation on inert substrates due to an increase in the nucleation rate and a decrease in the radius of the zero nucleation zones. As a result of this, a compact surface metal film is formed with a smaller quantity of electrodeposited metal, and its coarseness and porosity decrease with a decreasing exchange current density. On the other hand, at sufficiently negative equilibrium potentials and low hydrogen overpotential for an inert substrate, decreasing the exchange current density of the deposition process can produce a porous deposit due to hydrogen co-deposition.

3.1.5 Deposition from complex salt solutions

The silver deposits obtained from a nitrate bath at overpotentials corresponding to an initial current densities of $0.5j_L$ and from the ammonium complex salt bath at an overpotential corresponding to the current density of $0.3j_L$, are presented in Fig. 3.11[11]. The solutions are the same as in section 3.1.2.2. It can be seen that the deposit obtained from the nitrate bath consists of a small number of nuclei and that boulders are formed even at $0.5j_L$, which leads to the formation of a non-compact deposit.

On the other hand the deposit obtained from the ammonium complex salt bath is microcrystalline.

The poor microthrowing power of deposits obtained from nitrate solutions at smaller current densities can be explained in the following way.

For $j_0 >> j_L$, the deposition overpotential is given by:

$$\eta = \varphi_0 \ln\left(1 - \frac{j}{j_L}\right) \tag{3.13}$$

according to Eq. 2.41, where:

$$\varphi_0 = \frac{RT}{nF} \tag{3.14}$$

and, for $j << j_L$, $\eta \to 0$ and $N \to 0$ from Eqs. 3.11 and 3.12.

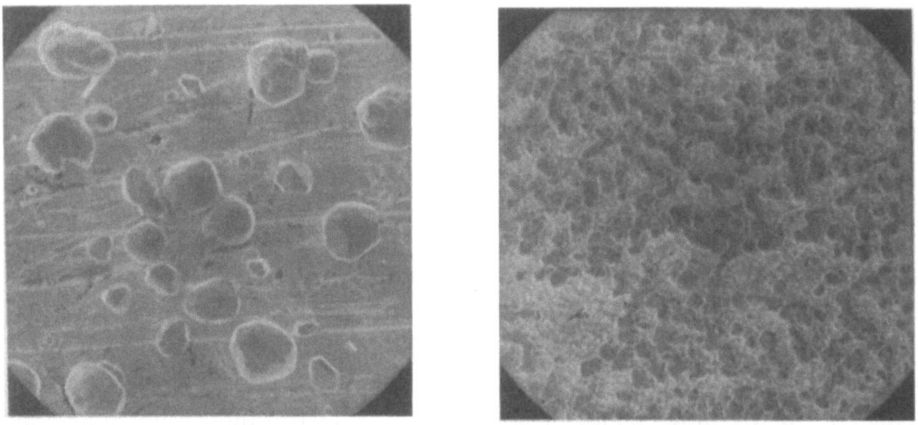

37 µm a) 37 µm b)

Figure 3.11. Silver deposits obtained from a) a nitrate bath at an overpotential of 18 mV (corresponding to an initial current density of $0.5j_L$) (90 mA min cm^{-2}), and b) an ammonium complex salt bath at an overpotential of 65 mV (corresponding to an initial current density of $0.3j_L$ (90 mA min cm^{-2}).[11] (Reprinted with permission from Elsevier Science).

Thus, at low current densities, poor microthrowing power is expected. In the ammonium complex salt solution, $j_0 << j_L$. For $j_0 << j$ and $j << j_L$, Eqs. 3.11, 3.12 and 2.37 are valid which means that $N > 0$. Hence, for deposition at low current densities, decreasing j_0 lead to increasing coverage of an inert substrate for a given quantity of deposited metal.

3.1.6 Deposition in the presence of adsorbed additives

3.1.6.1 Inorganic compounds

In order to illustrate the effect, silver was deposited onto Ag and Pt substrates from aqueous solutions containing 0.5 mol dm^{-3} $AgNO_3$ in 100 g dm^{-3} $NaNO_3$, Fig. 3.12a and 0.5 mol dm^{-3} $AgNO_3$ in 100 g dm^{-3} $NaNO_3$ + 6 g dm^{-3} H_3PO_4, Fig 3.12b. Galvanostatic and potentiostatic deposition conditions were applied in an open type cell[20-22].

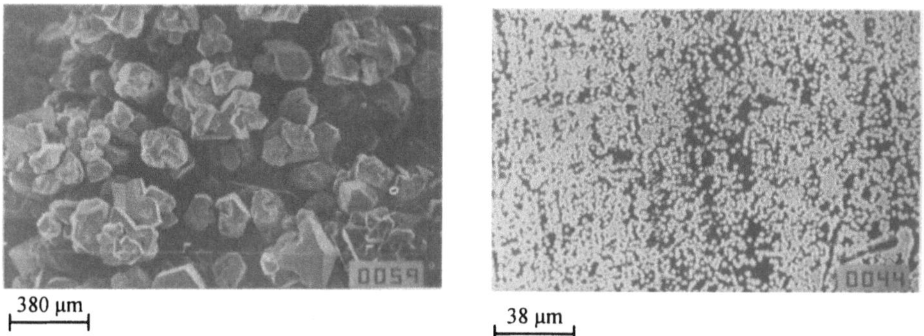

a) b)

Figure 3.12. a) Silver electrodeposits obtained potentiostatically from 0.5 M $AgNO_3$ in 100 g dm^{-3} $NaNO_3$ onto stationary wire electrodes. Quantity of electricity 100 mA h cm^{-2}. Deposition overpotential 120 mV. Exchange current density 26 mA cm^{-2}; [20] b) Silver electrodeposits obtained galvanostatically from 0.5 mol dm^{-3} $AgNO_3$ in 100 g dm^{-3} $NaNO_3$ + 6 g dm^{-3} H_3PO_4 onto stationary platinum wire electrodes I = 30 mA, t=2 s, exchange current density 5 mA cm^{-2}. [21]

Silver deposits formed from a PO_4^{3-}-ion containing electrolyte on to a Pt substrate with galvanostatic current pulses are shown in Fig. 3.12b. At a current density of 30 mA cm^{-2} (i.e. under the optimal film deposition conditions, as determined in Ref. 20), almost complete surface coverage was achieved even with a charge quantity of 2 mA h cm^{-2}.

This is probably due to the possibility of further nucleation occurring immediately next to the already existing nuclei, as a result of the smaller values of the radii of the nucleation exclusion zones. Obviously, this is due to the decrease of the exchange current density for the deposition process. For comparison, in phosphate-free nitrate solution, a compact Ag film had not been deposited even after 100 mA h cm^{-2} had been passed through the cell, as can be seen from Fig. 3.12 a. [20]

3.1.6.2 Organic compounds

The electrolyte used in all experiments was a solution of 0.25 mol dm^{-3} $CdSO_4$ in 0.5 mol dm^{-3} H_2SO_4 to which was added 3.3 g dm^{-3} poly(oxyethylene alkylphenol) (9.5 mol ethylene oxide)[23].

The overpotential-log(current density) plot is given in Fig. 3.13. A well-defined Tafel line characterised by $j_0 \approx 10^{-6}$ A cm^{-2} and b_c=160 mV dec^{-1} was observed at higher potentials also. This phenomenon is explained by the formation of a film of the organic additive which completely covers the cathode at sufficiently negative potentials[24,25]. Tafel linearity was also observed over a short overpotential range at low overpotentials. The values of $j_0 \approx 10^{-3}$ A cm^{-2} and b_c= 60 mV dec^{-1} obtained in this case are close to the values expected for deposition from a pure solution[26].

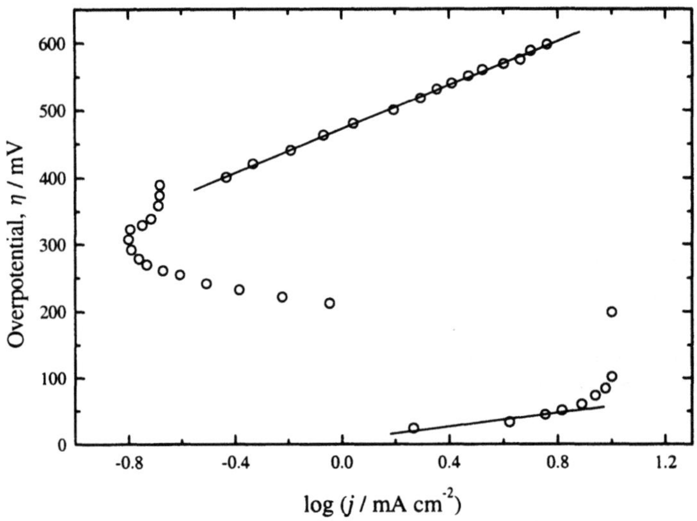

Figure 3.13. Overpotential *vs.* log (current density) for cadmium deposition in the presence of additive. [23] (Reprinted with permission from Elsevier Science).

It has recently been shown[27] that the optimum plating overpotential and current density are determined by the upper limit of the validity of the Tafel equation for the deposition process (see also section 3.2.1.3.1). In this case, as can be seen in Fig. 3.13, the optimum deposition overpotentials are about 530 mV and 40 mV in the presence and the absence of adsorption of the aditive respectively. Cadmium deposits 3 μm thick obtained at 40 mV and 530 mV are shown in Fig. 3.14. It can be seen that the deposits obtained at 40 mV have a large grain size, whereas those obtained at 530 mV are fine grained, due to the larger overpotential.

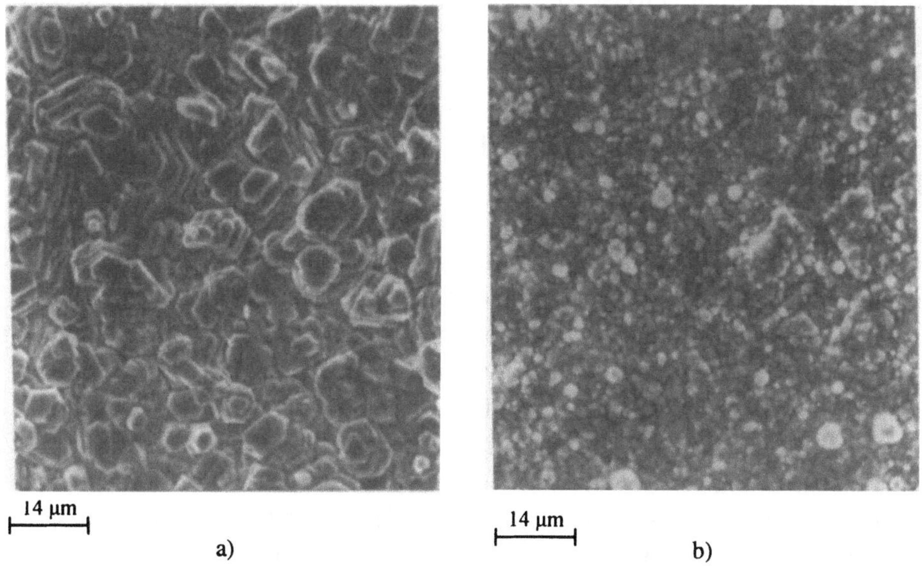

Figure 3.14. Cadmium deposit on the middle of a copper electrode, a) at 40 mV in the absence of adsiorption of additive; b) at 530 mV in the presence of adsorption of additive. The deposit thickness is 3 μm.[23] (Reprinted with permission from Elsevier Science).

3.1.7 Conclusions

A surface metal film on an inert substrate is formed by the coalescence of growing grains developed from corresponding nuclei, as is illustrated in Fig. 3.15, whereby the surface properties of the inert substrate are changed to those of the electrodeposited metal.

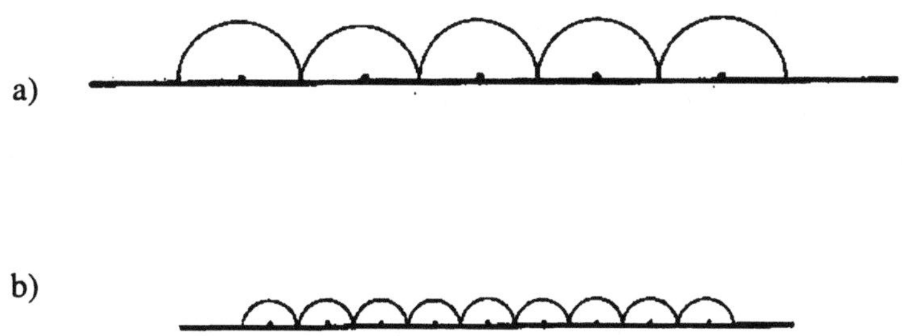

Figure 3.15. Schematic representation of the formation of the surface film: a) low nucleation rate, b) high nucleation rate.

It is obvious that the larger nucleus density, the thinner is the thickness of the metal film required to isolate the substrate from the solution. At the same time a thinner surface film will be less coarse than a thicker one. This means that a smoother and thinner surface film will be obtained at larger deposition overpotentials and nucleation rates, i.e. by electrodeposition processes characterized by high cathodic Tafel slopes and low exchange current densities.

It is obvious that the discussion concerning the effect of the value of exchange current density on the nucleation exclusion zone radius is not connected to the mechanism of surface film formation but to the mechanism of nucleation itself and the saturation nucleus number density. Papers dealing with three-dimensional growth and related phenomena[28] are mainly concerned with the determination of the mechanism of the formation of a surface film and are unimportant from the point of view of the estimation of the deposit thickness required to isolate a substrate from an electrolyte solution. For this purpose, the saturation nucleus density, or better to say the distribution of the distances between nearest neighbours is much more important. It is obvious that half of the largest distance between nearest neighbours[8,29] (as illustrated by Fig. 3.15) is the radius to which each grain must grow to produce a nonporous thin metal film. It is clear that the distribution of the distances between nearest neighbour crystallites is the most important dependence in the treatment of the thin metal film formation on an inert substrate. From the corresponding histograms, as shown by Milchev et al[8,29], it is possible to estimate the radius of the nucleation exclusion zones, as well as the maximum distance between nearest neighbour crystallites, which determines the thickness of a deposit required to isolate the substrate from the electrolyte solution, as illustrated in Fig. 3.15. If the distance between the nearest neighbour crystallites is smaller than the grain radius the deposit will overlap resulting in coarse deposit growth initiation.

Apart from the nucleation density, the preferential orientation of the nuclei is important in surface metal film formation. As the deposit becomes free of the influence of the substrate structure on thickening, instead of the formation of a randomly oriented grain structure, a preferred crystal orientation can develop, which gives a definite texture to the cross section of the deposit[30]. Texture can be expressed in terms of degree of orientation of the grains constituting the deposit.

It is to note a the theoretical approach to the problems of deposit orientation was successfully developed by Pangarov et al[3-5]. Using this theory, it was possible to determine the preferred orientation as a function of overpotential from silver single crystals[4] to nickel and iron thin films[3,5].

3.2 THICK COMPACT METAL ELECTRODEPOSITS

After formation of a thin surface metal film on an inert substrate further deposition takes place in the same way as on a massive electrode of the metal, the ions of which were reduced to form the electrodeposit. The final thickness of the electrodeposited surface layer varies from the order of ten micrometers in electroplating, to many times larger in electrowinning and refining of metals. The surface of deposits obtained in electroplating must be smooth and bright, in the other cases the surfaces of the deposit surface has to be as smooth as possible. How this can be achieved will be revealed in this section.

3.2.1 Coarse surfaces

3.2.1.1 Mathematical models

Any solid metal surface that represents a substrate for metal deposition possesses a certain roughness. In addition, it may appear coarse or smooth, and this is not necessarily related to the roughness. Fig. 3.16 shows cases of surfaces with a) the same roughness and profoundly different coarseness and b) vice versa.

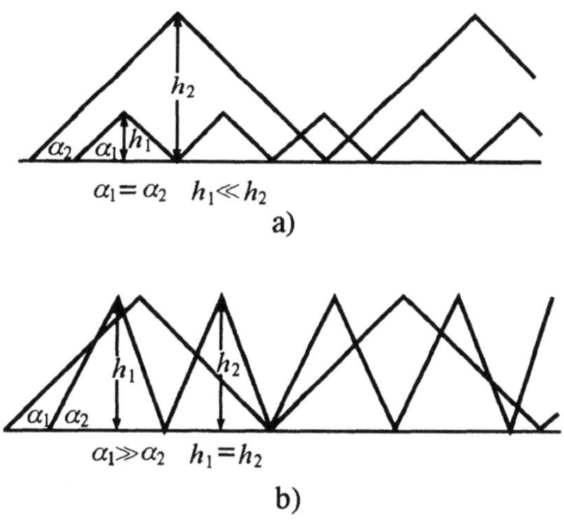

Figure 3.16. Models of surfaces with a) the same surface roughness and different coarseness and b) vice versa.[31]

It is the level of coarseness which determines the appearance of metal deposits, while even with considerable roughness, if below the visual level, the surface may appear smooth.

It is convenient to define the surface coarseness as the difference in thickness of the metal at the highest and lowest points above an arbitrary reference plane facing the solution. In early models used to describe the surface by periodic functions this is equal to twice the amplitude of the function[31,32].

Historically, it was first established that under certain conditions of dissolution the surface coarseness tends to decrease[33]. Krichmar[34] was the first to point out that in some cases of deposition, under conditions somewhat analogous to those in which the coarseness decreases, the opposite effect occurs; i.e., in prolonged cathodic reduction, under conditions at which, the process is close to being under complete diffusion control, amplification of the surface coarseness occurs.

Taking a sinusoidal profile for the electrode surface:

$$H = H_0 \sin\left(\frac{2\pi x}{a}\right) \tag{3.15}$$

Krichmar[34] obtained the relationship:

$$H_0(t) = H_0 \exp\left(\frac{j}{j_L}\frac{Q}{Q_{0,K}}\right) \tag{3.16}$$

for $H_0(t) << a$, where

$$Q = jt \tag{3.17}$$

and

$$Q_{0,K} = \frac{nFa}{2\pi V}\tanh\left(\frac{2\pi\delta}{a}\right) \tag{3.18}$$

In the above equations a is the wavelight of the sinusoidal profile, F is the Faraday constant, H is the local elongation, H_0 is the initial amplitude of the sinusoidal profile, $H_0(t)$ is the amplitude of the sinusoidal profile at time t, j is the current density, j_L is the limiting diffusion current density, V is the molar volume of metal, n is the number of electrons, Q is the quantity of electricity, t is the time, x is the co-ordinate normal to the plane of the electrode and δ is the thickness of the diffusion layer.

Simpler mathematics were used in another, independently derived theory of the same phenomenon put forward by Despić et al[35], and Diggle et al[36]. A somewhat simplified treatment will be given here.

Consider the model of the surface irregularity shown in Fig. 3.17. The surface irregularity is buried deep in the diffusion layer, which is characterized by a steady linear diffusion to the flat portion of the surface. The current densities at the various parts of the surface are as follows.

a) At the flat part of the surface, the limiting diffusion current density is that for steady-state linear diffusion, i.e.,

$$j_L = \frac{nFDC_0}{\delta} \tag{2.29}$$

where D is the diffusion coefficient and C_0 is the bulk concentration of the depositing ions.

b) At the side of an irregularity, even when a possible lateral diffusion flux supplying the depositing ions is neglected, the current density, j_s, at any point of height h_s must be larger than the current density, j, at the flat part of surface.

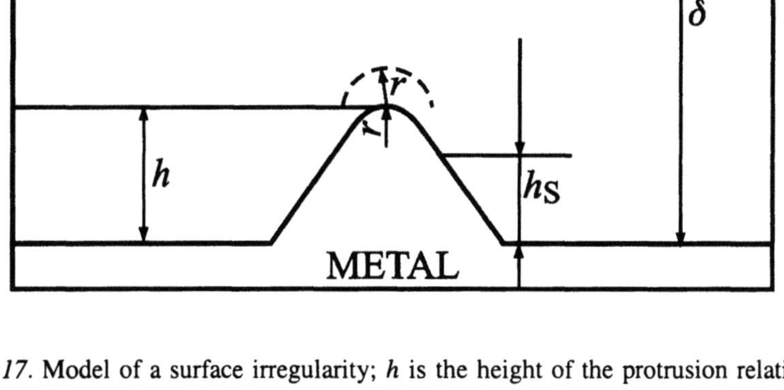

Figure 3.17. Model of a surface irregularity; h is the height of the protrusion relative to the flat portion of the surface, h_s is the corresponding local side elongation, r is the radius of the protrusion tip, and δ is the thickness of the diffusion layer. [32]

This is because the point is closer to the diffusion-layer boundary; i.e., the effective diffusion layer is thinner, and hence the diffusion flux and resulting current density are larger. Obviously, this is valid if the protrusion height does not affect the outer limit of the diffusion layer. i.e. if $\delta >> h_s$. The limiting diffusion current density, $j_{L,s}$, is given as:

$$j_{L,s} = \frac{nFDC_0}{\delta - h_s} = j_L \frac{\delta}{\delta - h_s} \tag{3.19}$$

c) At the tip of an irregularity, there is an additional reason for an increased current density. The lateral flux cannot be neglected, and the situation can be approximated by assuming a spherical diffusion current density, $j_{L,tip}$, given by:

$$j_{L,tip} = \frac{nFDC^*}{r} \tag{3.20}$$

where C^* is the concentration of the diffusing species at a distance r from the tip, assuming that around the tip a spherical diffusion layer having a thickness equal to the radius of the protrusion tip is formed[37,38]. If deposition to the macroelectrode is under full diffusion control, the distribution of the concentration C inside the linear diffusion layer is given by[36]

$$C = C_0 \frac{h}{\delta} \tag{3.21}$$

where $0 \le h \le \delta$. Hence,

$$C^* = C_0 \frac{h + r}{\delta} \tag{3.22}$$

and from of Eqs. 2.29, 3.20, and 3.22[39], it follows:

$$j_{L,tip} = j_L \left(1 + \frac{h}{r}\right) \tag{3.23}$$

The general equation of the polarisation curve for a flat surface is given by Eq. 2.28:

$$j = \frac{j_0(f_c - f_a)}{1 + \dfrac{j_0 f_c}{j_L}} \tag{2.28}$$

The current densities j_s and j_{tip} to different points of the electrode surface can then be obtained by substitution of j_L in Eq. 2.28 by appropriate values from Eqs. 3.19 and 3.23 for the side and the tip of the protrusion, around which spherical diffusion layer is formed, respectively. Hence:

$$j_s = \frac{j_0(f_c - f_a)}{1 + \left(\dfrac{j_0 f_c}{j_L}\right)\left(\dfrac{\delta - h_s}{\delta}\right)} \tag{3.24}$$

and

$$j_{tip} = \frac{j_0(f_c - f_a)}{1 + \left(\dfrac{j_0 f_c}{j_L}\right)\left(\dfrac{r}{r + h}\right)} \tag{3.25}$$

If a diffusion layer around the tip of a protrusion can not be formed (see 3.2.1, 3.2):

$$j_{tip} = \frac{j_0(f_c - f_a)}{1 + \left(\dfrac{j_0 f_c}{j_L}\right)\left(\dfrac{\delta - h}{\delta}\right)} \tag{3.26}$$

The effective rate of growth of the side elevation is equal to the rate of motion of the side elevation relative to the rate of motion of the flat surface[31]. Hence:

$$\frac{dh_s}{dt} = \frac{V}{nF}(j_s - j) \tag{3.27}$$

Substitution of j_s from Eq. 3.24 and j from Eq. 2.28 in Eq. 3.27 and further rearrangement gives[39]:

$$\frac{dh_s}{dt} = \frac{j^2 V h_s}{j_L n F \delta} \tag{3.28}$$

if, $\delta >> h_s$ and $f_c, >> f_a$, or in the integral form:

$$h_s = h_{0,s} \exp\left(\frac{j}{j_L}\frac{Q}{Q_{0,D}}\right) \quad (3.29)$$

where $h_{0,s}$ is the initial height of the local side elevation just as in the previous case (Eq. 3.16), Q is given by Eq. 3.17, and:

$$Q_{0,D} = \frac{nF\delta}{V} \quad (3.30)$$

According to both mechanisms (Eqs. 3.16 and 3.29), an increase in the surface coarseness can be expected with increasing quantity of deposited metal for the same deposition current density, as well as with increasing current density for the same quantity of electrodeposited metal.

In the same way, the propagation rate of the protrusion tip can be obtained by substituting j_{tip} from Eq. 3.25 and j from Eq. 2.28 into Eq. 3.27, where j_s and h_s are substituted by j_{tip} and h, on further rearrangement the following expression is obtained:

$$\frac{dh}{dt} = \frac{Vjj_{tip}h}{nFj_L(r+h)} \quad (3.31)$$

It should be noted that Eq. 3.31 is valid only if the radius of the protrusion tip is sufficiently large to make the surface energy term negligible[37].

It is obvious from Eqs. 3.28 and 3.31 that

$$\frac{dh}{dt} > \frac{dh_s}{dt} \quad (3.32)$$

because $j_{tip} > j$ and $h/(r+h) > h_s/\delta$, which means that the tip propagation protrusion will be larger under spherical diffusion control.

3.2.1.2 Physical Simulation

To test the validity of the above equations, Despić and Popov[40,41] carried out experiments on diffusion-controlled metal electrodeposition on a well-defined, triangularly shaped surface profile, through a diffusion layer of well-defined thickness $\delta >> h$. A phonograph disk negative was used as the substrate upon which a layer of an agar-containing copper sulfate-sulfuric acid solution was placed and left to solidify, as illustrated in Fig. 3.18.

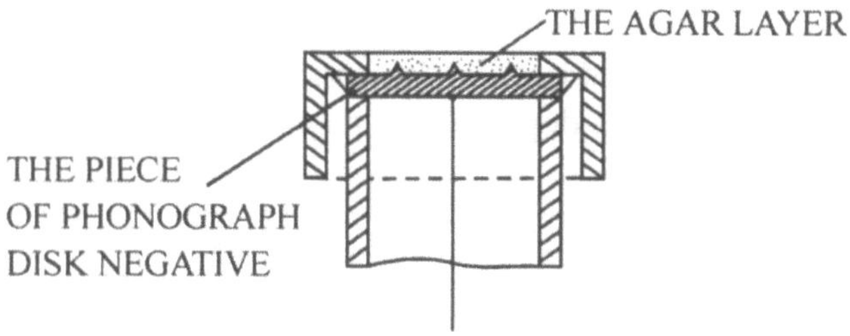

Figure 3.18. The model electrode for diffusion-controlled electrodeposition of metals.[40] (Reprinted with permission from the Serbian Chemical Society, Belgrade, Yugoslavia).

As current was passed and the layer was depleted of copper ions, an increase in the height of the triangular ridges was observed. Metallographic samples were made in wax and the cross sections of the deposit were photographed under a microscope, Fig. 3.19 was thus obtained[41].

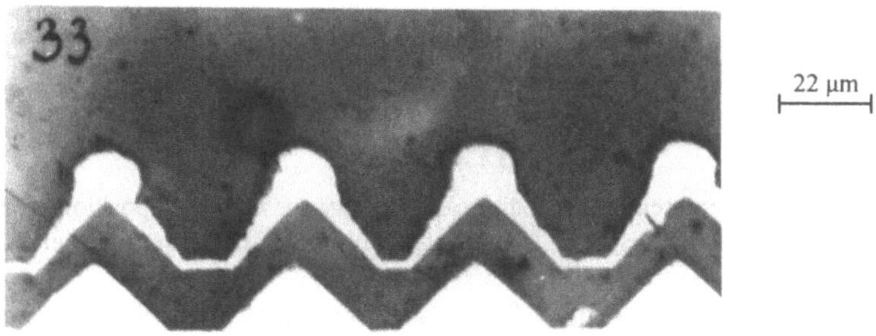

Figure 3.19. Cross-section of a copper electrodeposit in the system from Fig. 3.18 from 0.5 mol dm^3 $CuSO_4$ in 0.5 mol dm^{-3} H_2SO_4. The thickness of the agar diffusion layer was 1.0 mm. Deposition overpotential, 300 mV; deposition time, 120 min. The substrate is a piece of a phonograph disk negative.[41]

In accordance with the discussion presented in Section 3.2.1.1, three parts of the surface can be seen in Fig. 3.19: the flat part of the electrode and the sides and the tips of irregularities, providing an excellent physical illustration of the mathematical model.

The effect of deposition time at a given current density, i.e., the effect of the quantity of electrodeposited metal, on the protrusion height is obvious, the larger the deposition time, the larger is the surface coarseness[41].

3.2.1.3 Real Systems

3.2.1.3.1 Optimum current density for compact metal deposition

The Tafel plot for copper deposition is given in Fig. 3.20.

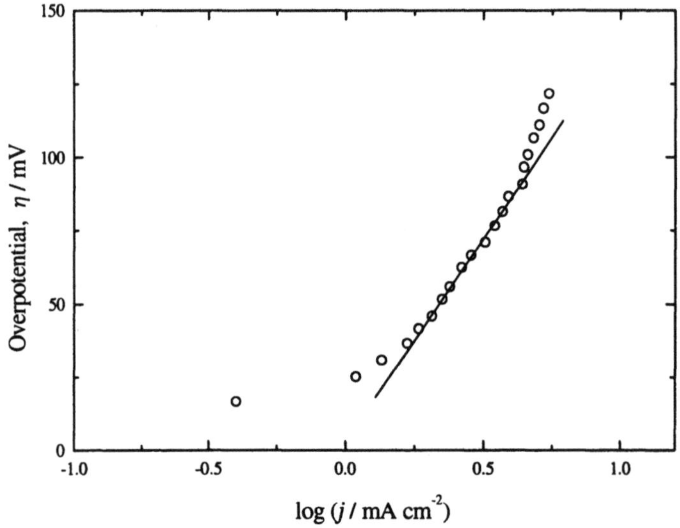

Figure 3.20. Overpotential vs. log current density for copper deposition from 0.1 mol dm^{-3} $CuSO_4$ in 0.5 mol dm^{-3} H_2SO_4.[43] (Reprinted with permission from Elsevier Science).

The surface coarseness for a fixed quantity of electrodeposited metal in mixed activation–diffusion controlled deposition increases strongly with increasing current density[42,43].

Activation-controlled deposition of copper produces large grains with relatively well-defined crystal shapes. This can be explained by the fact that the values of the exchange current densities on different crystal planes are quite different, whereas the reversible potential is approximately the same for all planes[44]. This can lead to preferential growth of some crystal planes, because the rate of deposition depends only on the orientation, which leads, to the formation of a large-grained rough deposit. However, even at low degrees of diffusion control, the formation of large, well-defined grains is not to be expected, because of irregular growth caused by mass-transport limitations. Hence, the current density which corresponds to the very beginning of mixed control (a little larger than this at the end of the Tafel linearity) will be the optimum one for compact metal deposition, as follows from Fig. 3.20.

All the above facts are illustrated in Fig. 3.21[43].

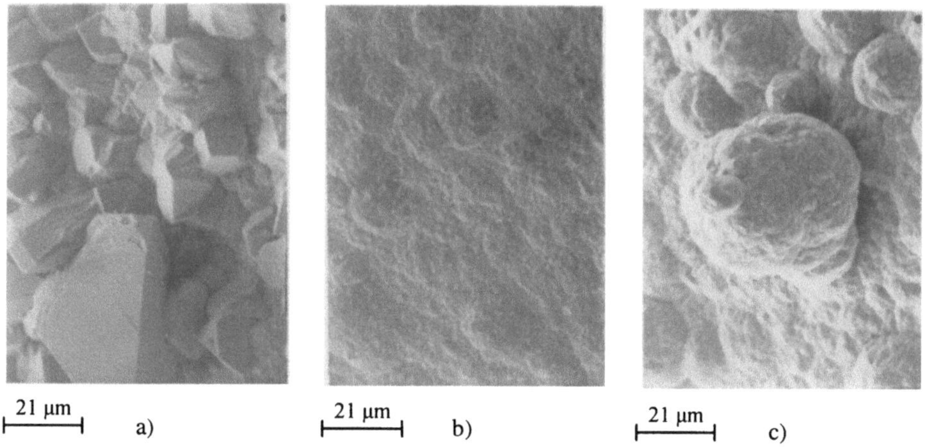

Figure 3.21. Copper deposits obtained from 0.1 mol dm^{-3} $CuSO_4$ in 0.5 mol dm^{-3} H_2SO_4. Quantity of electricity, 20 mAh cm^{-2}. a) Activation-controlled deposition; deposition overpotential, 90 mV; initial current density 3.3 mA cm^{-2}. b) Electrodeposition under mixed activation-diffusion control; deposition overpotential 140 mV; initial current density 4.2 mA cm^{-2}. c) Electrodeposition under mixed activation-diffusion control; deposition overpotential 210 mV; initial current density 6.5 mA cm^{-2}. The substrate was a copper wire electrode.[43] (Reprinted with permission from Elsevier Science).

3.2.1.3.2 Cauliflower like forms

It can be seen from Fig. 3.21c that the surface protrusions are globular and cauliflower-like. If the initial electrode surface protrusions are ellipsoidal, they can be characterized by the base radius R_0 and the height h as shown in Fig. 3.22a.

The tip radius is then given by:

$$r = \frac{R_0^2}{h} \tag{3.33}$$

The initial electrode surface protrusion is characterized by $h \to 0$ and $r \to \infty$ if $R_0 \neq 0$. In this situation, a spherical diffusion layer cannot be formed around the tip of the protrusion if $r < \delta - h$, and linear diffusion control occurs, leading to an increase in the height of the protrusion relative to the flat surface according to Eq. 3.29. When h increases, r decreases, and spherical diffusion control can be operative around the whole surface of protrusion, if it is sufficiently far from the other ones, as illustrated by Fig. 3.22b. In this situation, second-generation protrusions can grow inside the diffusion layer of first-generation protrusions in the same way as first-generation protrusions grow inside the diffusion layer of the macroelectrode, and so on.

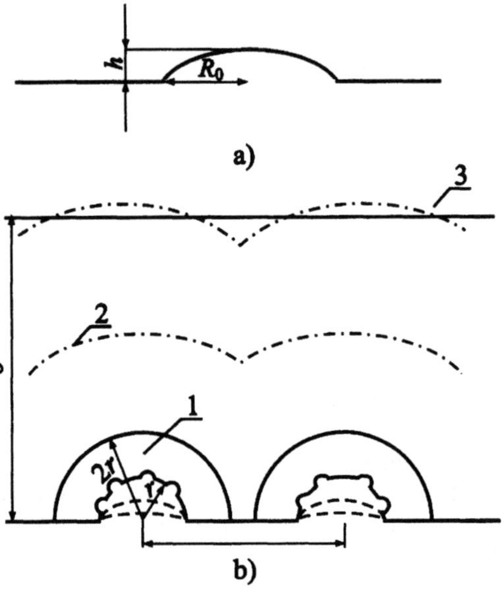

Figure 3.22. Schematic representation of: a) the initial electrode surface protrusion, b) the establishment of spherical diffusion layers around independently growing protrusions (1. $r < (\delta - h)$ and $r < 1/4l$ spherical diffusion zones are formed; 2. $r < (\delta - h)$ and $r > 1/4l$ spherical diffusion zones overlap); 3. $r > (\delta - h)$, spherical diffusion zones are not formed.[39] (Reprinted with permission from the Serbian Chemical Society, Belgrade, Yugoslavia).

A cauliflower deposit is formed under such conditions, as is shown in Fig. 3.23. It can be seen from Fig. 3.23a that the distance between the cauliflower grains is sufficiently large to permit the formation of spherical diffusion zones around each of them. At the same time, second-generation protrusions grow in all directions, as shown in Fig. 3.23b and c. This confirms the assumption that the deposition takes place in a spherically symmetric fashion.

To a first approximation, the rate of propagation can be taken to be practically the same in all directions, meaning that the cauliflower-type deposit formed by spherically symmetric growth inside the diffusion layer of the macroelectrode will be hemispherical, as is illustrated in Fig. 3.23a-c.

This type of protrusion is much larger than that formed by linearly symmetric growth inside the diffusion layer of the macroelectrode (Fig. 3.23a-c), as is predicted by Eq. 3.32.

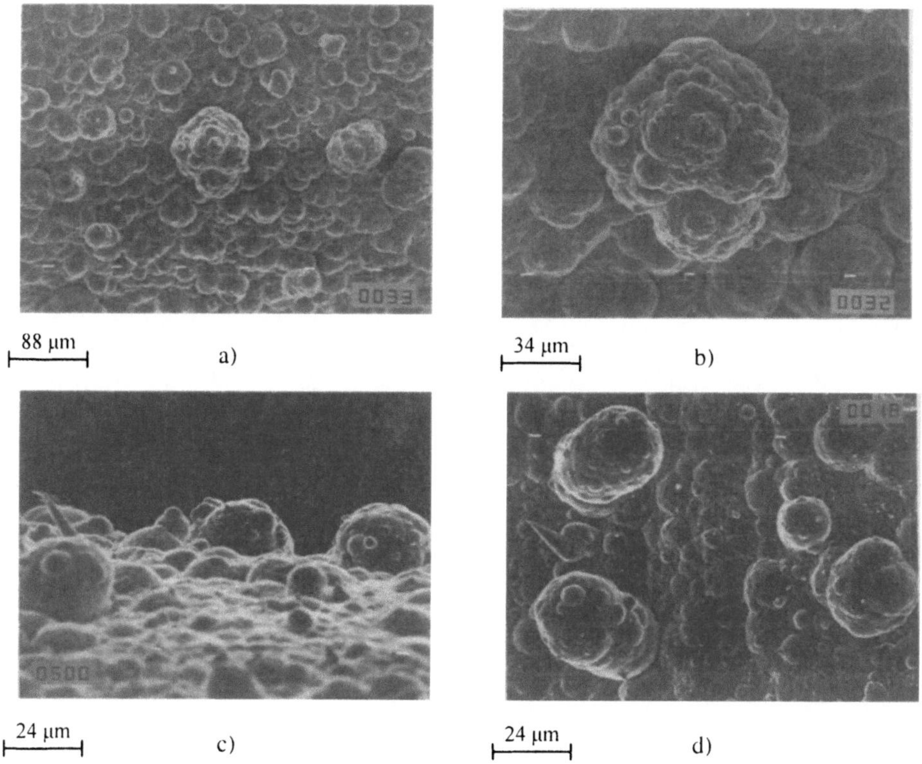

Figure 3.23. Copper deposits obtained from 0.3 mol dm^{-3} $CuSO_4$ in 0.5 mol dm^{-3} H_2SO_4 by electrodeposition under mixed activation-diffusion control. Deposition overpotential, 220 mV. a) Quantity of electricity 40 mAh cm^{-2}. b) The same as in a). c) and d) quantity of electricity 20 mAh cm^{-2}. The substrate was a copper wire electrode.[39] (Reprinted with permission from the Serbian Chemical Society, Belgrade, Yugoslavia).

This is because a spherical diffusion layer cannot be formed around closely packed protrusions, their diffusion fields overlap and they grow in the diffusion layer of the macroelectrode.

3.2.1.3.3 Carrot like forms

It can also be seen from Fig. 3.23c,d and 3.24 that the growth of some protrusions produces carrot-like forms, another typical form obtained in copper deposition under mixed activation-diffusion control. This happens under the condition $r/h << 1$, when spherical diffusion control takes place only around the tip of the protrusion, as is illustrated in Figs. 3.17 and 3.24. In this case Eq. 3.25, can be rewritten, if the surface energy effect is neglected, in the form:

$$j_{\text{tip}} = j_0(f_c - f_a) \tag{3.34}$$

meaning that deposition on the protrusion tip can be under pure activation control at overpotentials lower than the critical one for the initiation of dendritic growth (see section 3.3.2).

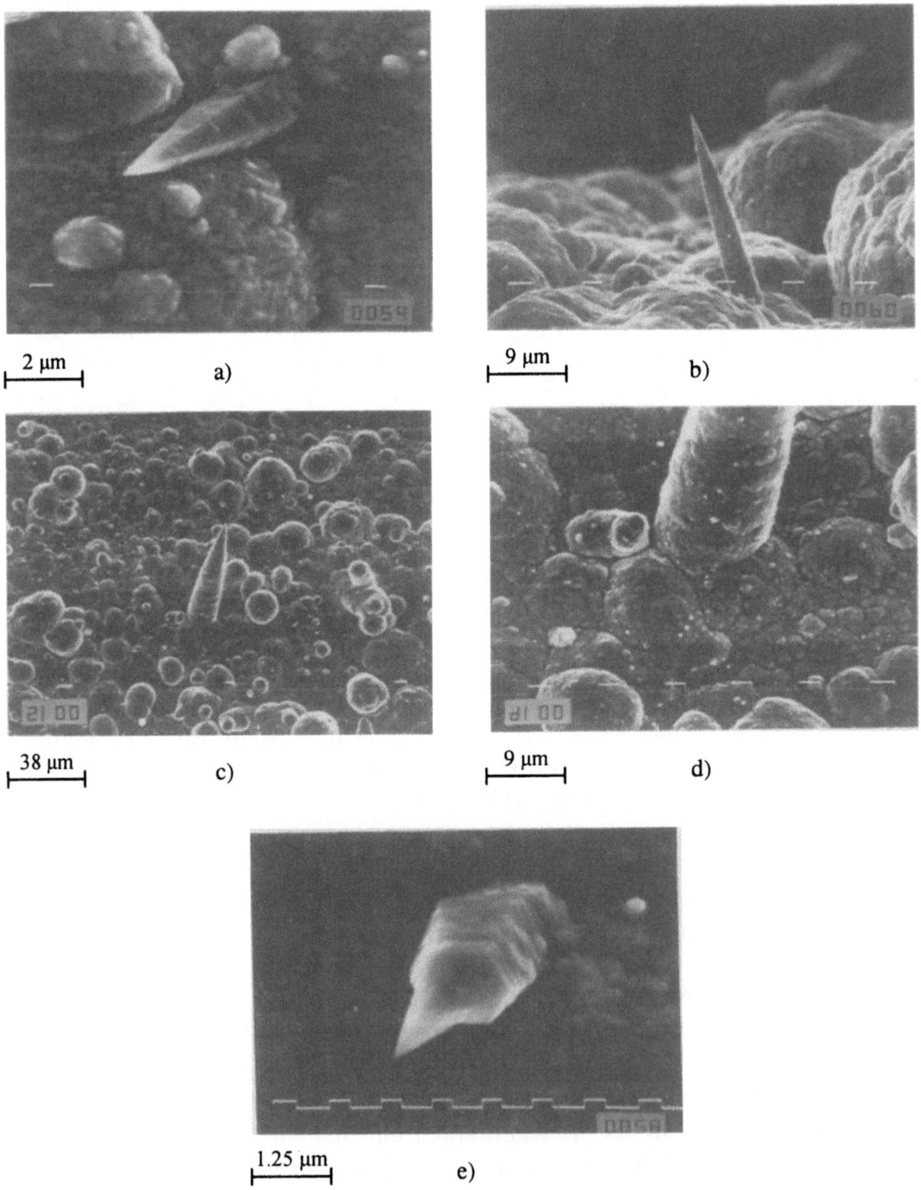

Figure 3.24. Copper deposits obtained from 0.3 mol dm^{-3} $CuSO_4$ in 0.5 mol dm^{-3} H_2SO_4 by electrodeposition under mixed activation-diffusion control. Deposition overpotential, 220 mV. Quantity of electricity: a) 10 mAh cm^{-2}, b) 40 mAh cm^{-2}; c) 20 mAh cm^{-2}; d) the root of the carrot from c) and e) 10 mAh cm^{-2}. The substrate is a copper wire electrode. [45] (Reprinted with permission from the Serbian Chemical Society, Belgrade, Yugoslavia).

This happens if the nuclei have a shape like that in Fig. 3.24a. The assumption that the protrusion tip grows under activation control is confirmed by the regular crystallographic shape of the tip[45] just as in the case of grains growing on the macroclectrode under activation control (see Fig. 3.21a).

The maximum growth rate at a given overpotential corresponds to activation-controlled deposition. As a result, the propagation rate at the tip will be many times larger than that in other directions, resulting in protrusions like that in Fig. 3.24b. The final form of the carrot-like protrusion is shown in Fig. 3.24c. It can be concluded from the parabolic shape that such protrusions grow as moving paraboloids in accordance with the Barton-Bockris theory[37], the tip radius remaining constant because of the surface energy effect. It can be concluded from Fig. 3.24d that thickening of such a protrusion is under mixed activation-diffusion control because the deposit is seen to be of the same quality as that on the surrounding macroelectrode surface. It can be seen from the Fig. 3.24e that activation control takes place only at the very tip of the protrusion.

Some of the new nuclei are precursors of carrot-like protrusions, depending on their crystal orientation and position relative to the already growing protrusion[32]. In this case, they are in the form of small hexagonal pyramids, as shown in Fig. 3.24e. Based on their morphology and because copper has a face-centred cubic crystal structure, it is reasonable to assume that they are truncated by a high-Miller-index plane. According to Pangarov et al[3-5], the orientation of nuclei is related to the applied overvoltage. It is reasonable to expect that the appearance of precursors of carrot protrusions have its own overvoltage range. Obviously, such kind of protrusions can produce short cuts and deposition of copper must be carried out at overpotentials lower than this at which carrot like protrusion can be formed.

3.2.2 Smooth surfaces

3.2.2.1 Basic Facts

It is well known[31] that in electrolytes containing specific substances as additives, a phenomenon opposite to the ones described so far can occur, i.e. a more rapid metal deposition at recessed points of the surface than at elevated points. This causes levelling of the surface irregularities as is illustrated in Fig. 3.25.

The fact that this phenomenon is only observed at microprofiles not exceeding 100 μm in amplitude necessitated the introduction of the concept of "microthrowing power" as a category different from ordinary throwing power[31]. The latter is used in technical literature to describe the quality of electrolytes in plating on macroprofiles, at which a similar effect is never observed. The difference between a microprofile and a macroprofile can be seen from Fig 3.26.

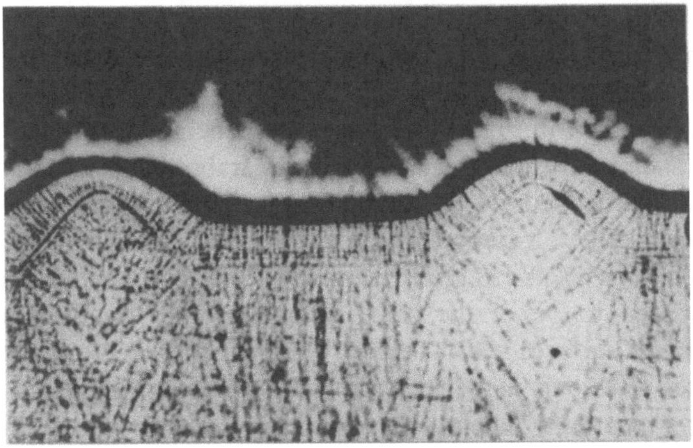

Figure 3.25. Semibright nickel at low levelling-agent concentration; height at peak, 0.00275 cm.[46] (Reprinted by permission of John Willey & Sons, Inc.).

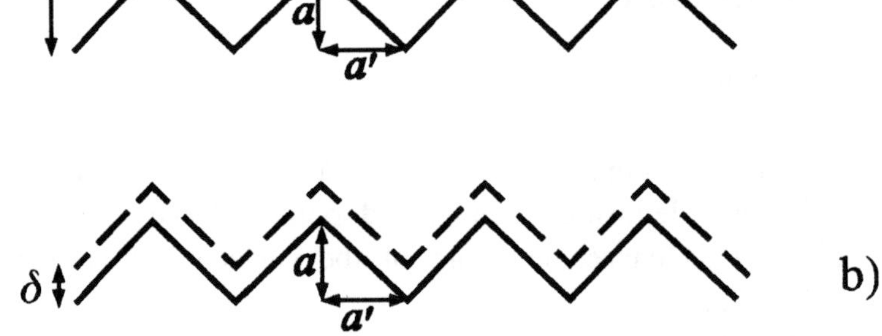

Figure 3.26. Serrated surface profile with diffusion layer, showing the difference between a) microprofile ($a \ll \delta$) and b) macroprofile ($a \gg \delta$).[47]

Detailed surveys of the literature on levelling are available in Refs. 31, 46, 48 and 49.

3.2.2.2 Model of leveling

All the experimental evidence points to the conclusion that leveling takes place under conditions when the supply of the substance causing inhibition of the electrode process is under diffusion control. It was already clear to early investigators that the explanation should be sought in the local variations in the supply of the leveling agent over the surface profile. Peaks at the surface receive larger amounts of an additive than the recesses. This results in an increase in inhibition and a decrease in the local current density of deposition at the protrusion relative to less-exposed parts of the surface. Thus, leveling is directly related to differences in the surface concentration of the additive which leads to differences in the local current density of deposition[31].

The deposition current density of metal ions must be close to the end of the Tafel linearity, i.e. it can be treated as the current density in activation controlled deposition, being independent on the geometry of the system. Hence, the current density at the tip of a protrusion, will be equal to the current density on the flat surface in the absence of additive. It should be noted that under such condition "geometric leveling" occurs, but true levelling requires the presence of an additive. If the additive is consumed at the electrode by the reaction

$$\text{Additive} + n^{\bullet}\text{e}^{-} \rightarrow \text{Product} \tag{3.35}$$

the limiting diffusion current density of the additive, $j^{\bullet}_{\mathrm{L,tip}}$, to the tip of the protrusion from Fig. 3.17, if spherical diffusion can be neglected, is given by:

$$j^{\bullet}_{\mathrm{L,tip}} = \frac{n^{\bullet} F D^{\bullet} C_0^{\bullet}}{\delta - h} \tag{3.36}$$

and that to the flat part of the electrode, $j^{\bullet}_{\mathrm{L,f}}$, by:

$$j^{\bullet}_{\mathrm{L,f}} = \frac{n^{\bullet} F D^{\bullet} C_0^{\bullet}}{\delta} \tag{3.37}$$

where $n^{\bullet}$ is the number of electrons in Eq. 3.35, $D^{\bullet}$ and $C_0^{\bullet}$ are the diffusion coefficient and concentration of the additive, respectively.

Assuming that the overall current density, j, at each point of the electrode surface is equal, the effective current density, j_{eff}, of metal deposition at the tip of a protrusion is given by:

$$j_{eff1} = j - \frac{n^{\bullet} F D^{\bullet} C_0^{\bullet}}{\delta - h} \tag{3.38}$$

and on the flat part of the surface by:

$$j_{eff2} = j - \frac{n^{\bullet} F D^{\bullet} C_0^{\bullet}}{\delta} \tag{3.39}$$

and following the same procedure as in the treatment of the increase of coarseness (see section 3.2.1), one can write:

$$\frac{dh}{dt} = \frac{V}{nF}\left(j_{eff1} - j_{eff2}\right) \tag{3.40}$$

or in the integral form:

$$h = h_0 \exp\left(-\frac{t}{\tau}\right) \tag{3.41}$$

where:

$$\tau = \frac{n\delta^2}{n^{\bullet} V D^{\bullet} C_0^{\bullet}} \tag{3.42}$$

where V (molar volume of metal) and n correspond to the electrodeposited metal. This is the simplest mathematical model of the leveling process[50]. Despite this, it elucidates the physical essence of the phenomenon under consideration well.

Obviously, for this model to be operative, two conditions must be satisfied: (a) the additive must be consumed in some manner at the electrode, so that it must be continuously supplied in order to maintain a certain surface concentration, and (b) the diffusion layer must not follow the microprofile but must have a smoother outer boundary, so that variations in its thickness arise, which cause variations of the diffusion flux of the additive.[31]

The first condition is fulfilled with all good levelling agents. Most of them undergo sufficiently strong adsorption to remain long enough at the metal surface to be surrounded by depositing atoms and be incorporated into the deposit. It is the balance between the rate of incorporation and that of diffusion of the substance from bulk of solution which maintains a given surface concentration of the additive. The larger the diffusion flux, the higher is the steady-state surface concentration of the additive. Conversely, higher rates of metal deposition cause a lowering of the latter.[31]

There is an optimal range of additive concentration and current density of deposition at which the differences in inhibition of deposition between the peaks and recesses, and hence the effect of levelling, are maximal. At too low surface concentrations of the additive, i.e., low bulk concentration and high current density of deposition, the process is practically uninhibited and little difference in the local current density of deposition can arise. This explains the decrease in the levelling effect with increasing current density.[31]

At somewhat higher bulk concentrations and lower current densities, linearity exists between the bulk and the surface concentration. This is the range of maximum difference in inhibition.[31]

However, at still higher concentrations, an adsorption/desorption equilibrium tends to be approached leading to a Langmuir-type relationship. Eventually, in spite of incorporation, saturation of the surface is reached and the surface concentration is no longer sensitive to local changes in the diffusion flux of the aditives. Hence, differences in inhibition vanish and leveling is lost.[31]

One should appreciate that some time is needed for the diffusion layer to develop to the extent that it separates from the surface microprofile and provides for local differences in the diffusion flux of the aditives. Hence, an induction time should be expected before the leveling effect appears. This is demonstrated by the observed sensitivity of the process to current interruptions[48].

3.2.2.3 Quantitative Treatment

Krichmar[51] made an attempt at a comprehensive quantitative consideration of the problem. He proposed that additives adsorbed to the surface of an electrode are incorporated into the deposit at a rate proportional to the surface coverage and current density.

For a sinusoidal profile of the electrode surface, Krichmar[51] obtained an exponential decrease of the amplitude, $H_0(t)$ with time:

$$H_0(t) = H_0 \exp\left(-\frac{t}{\tau}\right) \tag{3.43}$$

where τ is a time constant given by:

$$\tau = \frac{a}{A[B(C_0^\bullet / \delta) - j]} \coth\left(\frac{2\pi\delta}{a}\right) \tag{3.44}$$

where:

$$A = \frac{2\pi V}{nF} \tag{3.45}$$

$$B = \frac{K\alpha FnD^{\bullet}C^{\bullet}_{0,a}}{RT\delta V} \tag{3.46}$$

K and α in the above equations are constants and other symbols have meaning as in the Eqs. 3.18, 3.36 and 3.37.

An exponential decrease of the amplitude of the surface profile was found experimentally by Krichmar and Pronskaya[52].

Regardless, a model of the current distribution and numerical procedure for the calculation of the change in shape of an electrode, for electrodeposition followed by diffusion-controlled reduction or incorporation of a leveling agent, has been developed[49,53-55], the approach of Krichmar[51] seems to be the most important one for understanding the leveling process.

3.2.3 Bright surfaces

3.2.3.1 Silver mirror

The surface of a silver mirror can be taken as the reference level for bright surfaces[56]. It can be seen from Fig. 3.27, that a mirror surface consists of parts parallel to the base and flat on the atomic level with low steps heights between them, as is shown in Fig 3.28. Hence it can be expected that bright metal surfaces must be similar to the surface of the mirror.

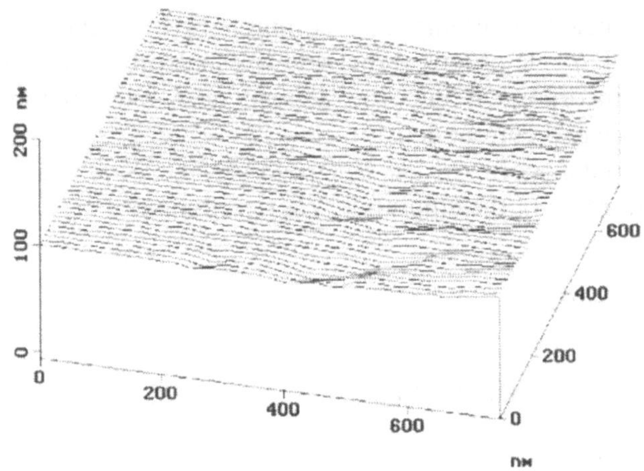

Figure 3.27. 3D STM image of a silver mirror.[56b] (Reprinted with permission from Serbian Chemical Society, Belgrade, Yugoslavia).

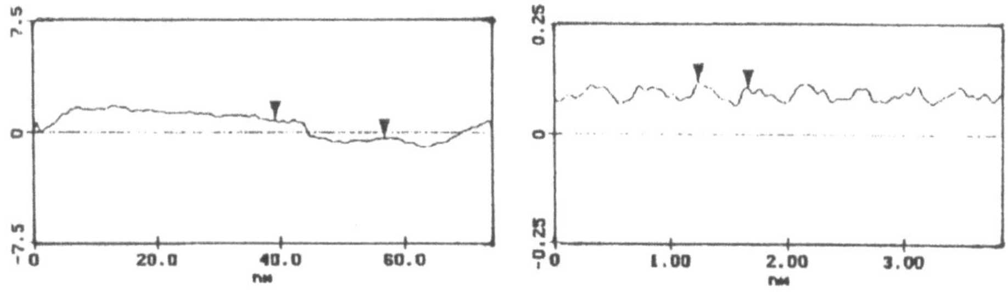

Figure 3.28. A typical surface profiles of a silver mirror.[56b] (Reprinted with permission from the Serbian Chemical Society, Belgrade Yugoslavia).

3.2.3.2 Electropolished surfaces

The phenomenon of decreasing the surface coarseness of a metal upon anodic dissolution under certain conditions is defined as electropolishing. In cases when polishing occurs, the current-voltage curve was found to exhibit a plateau characteristic for diffusion control of the dissolution process. Some facts point to the complex nature of the phenomenon of electropolishing.

It was also found that systems undergoing electropolishing exhibit a significant photoelectrochemical effect. This corresponds to the region of limiting current densities and also to the maximum polishing effect.

This suggests to the existence of a photosensitive semiconducting film at the surface and to a possible role of this film in the electropolishing process. Subsequent measurements of the capacitance and resistance of the double layer as functions of potential have shown that this film must be a very thin and a well-conducting one.

In spite of all this, considerable evidence has accumulated justifying the treatment of the electropolishing process as an essentially transport-controlled phenomenon; this film could be related to the effect of brightening[31].

A quantitative model was suggested by Edwards[57] and elaborated by Wagner[33]. According to this model the metal ions produced are complexed by a component of the electropolishing solution (e.g., phosphate ions or water molecules). Hence, for the reaction to be completed, not only must ions be formed, but also acceptor species have to diffuse to the surface from the bulk of the solution in order to form a complex reaction product. Diffusion of the acceptor from the bulk of the solution to the surface determines the overall rate of reaction.

Hence, differences in acceptor fluxes of different points at the surface arise. The slower diffusion of the acceptor to the recessed parts could cause an increased concentration of free metal ions. This would have many possible consequences such as: increasing the cathodic partial current which reduces the net dissolution current, producing changes in the reaction layer such as the formation of an oxide film by hydrolysis, making room for an additional phenomenon observed in electropolishing-brightening. This could be related to the dissolution of facets and other crystallites. Brightening seems to occur when the surface becomes covered by a protective film, which controls the rate of dissolution and makes it a random process, the energetic advantages of atoms at facets and dislocations being lost. Thus, it could be concluded that the Edwards-Wagner model seems to provide a reasonable basis for development of a comprehensive theory of the electropolishing process.[31]

The reaction between a metal ion and a ligand is usually sufficiently fast but it is the insufficient supply of the latter, which could cause the inhibition of this step and make it the rate-determining step.

The anodic partial current is independent of the presence of the ligand. However, if the ligand becomes scarce, the concentration of the free metal ions increases and, as a result, the cathodic partial current is increased as well. In an ideal case, when diffusion of free metal ions away from the electrode is negligible, the difference between the two partial currents, i.e., the net dissolution current, must be proportional to the flux of the ligand. This case was considered in detail by Wagner[33] who assumed a molecular diffusion mechanism of supply through a fixed hydrodynamic boundary layer of thickness δ, much larger than wavelength and amplitude of the surface profile.

For a sinusoidal profile of the electrode surface, Wagner[33] gives:

$$H_0(t) = H_0 \exp\left(-\frac{t}{\tau}\right) \tag{3.43}$$

where H_0 is the initial surface amplitude, and τ is given by

$$\tau = \frac{a\delta}{2\pi DVC^0} \tag{3.47}$$

a is the wavelength of the profile and C^0 is the concentration of the acceptor.

Equation 3.47 shows that the electropolishing process is faster if the thickness of the diffusion layer is smaller and if the bulk concentration of the ligand is larger. Also, it is faster the smaller the wavelength of the surface profile is. The latter reflects the radius of curvature of the elevations and recesses and the result indicates that "microroughnees" will disappear more rapidly than "macroroughness", which is in accordance with experimental experience (see also Fig. 3.30). The exponential time dependence of the height of the elevations given by Eq. 3.43 is in agreement with the findings of Krichmar and Pronskaya[58]. The electropolished metal surfaces are characterized by a specular reflection of light.

Mirror reflection in a desired direction can only be obtained from a suitably oriented flat metal surface, as one shown in Fig. 3.29.

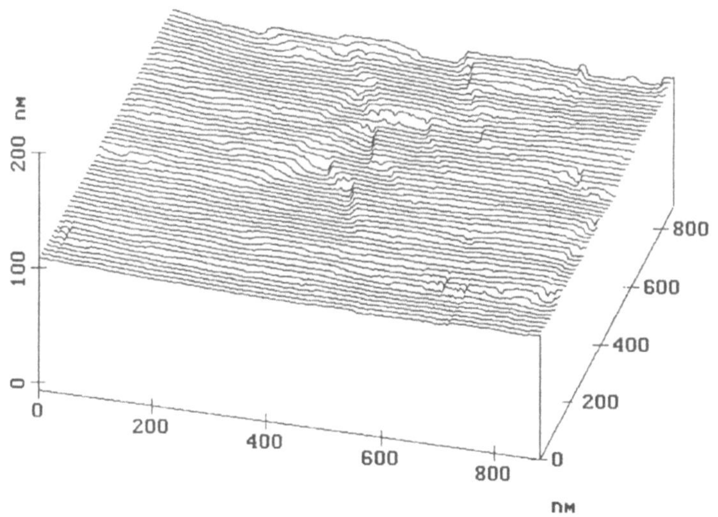

Figure 3.29. 3D STM image of a mechanically and electrochemically polished Cu surface.[59] (Reprinted with permission from the Serbian Chemical Society, Belgrade Yugoslavia).

In order to determine which structural features determine brightness, the flat parts of the profile, which were parallel to the base, were examined under different magnification. A part of the flat surface cross section was investigated first at lower magnifications and then at increasingly higher ones. Following this procedure Fig. 3.30 was obtained.

In the case of the mechanically polished surface, the amplitude of roughness was several atomic diameters of Cu (Figs. 3.30a and 3.30b) whereas the electrochemically polished surface exhibited smoothness on the atomic level (Figs. 3.30c and 3.30d). The increase in the specular reflection of 20–25% is due to this fact. It is interesting to note (Fig. 3.31) that the structure of the bright surface was not oriented. This is in accordance with the assumption that the dissolution process under polishing conditions is a random process.

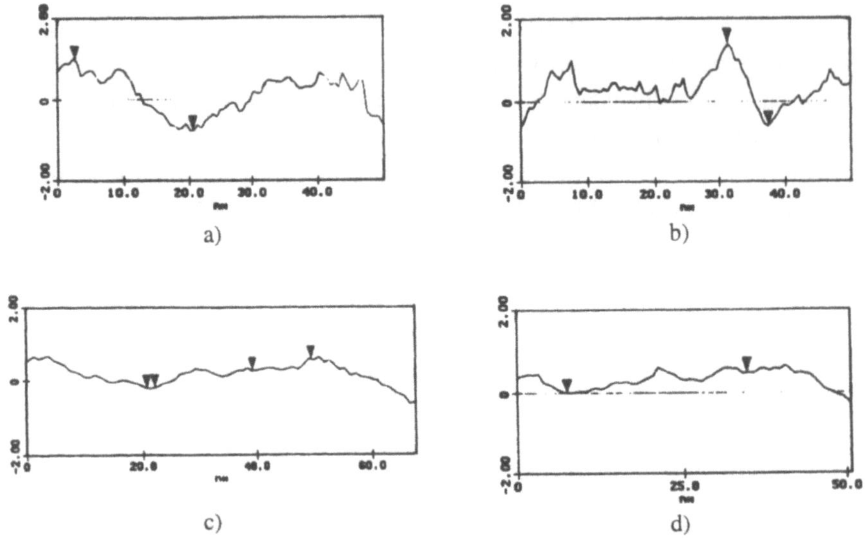

Figure 3.30. Typical STM images of the surface profile of the flat parts from a 50 x 50 nm^2 samples areas. Mechanically (a-b) and electrochemically (c-d) polished Cu surface. The vertical distances between the labels of the maximum and minimum points are: (a-b) $\Delta l = 2.0$ nm, (c-d) $\Delta l = 0.51$ nm. [59] (Reprinted with permission from the Serbian Chemical Society, Belgrade Yugoslavia).

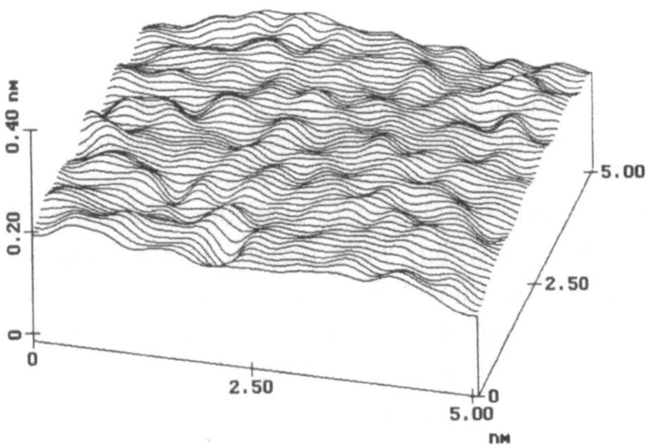

Figure 3.31. 3D STM image of a mechanically and electrochemically polished Cu surface. [59] (Reprinted with permission from the Serbian Chemical Society, Belgrade Yugoslavia).

Hence, it can be concluded that the condition for mirror brightness of a metal surface is smoothness on the atomic level of a suitably oriented flat part of the metal surface with low steps heights between, as in the case of a silver mirror.

3.2.3.3 Electrodeposited surfaces

The same situation appears in the case of bright electrodeposits as is illustrated in Fig. 3.32 and 3.33. The flat parts appear because the growth of the deposit in the vertical direction is suppressed by the adsorption of the additive[60] and mirror-like surface is formed.

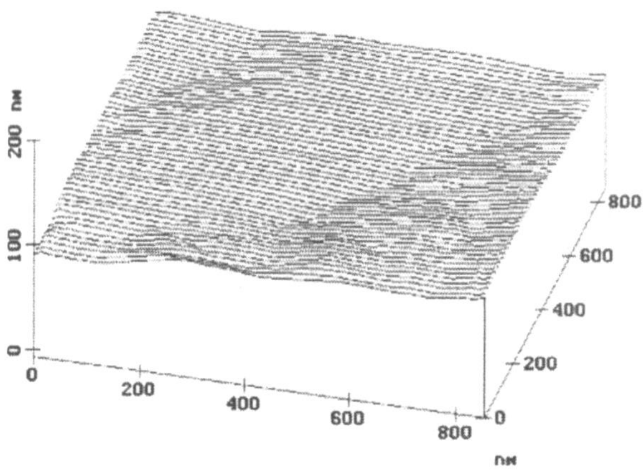

Figure 3.32. 3D STM image of electrodeposited bright copper. (Nikolić, Rakočević and Popov, unpublished results)

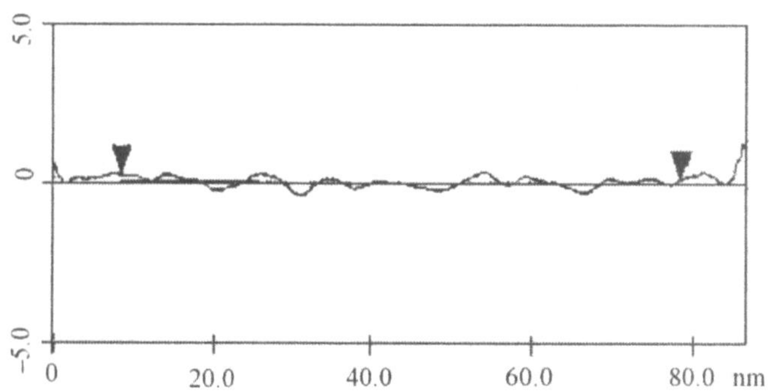

Figure 3.33. Typical surface profile of electrodeposited bright copper. (Nikolić, Rakočević and Popov, unpublished results)

This happens if the organic additive adsorbs strongly on the top of the copper crystallites and inhibits the formation of new growth centres. Deposition then occurs dominantly at the edges of the crystallite, resulting in lateral expansion of the layers, and the formation of smooth terraces on the atomic level. The growth of a new layer starts from defect arising at the interface between two adjanced crystallites and surfaces like those in Fig. 3.29 and 3.32 are formed.

3.2.3.4 Conclusions

Finally, it can be concluded that increasing the overpotential at one and the same current density leads to less coarse deposits with prolonged deposition, because of lower grains formed during the formation of the surface metal film. This conclusion is valid if $j_0 < j_L$. In the opposite case, a rough deposit is often not formed because of disperse or granular deposit formation. Hence, the first condition that must be satisfied in thick metal film deposition is that the exchange current density must be considerably lower than the limiting diffusion current density in the system under consideration. The second conditions is that the deposition current density must be a little larger than the one corresponding to the end of the Tafel linearity. In this way, the formation of large and well-defined crystal grains due to deposition under activation control will be prevented. At the same time, the increase of the surface coarseness due to the deposition in mixed activation-diffusion control will be minimal and the formation of carrot-like protrusion will be avoided.

On the other hand smooth and bright deposits can be obtained in the presence of organic additives only in the presence of additives which are incorporated into the deposits or undergo electrochemical reaction, or, in the presence of additives which are adsorbed on the flat part of the surfaces, permitting the deposition at the steps. In fact, bright deposits are obtained by the synergetic effect of the above two type of additives.

3.3 DISPERSE DEPOSITS

It was shown in the previous sections how compact electrodeposits are obtained during electrodeposition at low degrees of diffusion control. In condition close to complete diffusion control, disperse deposits are formed by the mechanisms discussed below.

3.3.1 Spongy deposits

3.3.1.1 Mathematical Model

It follows from Eq. 2.28

$$j = \frac{j_0(f_c - f_a)}{1 + \frac{j_0 f_c}{j_L}} \tag{2.28}$$

that deposition in systems with low exchange current densities comes under full diffusion control at sufficiently large overpotentials. On the other hand, if:

$$\frac{j_0}{j_L} >> 1 \tag{3.48}$$

deposition will be under complete diffusion control at all overpotentials if some other kind of control does not take place (e.g., for silver deposition on a well defined silver crystal grains at a silver electrode at low overpotentials two-dimensional nucleation is the rate-determining step[61]).

At low overpotentials a small number of nuclei are formed, and they can grow independently. The limiting diffusion current density to the growing nucleus $j_{L,N}$ is given by

$$j_{L,N} = \frac{nFDC_0}{r_N} \tag{3.49}$$

or

$$j_{L,N} = \frac{j_L \delta}{r_N} \tag{3.50}$$

where r_N is the tip radius of the nucleus. Hence, if $r_N \rightarrow 0$, the condition given by Eq. 3.48 is not satisfied and deposition is under activation or mixed control. Pure activation-controlled deposition is, thus, possible even at $j_0 >> j_L$ on very small electrodes such as nuclei on an inert substrate.

An increase in r_N leads to a decrease of $j_{L,N}$, and, at sufficiently large r_N, the deposition comes under mixed activation-diffusion control, i.e. when:

$$r_N > r_c \tag{3.51}$$

where r_c is the radius of a growing nucleus where the process comes under mixed control[9,32].

Under mixed control of the deposition, amplification of the surface irregularities on the growing nucleus occurs, leading to the formation of a spherical agglomerate of filaments. Thereby a spongy deposit is formed. The above reasoning is valid if spherical diffusion control can occur around growing grains, as in the case of cauliflower deposit growth. Assuming that around each grain with radius r_N, growing under spherical diffusion control, a diffusion layer of the same thickness is formed, then the initiation of spongy growth is possible if the number of nuclei per square centimetre, N, satisfies the condition

$$N \leq \frac{1}{(4r_c)^2} \tag{3.52}$$

On the basis of all the above facts, it can be concluded that the formation of a spongy deposit on an inert substrate may be caused by mass-transport limitations when the nucleation rate is low. Hence, suitable conditions for the formation of spongy deposits arise at low overpotentials in systems where $j_L < j_0$[9,32].

The validity of the condition, given by Eq. 3.48 can be easily tested using $CdSO_4$ solutions of widely varying concentrations. The j_0 values for cadmium deposition from sulfate solutions are estimated as 1 mA cm^{-2} for 5×10^{-3} mol dm^{-3} $CdSO_4$ (solution 1) and 10 mA cm^{-2} for 1 mol dm^{-3} $CdSO_4$ (solution 2). The corresponding limiting diffusion currents can be assumed to be 0.5 mA cm^{-2} and 100 mA cm^{-2}, respectively. Hence, $(j_0/j_L)_1 = 2$ and, $(j_0/j_L)_2 = 0.1$, where the subscripts 1 and 2 correspond to solutions 1 and 2. The cadmium deposit obtained from solution 1 is spongy as can be seen in Fig. 3.34a, meaning that the conditions given by Eq. 3.48 and Eqs. 3.51 and 3.52 are all satisfied. In the case of solution 2, Eq. 3.48 is not satisfied and so suitable conditions for the formation of a spongy deposit are not given. Hence, the grains grow under pure activation control until a complete surface film is formed, as illustrated in Fig. 3.34b, c[9,32].

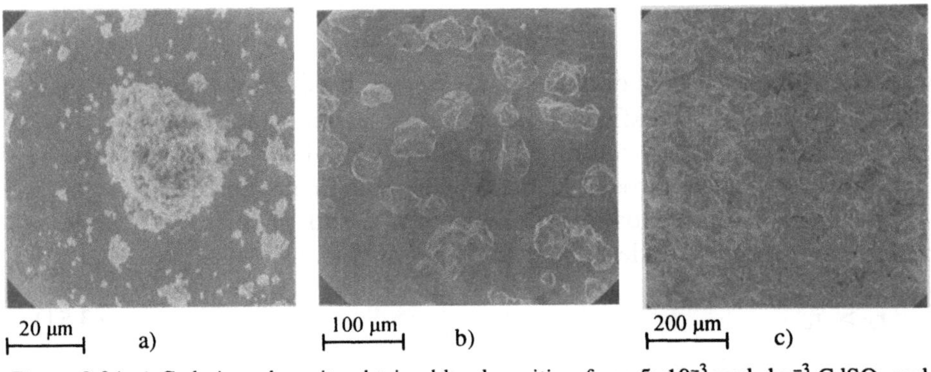

Figure 3.34. a) Cadmium deposits obtained by deposition from 5×10^{-3} mol dm^{-3} $CdSO_4$ and 0.4 mol dm^{-3} K_2SO_4 in 0.1 mol dm^{-3} H_2SO_4 solution, deposition overpotential. 10 mV; deposition time, 10 h. b) and c) cadmium deposits obtained by deposition from 1.0 mol dm^{-3} $CdSO_4$ in 0.5 mol dm^{-3} H_2SO_4 solution, deposition overpotential, 10 mV; deposition time: b) 24 min, c) 10 h. The substrate was a copper plane electrode.[9]

3.3.1.2 Physical Model

As was mentioned before, at a fixed value of the overpotential, the growth of a spongy deposit is possible if:

$$r_N > r_c \tag{3.51}$$

The situation in which spongy deposits can start to grow can easily be demonstrated[62]. Grains of the desired size and distribution can be grown at low overpotentials under conditions of activation-controlled deposition. This corresponds to growth of grains when $r < r_c$. The situation in which $r > r_c$ can be simulated by increasing the overpotential to a sufficiently high value to result in diffusion control around the growing grains and the amplification of surface irregularities. With increasing overpotential r_c decreases. This permits the simulation of the initial stage of spongy growth, as is illustrated in Fig. 3.35.

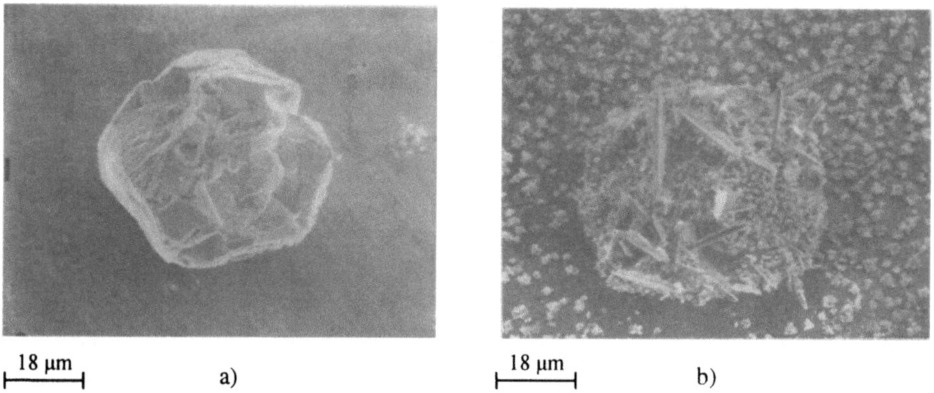

Figure 3.35. Cadmium deposits: a) from 1.0 mol dm^{-3} $CdSO_4$ in 0.5 mol dm^{-3} H_2S0_4 solution at 12 mV on a copper plane electrode; deposition time, 15 min; b) from 0.1 mol dm^{-3} $CdSO_4$ in 0.5 mol dm^{-3} H_2S0_4 solution at 120 mV on the substrate from a); deposition time, 45 s[62].

The growth of protrusions in all directions is a good proof that the deposition on the grain is under spherical diffusion control. At longer deposition times, the protrusions branch and interweave, as is shown in Fig. 3.36, causing the macroelectrode to have a spongy appearance.

Figure 3.36. Cadmium deposits obtained under the same deposition conditions as in Fig. 3.35b but after a deposition time of 120 s.[62]

3.3.1.3 Real Systems

Typical spongy electrodeposits are formed during zinc and cadmium electrodeposition at low overpotentials[9,32]. Scanning electron microscopy images of zinc deposited at an overpotental of 20 mV onto a copper electrode from an alkaline zincate solution are shown in Fig. 3.37.

The increase in the number of nuclei formed with increasing deposition time can be seen in Fig. 3.37a and b, and a spongy deposit is formed as can be seen in Fig. 3.37b. The spongy growth takes place on a relatively small number of nuclei, as is shown in Fig. 3.37b and c.

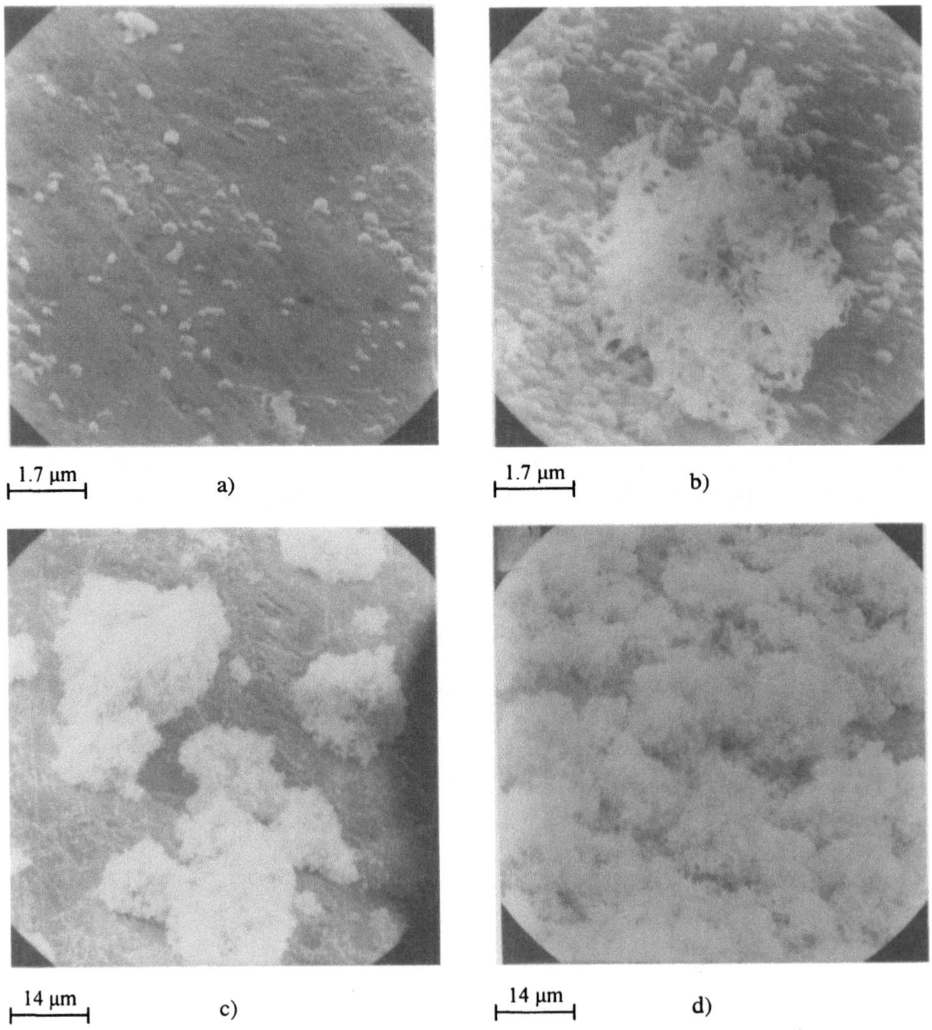

Figure 3.37. Zinc deposits obtained by deposition at 20 mV from 0.1 mol dm^{-3} zincate and 1.0 mol dm^{-3} KOH solution. Deposition time: a) 10 min. b) 20 min. c) 30 min. d) 60 min. The substrate is a copper plane electrode.[9]

The initiation of spongy growth at a fixed overpotential is possible if the condition $r_N > r_c$ (Eq. 3.51) is satisfied, which is the case after some time. On the other hand, increasing the deposition time leads to the formation of a larger number of nuclei, and so the condition given by Eq. 3.52 is not satisfied over a large part of the electrode surface. Regardless of this, the coverage of the electrode surface by spongy deposits increases with increasing deposition time up to full coverage, as can be seen in Fig. 3.37d, in the same way as was illustrated earlier by Fig. 3.36.

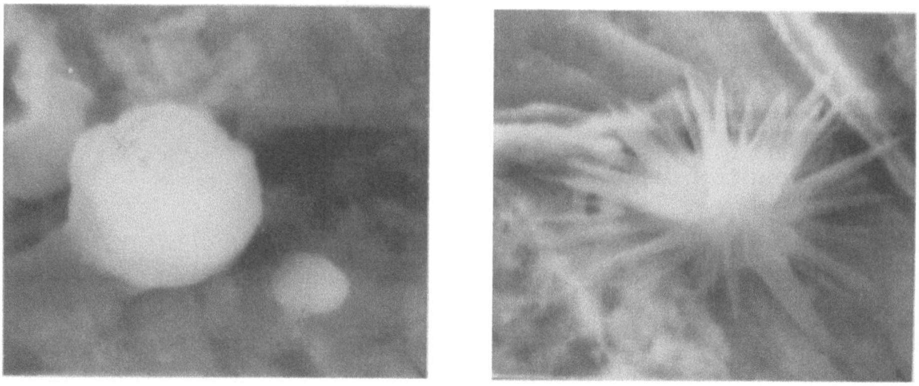

0.8 μm a) 1.3 μm b)

Figure 3.38. Zinc deposits obtained by deposition at 35 mV from 0.1 mol dm^{-3} zincate solution in 1.0 mol dm^{-3} KOH solution. Deposition time: a) 7 min. b) 15 min. The substrate is a copper plane electrode.[63] (Reprinted with permission from the Serbian Chemical Society, Belgrade, Yugoslavia.)

Spongy growth can start on the growing nucleus if the conditions given by Eqs. 3.51 and 3.52 are both satisfied simultaneously.

In the first stage of deposition, the formation of nuclei having a regular crystal shape can be expected because the deposition is activation-controlled. After r_c is reached, the system comes under mixed control, producing polycrystalline grains like those shown in Fig. 3.38a, just as in the case of mixed control of copper deposition[39], Fig. 3.21c. In this situation, amplification of the surface irregularities on the growing grains occurs, and spongy growth is initiated.

An ideal spongy nucleus obtained in a real system is shown in Fig. 3.38b which illustrates the above discussion and physical simulation well[63]. The agglomerate of filaments in Fig. 3.37b is obviously formed by further growth of nuclei like that in Fig. 3.38b.

Hence, it can be concluded that at low overpotentials the initiation of spongy growth is due to the amplification of surface protrusions directly inside the spherical diffusion layer formed around each independently growing grain, as in the case of the formation of cauliflower deposits. The growth of protrusions in all directions is good proof that the initial stage of deposition on the grain is under spherical diffusion control, while further growth takes place in the diffusion layer of the macroelectrode. In less ideal situations, non-ideal spongy nuclei are formed, which, however, after further deposition result in a macroelectrode with the same appearance.

It should be noted that some other possible mechanisms of spongy deposit formation have been considered in a qualitative way, as reviewed in Refs. 64 and 65, but the mechanism presented above seems to be the most probable[65]. However, the mechanism of formation of a spongy deposit over an initial coating, which is not seen in the case of cadmium but occurs in zinc deposition[9,63] requires clarification. For instance, the mechanism of spongy growth initiation in this case has not been elucidated.

3.3.2 Dendritic deposits

3.3.2.1 Mathematical Model

Two phenomena seem to distinguish dendritic from carrot-like growth[31,32]:

1. a certain well-defined critical overpotential value appears to exist below which dendrites do not grow,
2. dendrites exhibit a highly ordered structure and grow and branch in well-defined directions. According to Wranglen[66], a dendrite is a skeleton of a monocrystal and consists of a stalk and branches, thereby resembling a tree.

It is known that dendritic growth occurs selectively at three types of growth sites[31]:

1. dendritic growth occurs at screw dislocations. Sword-like dendrites with pyramidal tips are formed by this process[31,36].
2. many investigations of the crystallographic properties of dendrites have reported the existence of twin structures[67-69]. In the twinning process, a so-called indestructible re-entrant groove is formed. Repeated one-dimensional nucleation in the groove is sufficient to provide for growth extending in the direction defined by the bisector of the angle between the twin plants[31,32].
3. it is a particular feature of a hexagonal close-packed lattice that growth along a high-index axis does not lead to the formation of low index planes. Grooves containing planes are perpetuated, and so is the chance for extended growth by the one-dimensional nucleation mechanism[70].

In all the above cases, the adatoms are incorporated into the lattice by repeated one-dimensional nucleation. On the other hand deposition to the tip of screw dislocations can be theoretically considered as deposition to a point; in the other two cases, the deposition is to a line.

From the electrochemical point of view, a dendrite can be defined as an electrode surface protrusion that grows under activation control, while deposition to the macroelectrode is predominantly under diffusion control[31,32,36]. The polarisation curve for a flat macroelectrode is given by Eq. 2.28

$$j = \frac{j_0(f_c - f_a)}{1 + \dfrac{j_0 f_c}{j_L}} \tag{2.28}$$

and that for the tip of a protrusion growing inside the macroelectrode diffusion layer is given by Eq. 3.25

$$j_{tip} = \frac{j_0(f_c - f_a)}{1 + \left(\dfrac{j_0 f_c}{j_L}\right)\left(\dfrac{r}{r+h}\right)} \tag{3.25}$$

It follows from Eq. 2.28 that $j = j_L$ if $j_0 f_c / j_L >> 1$ and $f_c >> f_a$. On the other hand, if $h >> r$ and $r/h << j_0 f_c / j_L$ activation-controlled deposition to the tip of the protrusion takes place, and Eq. 3.25 can be rewritten in the form:

$$j_{tip} = j_{0,tip} f_c \tag{3.53}$$

where $j_{0,tip}$ is the corrected value of the exchange current density. This is so because if $r/h << 1$, Eq. 3.22

$$C^* = C_0 \frac{h+r}{\delta} \tag{3.22}$$

can be rewritten in the form

$$C_{tip} = C_0 \frac{h}{\delta} \tag{3.54}$$

where C_{tip} is the concentration at the tip of a protrusion growing under activation control inside the diffusion layer of the macroclectrode if the deposition to it is under full diffusion control[36]. A similar situation can arise if cylindrical diffusion around the tip of a growing protrusion occurrs[71].

The exchange current density $j_{0,tip}$ at the tip of such a protrusion is given by[72]:

$$j_{0,\text{tip}} = \left(\frac{C_{\text{tip}}}{C_0}\right)^{\gamma} j_0 \tag{3.55}$$

or, taking into account Eq. 3.54,

$$j_{0,\text{tip}} = \left(\frac{h}{\delta}\right)^{\gamma} j_0 \tag{3.56}$$

where γ is a function of the symmetry factor β and j_0 is the exchange current density corresponding to the bulk concentration of the depositing ion. Assuming for the sake of simplicity that γ is approximately 1, Eq. 3.53 can then be rewritten in the form:

$$j_{\text{tip}} = j_0 \frac{h}{\delta} f_{\text{c}} \tag{3.57}$$

Dendrites grow faster than the flat electrode surface, and the condition for the initiation of dendritic growth is:

$$j_{\text{L}} \leq j_0 \frac{h}{\delta} f_{\text{crt}} \tag{3.58}$$

where f_{crt} corresponds to η_{crt}, the critical overpotential at which initiation of dendritic growth from the tip of a growing protrusion, whose height is h, is instantaneously possible after the steady-state concentration distribution inside the diffusion layer of the macroelectrode has been reached. Taking into account the relationship:

$$f_{\text{c}} = \exp \frac{2.3\eta}{b_{\text{c}}} \tag{2.19}$$

Eq. 3.58 can be rewritten in the form:

$$\eta_{\text{crt}} = \frac{b_{\text{c}}}{2.3} \ln\left(\frac{j_{\text{L}}}{j_0} \frac{\delta}{h}\right) \tag{3.59}$$

The minimum overpotential at which dendritic growth can be initiated, η_{i}, corresponds to $h=\delta$, and Eq. 3.59 becomes:

$$\eta_i = \frac{b_c}{2.3} \ln \frac{j_L}{j_0} \tag{3.60}$$

A different situation arises if $j_0 f_c / j_L >> 1$ and $r/h << j_0 f_c / j_L$ but $f_c > f_a$. The current density to the tip of the protrusion is then given by:

$$j = j_{0,tip}(f_c - f_a) \tag{3.61}$$

for $h=\delta$, $j_{0,tip}=j_0$, while the diffusion current density to the macroelectrode is given by

$$j_d = j_L\left(1 - \frac{f_c}{f_a}\right) < j_L \tag{3.62}$$

Assuming that j_L can be used instead of j_d and following the same reasoning as in the derivation of Eqs. 3.58 and 3.60, one obtains:

$$\eta_i = \frac{RT}{nF}\frac{j_L}{j_0} \tag{3.63}$$

It should be noted that all the above derivations are valid, if the protrusion tip radius is sufficiently large to make the effect of the surface energy term negligible[37], and if Eqn. 2.32 is also valid.

It follows from Eq. 3.63 that for systems with $j_0 \to \infty$, dendritic growth is possible at all overpotentials. Experimentally, some critical overpotential of dendritic growth initiation exists in all cases, being of the order of 10 mV[37,73,74]. Assuming that under complete diffusion and surface energy control ($j_0 \to \infty$,) the current density to the macroclectrode is given by[37]:

$$j = \eta \frac{(nF)^2 DC_0}{\delta RT} \tag{3.64}$$

and, assuming that Eq. 3.54 is valid, the current density on the tip of a dendrite growing inside the diffusion layer of a macroelectrode is given by:

$$j_d = \eta^2 \frac{(nF)^3 DC_0}{8\sigma VRT}\frac{h}{\delta} \tag{3.65}$$

then, it is possible to derive the relationships:

$$\eta_{crt} = \frac{8\sigma V}{nFh} \tag{3.66}$$

and

$$\eta_{i} = \frac{8\sigma V}{nF\delta} \tag{3.67}$$

using the same procedure as in the derivation of Eqs. 3.59 and 3.60[75].

3.3.2.2 Physical Simulation

The cross sections of the copper deposits obtained in a model system, described earlier (Fig. 3.18), are shown in Fig. 3.39.

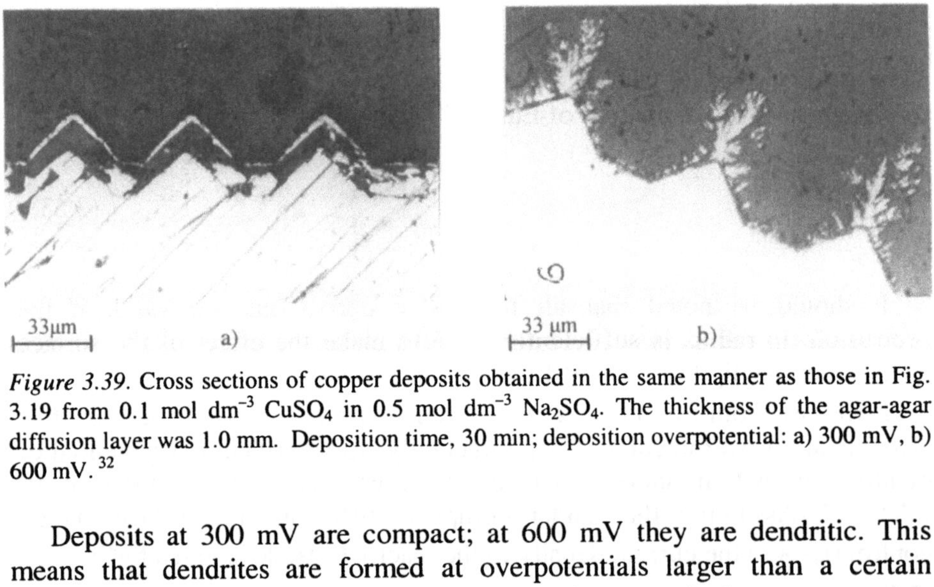

Figure 3.39. Cross sections of copper deposits obtained in the same manner as those in Fig. 3.19 from 0.1 mol dm^{-3} $CuSO_4$ in 0.5 mol dm^{-3} Na_2SO_4. The thickness of the agar-agar diffusion layer was 1.0 mm. Deposition time, 30 min; deposition overpotential: a) 300 mV, b) 600 mV.[32]

Deposits at 300 mV are compact; at 600 mV they are dendritic. This means that dendrites are formed at overpotentials larger than a certain critical value, as required by Eq. 3.60, because both overpotentials correspond to the plateau of the limiting diffusion current. It is seen that the current density to the tips of dendrites depends on the h/δ ratio (see Eq. 3.57), so that larger dendrites are produced at more elevated points of the electrode surface. This is because the effective height of the dendrite precursor in the modelled diffusion layer is equal to the sum of the height of the precursor and the height of the point at which nucleation took place relative to the flat part of the electrode surface. In the same way, for nuclei formed on the tip of a protrusion (Fig. 3.39b), η_{crt}, (see Eq. 3.59) is lower than for those formed on the flat surface, and a dendrite is formed at the tip of the protrusion while at the same overpotential dendrites are not formed on the flat part of the electrode.

The validity of Eq. 3.60 can be qualitatively tested by using the same solutions (1 and 2) as where used for the examination of spongy deposit formation. In this way, different j_L/j_0 ratios for the same deposition process can be obtained, while the surface energy and the crystallographic properties of the metal are kept the same. As expected, because of the lower j_L/j_0 ratio, dendrites appear at lower overpotentials from the more dilute solution than from the more concentrated one. This is illustrated in Fig 3.40.

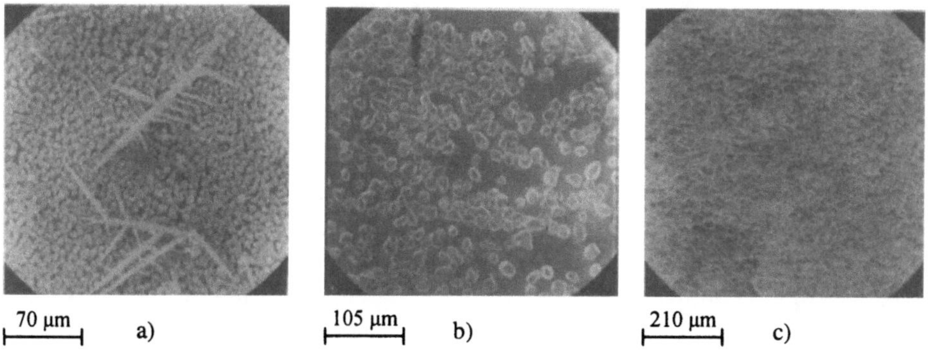

Figure 3.40. Cadmium deposits. a) From 5×10^{-3} mol dm^{-3} $CdSO_4$ in 0.4 mol dm^{-3} K_2SO_4 + 0.1 mol dm^{-3} H_2SO_4 solution. Deposition overpotential 40 mV; deposition time, 2 h: b) From 1.0 mol dm^{-3} $CdSO_4$ in 0.5 mol dm^{-3} H_2SO_4 solution. Deposition overpotential, 40 mV; deposition time, 4 min. c) Same conditions as in b) except the deposition time was 2 h. The substrate is a copper plane electrode.[9]

3.3.2.3 Real Systems

There is an induction period before the initiation of dendritic growth[31,32,37]. During this induction period, dendrite precursors are formed and become sufficiently high to satisfy Eq. 3.59 at a given overpotential, as illustrated in Fig. 3.41. The crosslike grains seen in Fig. 3.41a and b further develop into dendrite precursors (Fig. 3.41a, c).

The propagation of this structure by branching (Fig. 3.41d) produces dendrites as shown in Fig. 3.41e.

The initiation of dendritic growth is followed by a change in the slope of the current density-time curves[31,32], indicating a change in the growth mechanism of the deposit.

The slopes of these dependences are similar to each other and independent of the deposition overpotential during the non-dendritic amplification of the surface-coarsening.

The change of the slope of the current-time dependences due to the dendritic growth initiation will be treated here in somewhat simplified way.

The limiting diffusion current density to the tip of a surface protrusion, $j_{L,tip}$, is given by[75]:

Figure 3.41. SEM micrographs of copper deposits obtained by deposition from 0.3 mol dm^{-3} $CuSO_4$ in 0.5 mol dm^{-2} H_2SO_4 onto a copper wire electrode. Deposition overpotential, 550 mV. Quantity of electricity a) 2 mAh cm-^{2}; b) 2 mAh cm^{-2}; c) 5 mAh cm^{-2}; d) 10 mAh cm^{-2}; e) 10 mAh cm^{-2}.[76] (Reprinted with permission from the Serbian Chemical Society, Belgrade, Yugoslavia).

$$j_{\mathrm{L,tip}} = \frac{nFDC_0}{\delta - h} \tag{3.68}$$

if the spherical flux around the tip can be neglected, and:

$$j_{\mathrm{L}} = \frac{nFDC_0}{\delta} \tag{2.29}$$

to the flat part of the electrode.

Differentiation of Eq. 3.68 gives:

$$\frac{\mathrm{d}j_{\mathrm{L,tip}}}{\mathrm{d}t} = \frac{nFDC_0}{(\delta - h)^2}\frac{\mathrm{d}h}{\mathrm{d}t} = \frac{j_{\mathrm{L,tip}}}{\delta}\frac{\mathrm{d}h}{\mathrm{d}t} \tag{3.69}$$

and as in Eqn. 3.27

$$\frac{\mathrm{d}h}{\mathrm{d}t} = \frac{V}{nF}(j_{\mathrm{L,tip}} - j_{\mathrm{L}}) = \frac{VDC_0}{\delta^2}h \tag{3.70}$$

taking into account Eqs. 3.68 and 2.29 if $\delta >> h$.

Substitution of $\mathrm{d}h/\mathrm{d}t$ from Eqn. 3.70 in Eqn. 3.69 produces

$$\frac{\mathrm{d}(\ln j_{\mathrm{L,tip}})}{\mathrm{d}t} = \frac{VDC_0}{\delta^3}h \tag{3.71}$$

being independent on overpotential.

After initiation of dendritic growth, the slopes become dependent on the overpotential. A dendrite is a surface protrusion growing under mixed or activation control, while deposition to the flat part of the electrode surface is under complete diffusion control. The overpotential η and current density j_{tip} on the tip of a dendrite are related by:

$$j_{\mathrm{tip}} = j_0 \frac{h}{\delta} f_{\mathrm{c}} \tag{3.57}$$

Differentiation of Eq. 3.57 produces:

$$\frac{\mathrm{d}\,j_{\mathrm{tip}}}{\mathrm{d}t} = \frac{j_0}{\delta} f_{\mathrm{c}} \frac{\mathrm{d}h}{\mathrm{d}t} \tag{3.72}$$

and as in the derivation of Eqn. 3.71

$$\frac{\mathrm{d}h}{\mathrm{d}t} = \frac{V}{nF}(j_{\mathrm{L,tip}} - j_{\mathrm{L}}) \approx \frac{V}{nF} j_{\mathrm{L,tip}} \tag{3.73}$$

and

$$\frac{d(\ln j_{tip})}{dt} = \frac{V}{nF}\left(j_0\frac{f_c}{\delta}\right)^2 \tag{3.74}$$

Hence, the maximum overpotential at which the slope of the apparent current density-time dependence remains constant and equal to that in nondendritic amplification of the surface-roughness corresponds to η_i. The minimum overpotential at which this slope cannot be recorded corresponds to η_{crt}.

In this way η_i and η_{crt} can be estimated. It is known that the *j-t* dependence are different from case to case owing to different mechanisms of dendritic growth initiation and dendritic growth[31,32]. As a result of this, the analytical approach to the determination of η_i and η_{crt} must be specific for each system under consideration; the procedure for one particular case is as follows.

Typical log (current)-time dependences obtained for copper deposition from 0.2 mol dm^{-3} $CuSO_4$ in 0.5 mol dm^{-3} H_2SO_4 at overpotentials belonging to the limiting diffusion current plateau are shown in Fig. 3.42. According to the above discussion, it is clear that the intersection points of the two linear dependencies determines the induction time of dendritic growth initiation[75].

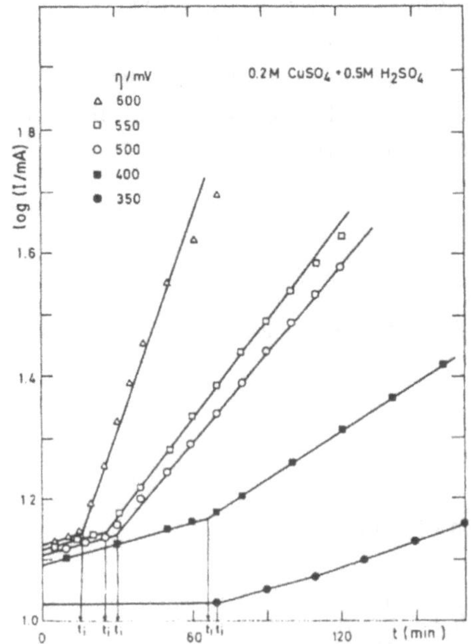

Figure 3.42. log *I* as a function of time for copper deposition.[75]

The induction times for dendritic growth initiation extracted from the graphs in Fig. 3.42 can be presented as a function of overpotential, and the critical overpotential for instantaneous dendritic growth can be obtained by extrapolation to zero induction time.

The critical overpotential of dendritic growth initiation can be determined by plotting the logarithm of the slopes of the straight lines from Fig. 3.42 as a function of overpotential and the intersection point of the two straight lines determines η_i. A similar procedure was followed for the deposition of cadmium from 0.1 mol dm^{-3} $CdSO_4$ in 0.5 mol dm^{-3} H_2SO_4.

The cross sections of the copper and cadmium deposits obtained at $\eta < \eta_i$, $\eta_i < \eta < \eta_{crt}$, and $\eta > \eta_{crt}$ are shown in Figs. 3.43a-c and 3.44a-c , respectively. It can be seen that there is no dendrite formation when $\eta < \eta_i$, both compact and dendritic deposits are formed when $\eta_i < \eta < \eta_{crt}$ and only dendritic metal is deposited when $\eta > \eta_{crt}$. This is in perfect agreement with findings of Calusaru[77] for the morphology of deposits of the same metals deposited at overpotentials corresponding to full diffusion control.

The η_i and η_{crt} of 260 mV and 660 mV for copper deposition (lower j_0 value) and 27 mV and 110 mV for cadmium deposition (larger j_0 value), are successfully determined using the above given procedure, being in perfect agreement with experimental findings as can bee seen from Figs 3.43 and 3.44.[75]

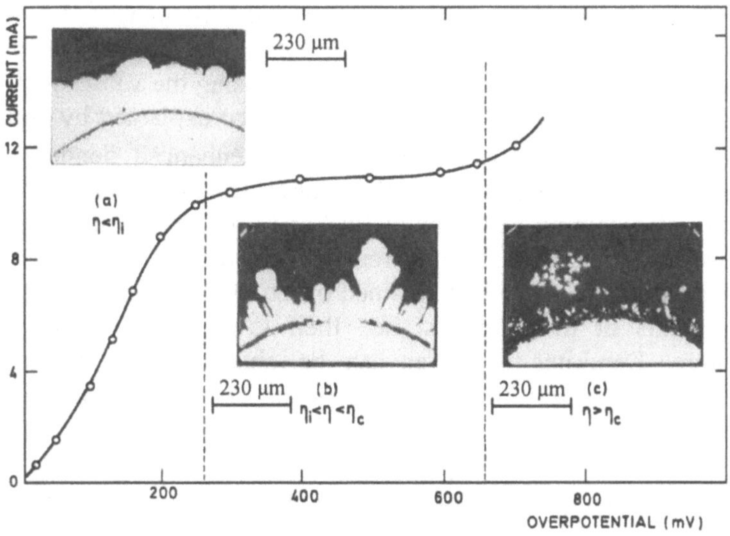

Figure 3.43. Polarisation curve for the potentiostatic deposition of copper from 0.2 mol dm^{-3} $CuSO_4$ in 0.5 mol dm^{-3} H_2SO_4 and the cross sections of copper deposits obtained on copper wire electrodes previously plated with nickel. a) overpotential 200 mV, deposition time 6 hours; b) overpotential 300 mV, deposition time 5 hours; c) overpotential 700 mV, deposition time 2 min.l[75].

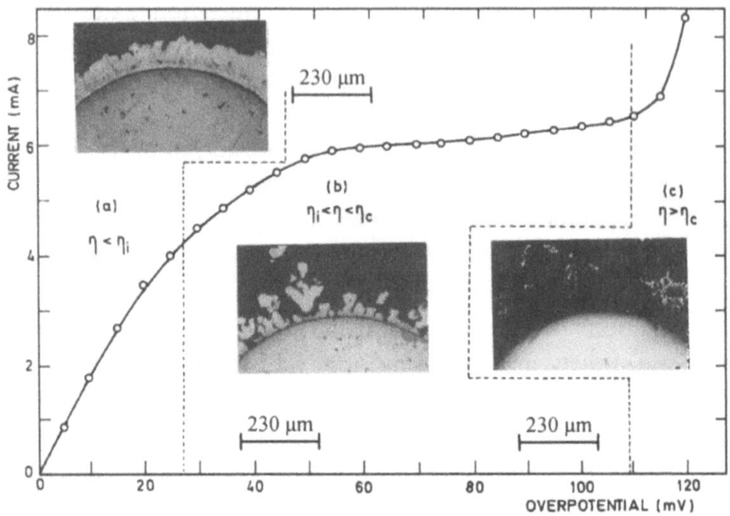

Figure 3.44. Polarisation curve for the potentiostatic deposition of cadmium from 0.1 mol dm^{-3} $CdSO_4$ in 0.5 mol dm^{-3} H_2SO_4 and the cross section of cadmium deposits obtained on copper wire electrode a) overpotential 20 mV, deposition time 8 hours; b) overpotential 40 mV, deposition time 2 hours; c) overpotential 120 mV, deposition time 9 min. [75]

It is known[78] that, apart from decreasing the concentration of the depositing ion, the formation of a dendritic deposit can also be enhanced by increasing the concentration of the supporting electrolyte, increasing the viscosity of the solution, decreasing the temperature, and decreasing the velocity of motion of the solution. Practically, all the above facts can be explained by Eqs. 3.60 and 3.63, assuming that a decrease in η_i means enhanced dendrite formation because of the lower electrical work required to produce the dendrites. The possibility of obtaining dendrites of Pb[73] and Sn[74] from aqueous solutions at lower overpotentials than required for the formation of dendrites of Ag from aqueous solutions can also be explained by Eq. 3.67 owing to the much lower melting points of these metals, i.e., their lower surface energy at room temperature. Dendrites of silver can be obtained from molten salts at overpotentials of a few millivolts[37], as in the case of Pb and Sn deposition from aqueous solutions[73,74], because the difference between the melting point of silver and the working temperature for deposition from molten salts is not very different from the difference between the melting point of lead or tin and room temperature. On the other hand, dendrites grow from screw dislocation and nuclei of higher indices or twinned ones only[31,32]. The probability of formation of such nuclei increases with increasing overpotential[79] and η_i, can also be defined as the overpotential at which they are formed. Regardless of this, Eqs. 3.60, 3.63 and 3.67 illustrate well the effect of different parameters on the initiation of dendritic growth.

It is obvious that the electrochemical conditions, as well as the crystallographic ones, under which dendritic deposits are formed can be precisely determined. One problem that still seems to remain unresolved is the question of what causes the dendrite precursors to appear at regularly spaced locations along the dendrite stem. Further investigations in this direction are necessary.[31]

3.3.3 Powdered deposits

A metal powder represents a dendritic deposit which can spontaneously fall or can be removed from the electrode by tapping or in a similar way.

All metals, which can be electrodeposited, exhibit a tendency to appear in the form of powders at current densities larger than a certain critical value j_p. This value is equal to the limiting diffusion current density in galvanostatic deposition, as was shown by Ibl[80]. Simultaneously it was observed that the product of the employed current density and the square root of the time of powder formation t_p is a constant quantity[31]. Such dependencies are characteristic for processes controlled by diffusion and the time of powder formation coincides with the transition time. The time for powder formation at current densities equal to or larger than j_p can be observed visually as the appearance of the electrode is seen to turn suddenly from lustrous to black.

It is known that increasing the overpotential leads to the formation of a more dispersed deposit characterised by decreased particle size, even at the same initial current density (and real current density in potentiostatic deposition) because increasing the overpotential means the increasing the electrical work, thus a powder with larger specific surface area is produced.

This is illustrated in Fig. 3.45, where copper particles obtained at different overpotentials are presented[81].

In the same way, the differences in the grain size of the powder particles of different metals can be explained assuming that their surface energies are similar. It can be seen that an increase in b_c and the j_L/j_0 ratio leads to an increase in η_i and, hence a decrease in the grain size of powder particles can be expected as is illustrated in Fig. 3.45a and Fig. 3.46a. In the same way, the different grain sizes of the same metal powder particles but obtained from different electrolytes can be explained, as is demonstrated in Fig. 3.46. It was shown earlier[11], that deposition of Ag from 0. 1 mol dm^{-3} $AgNO_3$ in 0.5 mol dm^{-3} HNO_3 is characterised by $j_0 >> j_L$, and the deposition of silver from 0.1 mol dm^{-3} $AgNO_3$ in 0.5 mol dm^{-3} $(NH_4)_2SO_4$ is characterized by $j_0 << j_L$, as in the case of copper. It is also noteworthy that in soft metal (low melting points) powder deposition agglomerates are formed due to the plasticity of the growing dendrites, as can be seen in Fig. 3.47.

The effect of deposition conditions on the grain size of powder particles can not be discussed using Eqs. 3.59 and 3.60 alone. Despite this, in all cases increasing the overpotential leads to the formation of smaller particles and to a narrower particle–size distribution curve. It was shown that changing concen-

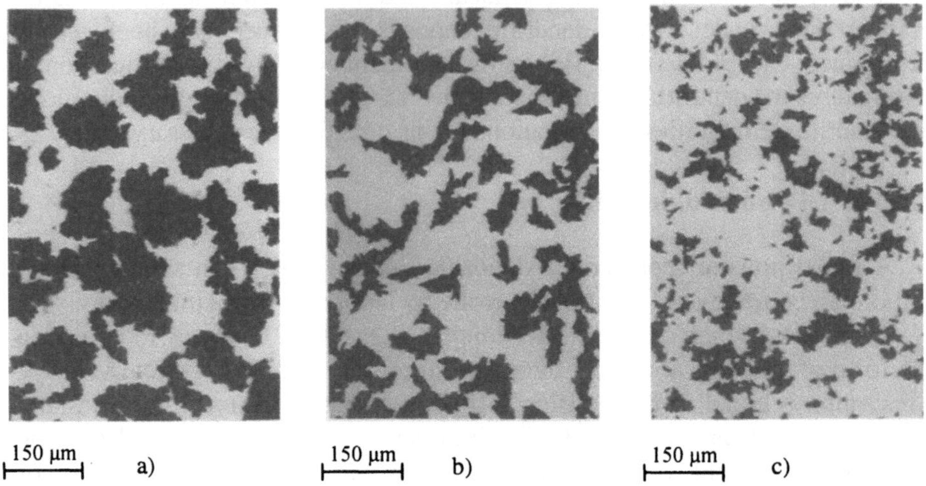

Figure 3.45. Copper powder particles obtained at different constant overpotentials from 0.1 mol dm^{-3} $CuSO_4$ in 0.5 mol dm^{-3} H_2SO_4 onto stationary platinum wire electrodes painted with shellac. Deposition time 15 min. at room temperature at deposition overpotentials of: a) 500 mV, b) 600 mV, c) 700 mV.[81]

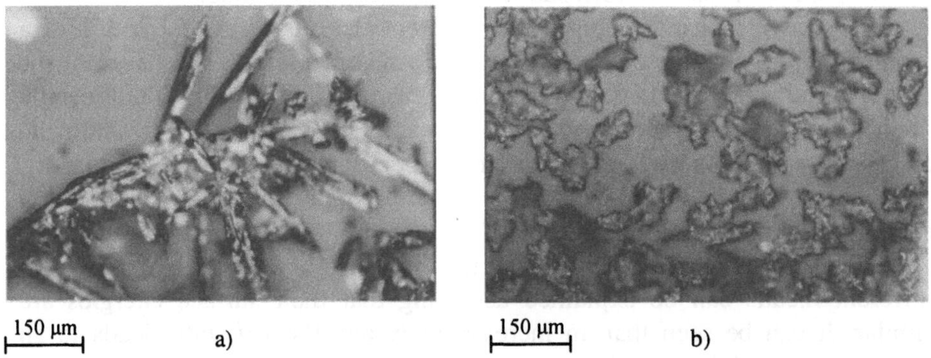

Figure 3.46. a) Silver powder particles obtained at a constant overpotential of 150 mV from 0.1 mol dm^{-3} $AgNO_3$ in 0.5 mol dm^{-3} HNO_3. Deposition time 15 min. b) Silver powder particles obtained at a constant overpotential of 600 mV from 0.1 mol dm^{-3} $AgNO_3$ in 0.5 mol dm^{-3} $(NH_4)_2SO_4$. Depositions time 15 min. The depositions were carried out on platinum electrodes at room temperature.[82] (Reprinted with permission from ISE).

tration of the electrolyte[83] or the stirring rate[84] do not affect appreciably the η_i and η_{crt} values. Also, increasing the temperature leads to an increase in both j_L and j_0, as dues increasing the concentration, and a significant effect on the value of η_i and η_{crt}, is not to be expected. In these cases, however, deposition at a similar overpotential means deposition at very different deposition current densities. Consequently, an increase in the particles grain size is to be expected, for the same deposition time, with increasing concentration, temperature, and decreasing concentration of the supporting electrolyte, as is illustrated in Fig. 3.48. Stirring, according to Pavlović et al[84] has the same effect as increasing the concentration, as well as increasing the deposition time.

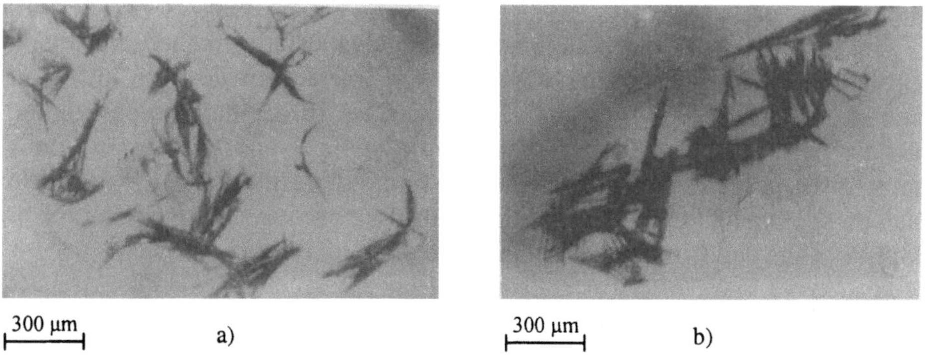

Figure 3.47. Tin powder particles obtained at a constant-overpotential from an electrolyte containing 20 g dm^{-3}, $SnCl_2 \times 2H_2O$ and 250 g dm^{-3} NaOH on a copper wire electrode at a deposition overpotential of 100 mV.[74] (Reprinted with permission from Elsevier Science).

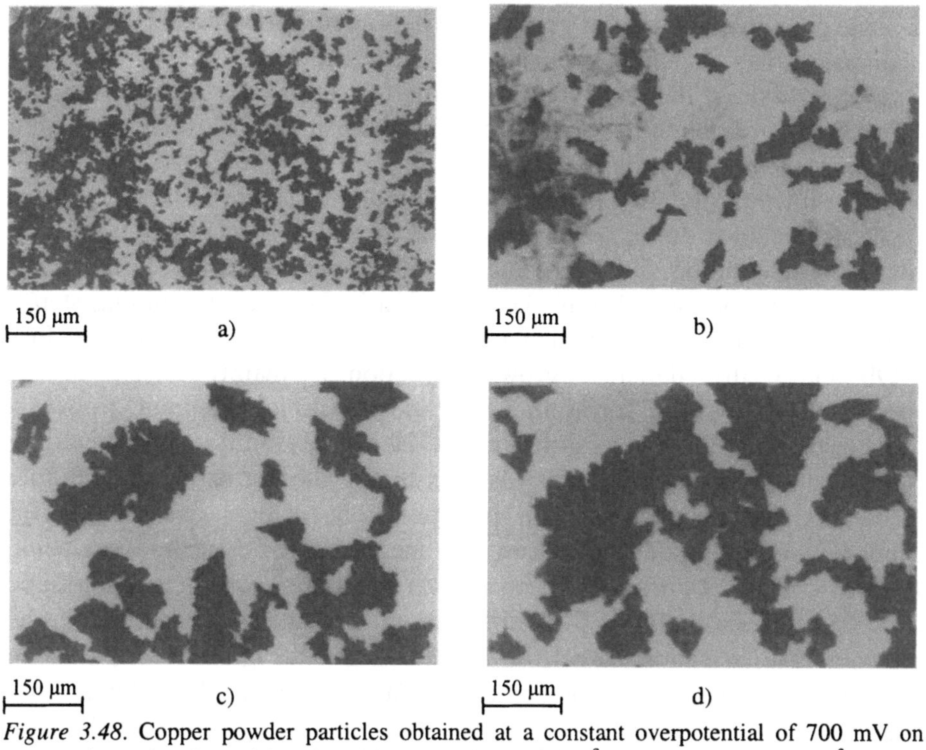

Figure 3.48. Copper powder particles obtained at a constant overpotential of 700 mV on copper electrodes. Deposition time 15 min. a) 0.1 mol dm^{-3} $CuSO_4$ in 0.5 mol dm^{-3} H_2SO_4, temperature 25° C; b) 0.1 mol dm^{-3} $CuSO_4$ in 0.1 mol dm^{-3} H_2SO_4, temperature 25° C; c) 0.5 mol dm^{-3} $CuSO_4$ in 0.5 mol dm^{-3} H_2SO_4, temperature 25° C; d) 0.1 mol dm^{-3} $CuSO_4$ in 0.5 mol dm^{-3} H_2SO_4, temperature 50° C.[82] (Reprinted with permission from ISE).

A large difference in the quality of a powdered deposit can arise in prolonged deposition due to different cathode materials[85] because of the different properties of the solid–solution interface, as well as in galvanostatic and potentiostatic cases.

The use of either the potentiostatic or galvanostatic methods result no substantial difference in the deposition of metal powders on different cathode materials. The only real difference lies in the morphology of the powder particles obtained by potentiostatic and galvanostatic deposition[86], i.e. that the particles obtained by galvanostatic deposition are less dendritic than those obtained by potentiostatic deposition, because the overpotential at the end of the deposition is less negative than in this case.

3.3.4 Granular deposits

Granular deposits consisting of independently growing grains with a highly developed surface area are obtained by deposition of metals in processes characterised by large values of the exchange current density[87-89]. In some cases they consist of grains growing independent of each other until a compact film[9,90] or spongy deposit on it is formed[9].

On the basis of the facts concerning the formation of a surface film and the initiation of dendritic growth, the mechanism of the formation of a granular deposit can easily be proposed.

It is shown in Fig. 3.49 that various crystallographic forms, some of them ideal, are obtained during silver deposition at low overpotentials from pure silver nitrate solution, due to the independent grain growth inside the zones of zero nucleation, probably if 2D nucleation is the rate determining step[61]. Dendritic growth starts at higher overpotentials. It can be seen from Fig. 3.49b and c that dendritic growth initiation is mainly related to the appearance of twinned forms in which indestructible re-entrant groove is formed, being the precursors of dendrite. Hence, a granular deposit will be formed at overpotentials lower than those for the formation of dendrite precursors and dendritic growth initiation. The critical overpotential of dendritic growth initiation increases with increasing ion concentration, being 120 mV in 0.5 mol dm^{-3} $AgNO_3$ on a silver substrate[20]. The grains deposited from this solution at 100 mV are different in size, and less ideal than in the previous case, as can be seen from Fig. 3.50a. From Fig. 3.50b and c it can be seen that higher protrusions block the further growth of the lower ones and the formation of a surface film becomes impossible. The granulae will grow towards the bulk solution because lateral flux can be neglected and granular deposits are formed[22]. This is due to the formation of depletion zones around the growing grains[91].

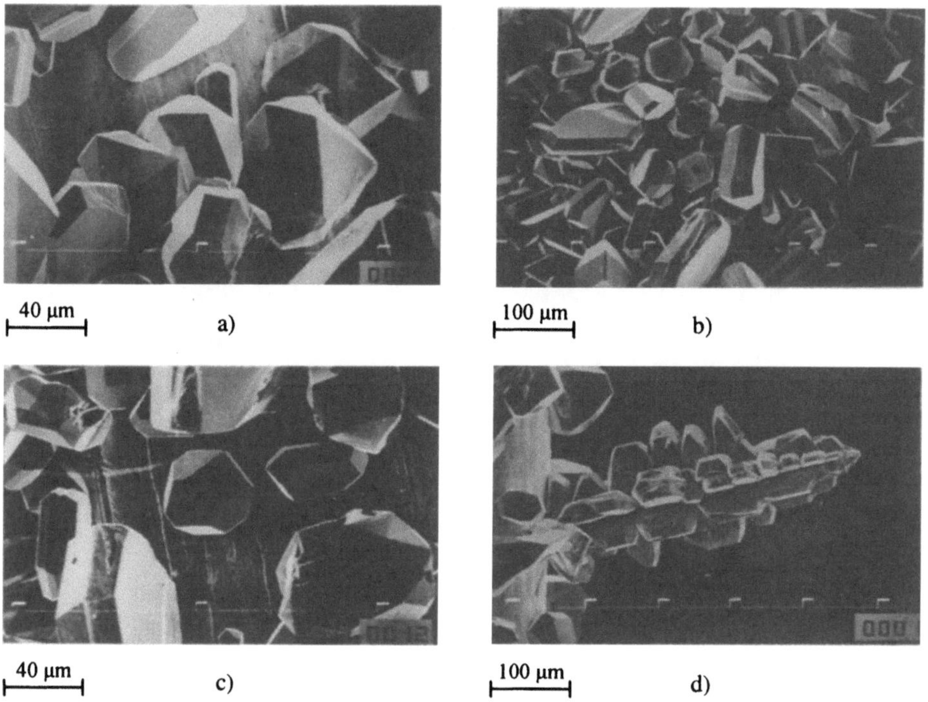

Figure 3.49. Silver deposits obtained potentiostatically on platinum electrode from 0.1 mol dm^{-3} $AgNO_3$ in 100 g dm^{-3} $NaNO_3$ at room temperature. Quantity of electricity 2 mAh cm^{-2}. Deposition overpotential: a) 60 mV; b) 60 mV; c) 70 mV, d) 70 mV[20]. (Reprinted with permission from Springer-Verlag).

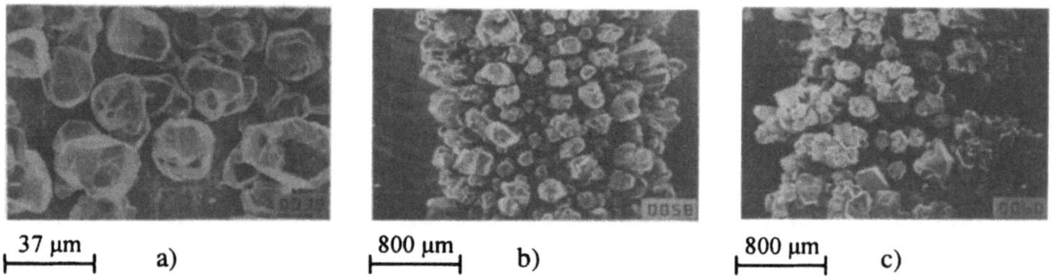

Figure 3.50. Silver deposits obtained potentiostatically on silver electrodes from 0.5 mol dm^{-3} $AgNO_3$ in 100 g dm^{-3} $NaNO_3$ at room temperature. Deposition overpotential 100 mV. Quantity of electricity: a) 2 mAh cm^{-2}, b) 2 mAh cm^{-2}, c) 20 mAh cm^{-2} and d) 60 mAh cm^{-2}. [22] (Reprinted with permission from Springer-Verlag,).

3.3.5 Whisker deposits

This form of crystal growth differs from that of dendrites in that (a) it tends to have a still larger ratio between the longitudinal and the lateral dimensions with an almost perfect preservation of the latter during the growth, and (b) it exhibits no tendency to sidebranching as can be seen from Fig 3.51. Impurities or additives in the electrolyte seem to be a prerequisite for its appearance[31,92].

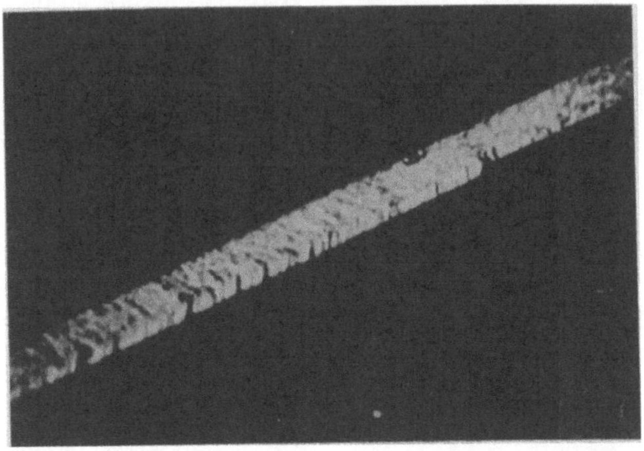

Figure 3.51. Photomicrograph of a silver whisker.[93] (Reprinted with permission from Elsevier Science).

Gorbunova et al[94,95] grew silver whiskers from fairly concentrated silver nitrate solutions (>0.3 mol dm^{-3}) containing oleic acid, gelatine, albumin, and heptyl, octyl, and nonyl alcohols.

A few more phenomena should be noted: a) while growing exclusively in one direction only, whiskers dissolve anodically at a practically uniform rate from all sides[92] and at an overpotential much smaller than that needed for growth; b) a higher overpotential is needed temporarily for the initiation of growth (or continuation after interruption) than for growth at a steady rate; c) if the growth is interrupted for a longer period of time, then it may continue at the tip, but usually assuming a new direction, or else it may be completely prevented and a new whisker started elsewhere. The minimum time required for complete cessation of further growth was found to depend on the concentration of the additive; (d) if a constant rate of growth is maintained, by a constant current flow through the cell to the individual whisker tip, fluctuations of overpotential are observed.

Finally, it should be noted that whiskers differ from other crystals of the same metal in two respects at least: they have an increased electrical resistivity (2-3 times that of crystals deposited in the absence of additives) and an increased tensile strength (3000-8000 lb in^{-2} compared to a few hundred lb in^{-2} observed in large, pure silver single crystals).[31]

A model of the growth mechanism was developed by Price et al[92] which gives a good account of most of the phenomena observed. The basic assumption of the model is that molecules of impurities or additives are strongly adsorbed at all but one crystal plane and at such a concentration as to completely block the deposition and extension of the lattice. On the one plane, however, the process of adsorption is competitive with that of metal deposition whereby the adsorbed molecules are buried and, at a steady state, a sufficiently low surface coverage of foreign molecules is maintained for growth to be possible. The latter is assumed to occur by continuous nucleation and movement of steps over the close-packed surface. Indeed, the appearance of some whiskers, as for example, the one shown in Fig. 3.51, suggests repeated one-dimensional nucleation of the type shown in Fig. 2.8, and the extension of the step in two direction to the edge of the crystal.

3.3.6 Conclusions

It is obvious from the above discussions that the formation of disperse deposits is required only in electrodeposition of metal powders. In this situation, the deposition must be under complete diffusion control. Other types of disperse deposits are undesired and their appearance can easily be avoided by decreasing the exchange current density of the deposition processes (by complexing depositing ions or by appropriate organic additives) and maintaining the deposition overpotential below the values of the critical overpotential for dendritic growth initiation. It seems that in the case of whiskers nothing can be said in advance about their appearance.

3.4 FURTHER READINGS

1. Gutzov I., Kinetics of electrolytic phase-formation under galvanostatic conditions (in Bulgarian). Izv. Inst. Fiz. Chim. Bulgar. Acad. Nauk. 1964; 4:69-88
2. Klapka V., To the problem of crystalization overvoltage during electrocrystallization of metals. Collection Czechoslov. Chem. Comun. 1970; 35:899-906
3. Pangarov N.A., Vitkova S.D. Prefered orientation of electrodeposited iron crystallites. Electrochim. Acta 1966; 11:1719-31
4. Pangarov N.A., Velinov V. The orientation of silver nuclei on a platinum substrate. Electrochim. Acta 1966; 11:1753-58
5. Pangarov N.A., Vitkova S.D., Uzunova I. Electronographic investigation of the degree of preffered orientation of nickel electrodeposits. Electrochim. Acta 1966; 11:1747-51
6. Markov I., Boynov A., Toshev S. Screening action and growth kinetics of electrodeposited mercury droplets. Electrochim. Acta 1973; 18:377-84
7. Štrbac S., Rakočević Z., Popov K.I., Pavlović M.G., Petrović R. The role of surface defects in HOPG on the electrochemical and physical deposition of Ag. J.Serb.Chem.Soc. 1999; 64:483-93
8. Milchev A., Kruijt W.S., Sluyters-Rehbach M., Sluyters J.H. Distribution of the nucleation rate in the vicinity of a groving spherical cluster. Part 1. Theory and simulation results. J. Electroanal. Chem. 1994; 362:21-31; Kruijt W.S., Sluyters-Rehbach M., Sluyters J.H., Milchev A. Distribution of the nucleation rate in the vicinity

of a growing spherical cluster. Part 2. Theory of some special cases and experimaental results. J. Electroanal. Chem. 1993; 371:13-26
9. Popov K.I., Krstajić N.V. The mechanism of spongy electrodeposits formation on inert substrate at low overpotentials. J. Appl. Electrochem. 1983; 13:775-82
10. Markov I. Saturation nucleus density in the electrodeposition of metal onto inert electrodes. I. Theory. Thin Solid Films. 1976; 35:11-20; Markov I., Stoycheva E. Saturation nucleus density in the electrodeposition of metal onto inert electrodes. II. Experimental. Thin Solid Films. 1976; 35:21-35
11. Popov K.I., Krstajić N.V., Popov S.R. Fundamental aspects of plating technology. II: Morphological aspects of metal electrodeposition from complex salt solutions. Surf. Technol. 1983; 20:203-08
12. Popov K.I., Grgur B.N., Stoilković E.R., Pavlović M.G., Nikolić N.D. The effect of deposition process exchange current density on the thin metal films formation on inert substrate. J. Serb. Chem. Soc. 1997; 62:433-42
13. Kaishew R., Mutafctschiew B. Electrolytic nucleation of mercury (in Germane). Electrochim. Acta 1965; 10:643-50
14. Erdey-Grúz T., Volmer Z. Overvoltage of metals (in German). Z. Phys. Chem. 1931;157A: 165-81
15. Fetter, Klaus , *Electrochemical Kinetics (in Russian).* Moscow: Khimiya, 1967.
16. Fleischmann M., Thirsk H.R. The potentiostatic study of the growth of deposits on electrodes. Electrochim. Acta 1959; 1:146-60
17. Kovarskii N.Ya., Lisov A.V. Periodicities in the surface structure of polycrystaline electrolytic deposits (in Russian). Elektrokhimiya 1984; 20:221-25
18. Kovarskii N.Ya., Lisov A.V. The reasons for the structure periodicity in the surfaces of electrolyte copper deposits (in Russian) Elektrokhimiya, 1984; 20:833-837
19. Kovarskii N.Ya., Arzhanova T.A. On the nature of the "no nucleation" zones in the electrocrystallization process (in Russian). Elektrokhimiya 1986; 20:452-58
20. Dimitrov A.T., Hadži-Jordanov S., Popov K.I., Pavlović M.G., Radmilović V. Electrodeposition of silver from nitrate solutions: Part I. Effect of phosphate ions on morphology. J. Appl. Electrochem. 1998; 28:791-96
21. Popov K.I., Pavlović M.G., Grgur B.N., Dimitrov A.T., Hadži-Jordanov S. Electrodeposition of silver from nitrate solutions: Part II. Mechanism of the effect of phospate ions. J. Appl. Electrochem. 1998; 28:797-801
22. Radmilović V., Popov K.I., Pavlović M.G., Dimitrov A.T., Hadži-Jordanov S. The mechanism of silver granular electrodeposits formation. J.Solid State Electrochem. 1998; 2:162-69
23. Popov K.I., Rodaljević Z.P., Krstajić N.V., Novaković S.D. Fundamental aspects of plating technology V: The effect of strongly adsorbed species on the morphology of metal deposit. Surf. Technol. 1985; 25:217-22
24. Meibuhr S., Yeger E., Kozawa A., Hovorka F. The electrochemistry of tin I. Effect of nonionic addition agents on electrodeposition from stannous sulfate solutions. J. Electrochem. Soc. 1963; 110: 190-202
25. Kabanov, Boris N. *Electrochemistry of Metal and Adsorption (in Russian),* Moscow: Nauka, 1966.
26. Lorenz W. Oscillographic overvoltage measurements (in German). Z. Electrochem. 1954; 58:912-18
27. Popov K.I., Krstajić N.V., Popov S.R. Fundamental aspect of plating technology I: The determination of the optimum deposition current density. Surf. Technol. 1983; 20:199-202

28. Scharifker B., Hills G. Theoretical and experimental studies of multiple nucleation. Electrochim. Acta 1983; 28:879-89
29. Milchev A. Role of the substrate state in electrochemical nucleation. Electrochim. Acta 1983; 28:947-53
30. Despić, Aleksandar. "Deposition and Dissolution of Metals and Alloys." In *Comprehensive Treatise of Electrochemistry,* Vol. 2, John O'M. Bockris, Brian E. Conway, Ernest Yeger, Ralph E. White, eds. New York, NY: Plenum Press, 1983.
31. Despić, Aleksandar; Popov, Konstantin. "Transport Controlled Deposition and Dissolution of Metals." *In Modern Aspects of Electrochemistry,* Vol. 7, Brian E. Conway, John O'M. Bockris, eds. New York, NY: Plenum Press, 1972.
32. Popov, Konstantin; Krstajić, Nedeljko; Čekerevac, Milan. "The Mechanism of Formation of Coarse and Disperse Electrodeposits." *In Modern Aspects of Electrochemistry,* Vol. 30, Ralph E. White, Brian E. Conway, John O'M. Bockris, eds. New York, NY: Plenum Press, 1996.
33. Wagner, C. Contribution to the theory of Electropolishing. J. Electrochem. Soc. 1954; 101:225-28
34. Krichmar S.I. Study of the negative leveling effect in the deposition of silver from iodine electrolyte (in Russian). Elektrokhimiya 1965; 1:609-12
35. Despić A.R., Diggle J.W., Bockris J.O'M. Mechanism of formation of zinc dendrites. J. Electrochem. Soc. 1968; 115:507-08
36. Diggle J.W., Despić A.R., Bockris J.O'M. The mechanism of the dendritic electrocrystallization of zinc. J. Electrochem. Soc. 1969; 116:1503-14
37. Barton J.L., Bockris J.O'M. The electrolytic growth of dendrites from ionic solutions. Proc. Roy. Soc. London 1962; A268:485-505
38. Hamilton D.R. A theory of dendritic growth in electrolytes. Electrochim. Acta 1963; 8:731-40
39. Popov K.I., Grgur B.N., Pavlović M.G., Radmilović V. The morphology of copper electrodeposits: I. The mechanism of copper cauliflower-like electrodeposit formation. J. Serb. Chem. Soc. 1993; 58:1055-62
40. Popov K.I., Despić A.R. A contribution of the study of surface roughness amplification in diffusion controlled metal deposition (in Serbian), Bull. Soc. Chim. Belgrade 1971; 36:173-77
41. Popov, Konstantin, *Deposition and Dissolution of Metals in Diffusion Control Conditions (in Serbian),* Ph.D. thesis, University of Belgrade, Belgrade, 1971.
42. Popov K.I., Pavlović Lj. J., Pavlović M.G., Čekerevac M.I. Electrode surface coarsening in potentiostatic copper electrodeposition. Surf. Coat. Technol. 1988; 35:39-45
43. Popov K.I., Pavlović M.G., Pavlović Lj. J., Čekerevac M.I., Remović G.Ž. Electrode surface coarsening in pulsating overpotential copper electrodeposition. Surf. Coat. Technol. 1988; 34:355-63
44. Damjanović A. On the mechanism of metal electrocrystallization. Plating 1965; 52:1017-26
45. Popov K.I., Radmilović V., Grgur B.N., Pavlović M.G. The morphology of copper electrodeposits.II. The mechanism of carrot-like electrodeposits formation. J.Serb.Chem.Soc. 1994;59:47-52
46. Kardos, Otto; Foulke Gardner. Applications of Mass Transfer Theory: "Electrodeposition on Small-Scale Profiles" *In Advances in Electrochemistry and Electrochemical Engineering,* Vol. 2., Paul Delahay, Charles W. Tobias, eds. New York, NY: Interscience, 1962.

47. Ibl, Norbert "Current distribution" *In Comprehensive Treatise of Electrochemistry,* Vol. 6, Ernest Yeger, John O'M. Bockris, Brian E. Conway, S. Sarangapani, eds. New York, NY: Plenum Press, 1983.
48. Kruglikov S.S., Kudriavtsev N.T., Vorobiova G.F., Antonov A.Ya. On the mechanism of leveling by addition agents in electrodeposition of metals. Electrochim. Acta 1965; 10:253-62
49. Dukovic J.O., Tobias C.W. Simulation of leveling in electrodeposition. J. Electrochem. Soc. 1990; 137:3748-55
50. Đorđević, Spasoje; Maksimović, Miodrag; Pavlović, Miomir; Popov, Konstantin, *Electroplating (in Serbian).* Beograd: Tehnička Knjiga, 1997.
51. Krichmar S.I. On the theory of the levelling action in the electrochemical behaviour of metals (in Russian). Elektrokhimiya 1965; 1:858-61; Krichmar S.I. Leveling mechanism in the cathodic deposition of nickel. (in Russian) Zh.Fiz.Khim. 1965; 39:602-03
52. Krichmar S.I., Pronskaya A.Y. Experimental investigation of the levelling effect in the cathodic deposition of nickel from coumarine containing electrolytes (in Russian). Zh.Fiz.Khim. 1965; 39:741-44
53. Jordan K.G., Tobias C.W. The effect of inhibitor transport on leveling in electrodeposition. J. Electrochem. Soc. 1991; 138:1251-59
54. Dukovic J.O. Feature-scale simulation of resist-pattered electrodposition. IBM J.Res.Develop. 1993; 37:125-41
55. Andricacos P.C., Uzoh C., Dukovic J.O. Horkans J., Deligianni H. Damascene copper electroplating for chip interconection. IBM J.Res.Develop. 1998; 42:567-74
56. Nikolić N.D., Rakočević Z., Popov K.I. The structure of bright metal electrodeposits. J. Electroanal. Chem., 2001; 514: 56-66, Nikolić N.D., Rakočević Z., Popov K.I. The STM analysis of a silver mirror surface. J. Serb. Chem. Soc., 2001; 66:723-7
57. Edwards J. The mechanism of electropolishing of copper in phosphoric acid solutions, I. Processes preceding the establishment of polishing conditions. J. Electrochem. Soc. 1953; 100:189c-94c. The mechanism of electropolishing of copper in phosphoric acid solutions, II. The mechanism of smothering. J. Electrochem. Soc. 1953; 100:223c-30c
58. Krichmar S.I., Pronskaya A.Ya. Study of the levelling effect in electrochemical polishing of metals (in Russian). Elektrokhimiya 1966; 2:69-73
59. Popov K.I., Pavlović M.G., Rakočević Z., Škorić D.M. The structure of bright copper surfaces. J. Serb. Chem. Soc. 1995; 60:873-78
60. Nichols R.J., Bach C.E., Meyer H. The effect of three organic additives on the structure and growth of electrodeposited copper: an in-sity scaning probe microscopy study. Ber.Bunsenges. Phys.Chem. 1993; 97:1012-20
61. Popov K.I., Krstajić N.V., Jerotijević Z.Đ., Marinković S.R. Electrocrystallization of silver from silver nitrate solutions at low overpotentials. Surf. Technol. 1985; 26:185-88
62. Popov K.I., Krstajić N.V., Popov S.R., Čekerevac M.I. Spongy electrodeposit formation. J. Appl. Electrochem. 1986; 16:771-774
63. Popov K.I., Krstajić N.V., Simičić M.V., Bibić N.M. The initial stage of spongy electrodeposit formation on inert substrate. J. Serb. Chem. Soc. 1992; 57:927-33
64. Jakšić M.M. Impurity effects on the macromorphology of electrodeposited zinc I: Theoretical consideration and a review of existing knowledge. Surf. Technol. 1985; 24:193-17
65. Murashova, Irina; Pomosov, B. "Electrodeposition of metals in dendritic shapes" (in Russian) *In Itogi nauki i tehniki, Seria Elektrokhimiya,* Vol 30., Yu. M. Polukarov, ed. Moscow: Acad. Sci , 1989.

66. Wranglen G. Dendrites and growth layers in the electrocrystallization of metals. Electrochim. Acta 1960; 2:130-44
67. Bechtoldt C.J., Ogburn F., Smith J. Structure and morphology of electrodeposited molybdenum dendrites. J. Electrochem. Soc. 1968; 115:813-16
68. Faust J.W., John H.F. Growth twins in F.C.C metals. J. Electrochem. Soc. 1963; 110:463-64
69. Faust J.W., John H.F. Germanium dendritic studies, I Studies of thin structures and the seeding mechanism. J. Electrochem. Soc. 1961; 108:855-860
70. Justinijanović I.N., Despić A.R. Some observation on the properties of zinc electrodeposited from alkaline zincate solutions. Electrochim. Acta 1973; 18:709-12
71. Popov K.I., Čekerevac M.I. Dendritic electrocrystallization of cadmium from acid sulphate solution II: The effect of the geometry of dendrite precursors on the shape of dendrites. Surf Technol. 1989; 37:435-40
72. Newman, John, *Electrochemical Systems*, N.J: Prentice-Hall INC, Engelwood Clifts, 1973.
73. Popov K.I., Krstajić N.V., Pantelić R.M.,Popov S.R. Dendritic electrocrystalisation of lead from lead nitrate solution. Surf. Technol. 1985; 26:177-83
74. Popov K.I., Pavlović M.G., Jovićević J.N. Morphology of Tin powder particles obtained in electrodeposition on copper cathode by constant and square - wave pulsating overpotential from Sn(II) alkaline solution. Hydrometallurgy 1989; 23:127-37
75. Popov K.I., Maksimović M.D., Trnjančev J.D., Pavlović M.G. Dendritic electrocrystallization and the mechanism of powder formation in the potentiostatic electrodeposition of metals. J. Appl. Electrochem. 1981; 11:239-46
76. Popov K.I., Radmilović V., Grgur B.N, Pavlović M.G. The morphology of copper electrodeposits. III..The disperse deposits formation, J.Serb.Chem.Soc., 1994; 59: 119-25
77. Calusaru, Aurelian, *Electrodepositon of Metal Powders*. Material Science Monography Vol. III. Amsterdam: Elsevier, 1979.
78. Popov, Konstantin; Pavlović, Miomir. "Electrodeposition of Metal Powders with Controlled Grain Size and Morphology." *In Modern Aspects of Electrochemistry,* Vol. 24, Ralph E. White, John O'M. Bockris, Brian E. Conway, eds. New York, NY: Plenum Press, 1993.
79. Pangarov N.A. Twining processes in the electrocrystallization of face-centred cubic metals. Phys. Stat. Sol. 1967; 20:371-77
80. Ibl, Norbert. "The Formation of Powdered Metal Deposits". In *Advances in Electrochemistry and Electrochemical Engineering*, Vol. II, Paul Delahay, Charles W. Tobias, eds. New York, NY: Intersience, 1962.
81. Popov K.I., Pavlović M.G., Maksimović M.D. Electrodeposition of copper powders. The 10th Meeting of Mines and Metallurgist, University of Belgrade, Bor, 1978.
82. Pavlović M.G., Popov K.I. Electrodeposition of mettal powders with controlled particle size and morphology. Proceedings of 43rd Meeting of ISE; 1992; Kordova.
83. Popov K.I., Pavlović M.G., Maksimović M.D. Comparison of critical conditions for initiation of dendritic growth and powder formation in potentiostatic and galvanostatic copper electrodeposition. J. Appl. Electrochem. 1982; 12:525-31
84. Pavlović M.G., Kindlova Š., Roušar I. The initiation of dendritic growth of electrodeposited copper on rotating disc electrode with changing copper concentration and diffusion layer thickness. Electrochim. Acta 1992; 37:23-27
85. Pavlović M.G., Maksimović M.D., Popov K.I. The effect of cathode material and electrolysis duration on the powder morphology and particle grain size in potentiostatic copper powder electrodeposition (in Serbian). Hem. Ind. 1978; 32:15-21

86. Popov K.I., Pavlović M.G., Maksimović M.D., Krstajić S.S. The comparison of galvanostatic and potentiostatic copper powder deposition on platinum and aluminum electrodes. J. Appl. Electrochem. 1978; 8:503-14
87. Bockris J. O'M., Nagy Z., Dražić D. On the morphology of zinc electrodeposited from alkaline solutions. J. Electrochem. Soc. 1973; 120:30-41
88. Jovićević J.N., Despić A.R., Dražić D.M. Studies of the deposition of cadmium on foreign substrates. Electrochim. Acta 1977; 22:577-87
89. Despić A.R., Dražić M.D., Mirjanić M.D. Granular growth of electrochemically deposited metals. araday Discussion of the Chemical Society, 1978; 12:126-35
90. Popov K.I., Čekerevac M.I., Nikolić Lj.N. The dendritic electrocrystallization of cadmium from acid sulphate solutions I: Granular cadmium substrate. Surf. Coat. Technol. 1988; 34:219-29
91. Sluyters-Rehbach M., Wijenberg J.H.O.J, Bosco E., Sluyters J.H. The theory of chronoamperometry for the investigation of electrocrystallization. Mathematical description and analysis in the case of diffusion-controlled growth. J. Electroanal. Chem. 1987; 236:1-20
92. Price P.B., Vermilyea D.A., Webb M.B., The growth and properties of electrolytic whiskers. Acta Mat. 1958; 6:524-31
93. Graf L., Weser W. The appearance of whiskers during electrocrystalization of silver (in German). Electrochim. Acta 1960; 2:145-64
94. Gorbunova K.M., Zhukova A.J. Crystallochemical and diffusion mechanism of electrocrystallization (in Russian). Zh.Fiz.Khim. 1949; 23:605-15
95. Gorbunova K.M., Pankov P.D. Regularities in the crystallization of thin silver filaments (in Russian). Zh.Fiz.Khim. 1949; 23:616-24

Chapter 4

THE CURRENT DISTRIBUTION IN ELECTROCHEMICAL CELLS

It is a known fact that different morphologies of electrodeposited metal can appear at the different positions of the electrode surface. This means that the local current density during electrodeposition of a metal varies from point to point on an electrode surface. This is due to the following factors[1,2]:

- the geometry of the system;
- the conductivity of the solution and electrodes;
- the activation overpotential;
- the diffusion overpotential;
- the hydrodynamics of the system.

Although all factors effect the current distribution simultaneously, there are three main types of current distribution on a macroprofile.

If the influence of overpotential is negligible, primary distribution is determined by the geometry of the system and the conductivity of the solution.

In the case of the secondary and tertiary distribution the activation overpotential and both the activation and diffusion overpotentials have to be taken into consideration, also.

Even for a simple electrode configuration, the calculation of the current distribution is a complex problem and the difficulties further increase with increasing complexity of the geometry, especially if the limiting diffusion current varies over the electrode due to the different geometric and hydrodynamic conditions. Because of this, analytical solutions can be found only for some cases (Wagner[3], Newman[4,5]), while in other cases numerical solutions are available[1]. If the complete calculation can not be performed, it is possible to estimate certain trends, using a dimensionless group called the Wagner number, W_a, given by

$$W_a = \frac{d\eta_c}{dj}\frac{\kappa}{l}$$

where $d\eta_c/dj$ is the slope of the cathodic activation overpotential-current density dependence, κ is the conductivity of solution and l is a characteristic length.

Before W_a, the parameter

$$k_c = \kappa \frac{d\eta_c}{dj}$$

was used, according to Kasper[6], Hoar and Agar[7].

The Wagner number represents the ratio of the polarisation resistance to the solution resistance. The larger it is, the more even is the current distribution in spite of non-uniform geometry. In general, the current distribution is more uniform if[1]:

- the smaller characteristic length of the system is,
- the larger the conductivity of the solution is, and
- the larger the slope of the activation overpotential-current density curve is.

Obviously, the Wagner number can be used only to compare the current distribution in the cell with non-uniform geometry, which contains different electrolytes.

The same situation appears if the ability of an electrolyte to uniformly distribute the current density is experimentally determined using the method of Haring and Blum[8].

The current distribution on a macroprofile is very important in technical metal electrodeposition. In electroplating, the current distribution determines the local variations in the thickness of the coating. In electrowinning and electrorefining of metals, a non-homogenous current distribution can cause a short circuit with the counter electrode, and the corner weakness effect in electroforming. This is very important in the three-dimensional electrodes, as well as in some storage batteries. In all the cited cases a uniform current density distribution over the macroprofile is required.

The aim of this chapter is to present the procedure, based on simple equations of electrode kinetics, by which the condition in which a desired current density distribution can be obtained, or an undesired one avoided, under the assumption that the limiting diffusion current density does not vary over the whole electrode surface, including the edges of flat and the tips of wire electrodes.

4.1 TWO EQUAL PLANE PARALLEL ELECTRODES ARRANGEMENT

The cell with two equal plane parallel electrodes represents the elementary cell of electrode arrangement in electrochemical refining and winning processes.

It is a well-known fact that in a cell with parallel electrodes (if the electrode edges do not touch the side walls of the cell), the current density is higher at the edges than at the centre of the electrode[1,9]. This is because the current flow passes partially around the rectangular space between the electrodes. The increased current density at the edges of the electrodes can be easily noticed by observing the quality of the metal electrodeposit at the cathode. In some cases the deposit in the central part of the cathode may be compact and flat whereas the occurrence of dendrites is observed at the edges. The appearance of dendrites at the edges of the cathodes in such situations is the most important problem of the current density distribution, because the growing dendrites could cause short circuits followed by a decrease in the current efficiency, or even damage the power supply.

The aim of this section is to show in which way dendritic growth at the cathode edges can be avoided in electrowinning and refining processes.

4.1.1 Ohmic resistance of the cell

The current density distribution in a rectangular electrolytic cell in which parallel electrodes cover only part of the wall is illustrated in Fig. 4.1.

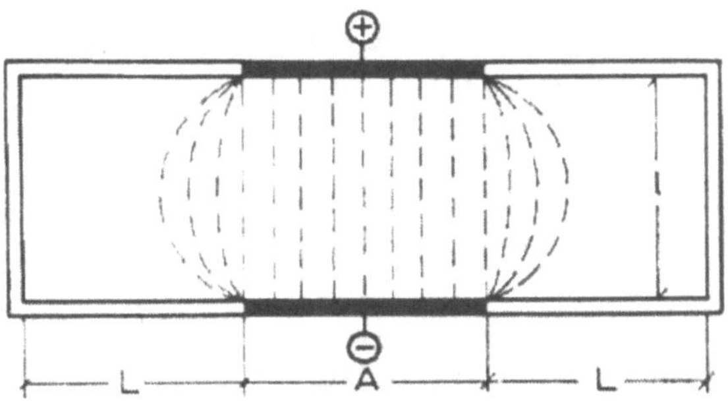

Figure 4.1. Current distribution in a parallel plate electrode geometry (*A* is the electrode width, *L* is the distance between the edge of the electrode and the side walls and *l* is the distance between the electrodes.[9] (Reprinted with permission from the Serbian Chemical Society, Belgrade, Yugoslavia)

The linear approximation of the current distribution in the cell with plane parallel electrodes shown in Fig. 4.1 is presented schematically in Fig. 4.2.

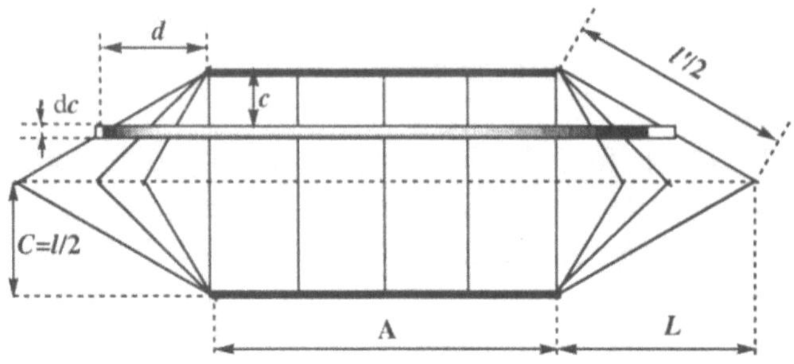

Figure 4.2. The linear approximation model showing the current flow passing around the space between the plane parallel electrodes.[10] (Reprinted with permission from Elsevier Science).

The analysis performed here for the current distribution between the electrode edges and the cell side walls is obviously valid also for the situation in which there is the distance between the upper edges of the electrodes and free surface of ssolution and lower edges to the bottom of the cell. In the case under consideration these two distances are zero.

The resistance dR of a section of the electrolyte of thickness dc is given by:

$$\mathrm{d}R = \frac{\rho}{B} \frac{\mathrm{d}c}{A + 2d} \tag{4.1}$$

where B is the height of the electrode and ρ is the specific resistance of the electrolyte. From the linear approximation:

$$d = \frac{L}{C} c \tag{4.2}$$

is obtained. The parameters d and c are indicated in Fig. 4.2.

The resistance of the whole electrolyte is then given by[10]:

$$R = \frac{\rho C}{BL} \ln\left(\frac{A + 2L}{A} \right) \tag{4.3}$$

and for $L \rightarrow 0$, by:

$$\lim_{L \to 0} R = \frac{2\rho C}{BA} = \frac{\rho l}{BA} = R_h \tag{4.4}$$

where R_h corresponds to the resistance of a system with a homogeneous current density distribution (the side walls touch the edges of the electrodes). For $0 \le L << \infty$, L can be related to A by a linear coefficient k as follows:

$$L = k\,A \tag{4.5}$$

which transforms Eq. 4.3 to:

$$R = \frac{R_h}{2k} \ln (1 + 2k) \tag{4.6}$$

and

$$l_{eff} = \frac{l}{2k} \ln(1+2k) \tag{4.7}$$

taking into account Eq. 4.4, where l_{eff} represents the interelectrode distance in a cell with L=0, the resistance of which is equal to the resistance of a cell in which the interelectrode distance is l and $L > 0$

The cell for the determination of the equivalent resistance is shown in Fig. 4.3.

Figure 4.3. Schematic representation of the cell used for the determination of the equivalent resistance.[10] (Reprinted with permission from Elsevier Science).

The side screens enabled the distance L between the edges of the electrodes and the side walls to be varied for given values A and $2C$. The resistance of the system for various adjusted values of L was measured by the bridge method using platinum electrodes in a 0.02 mol dm^{-3} KCl solution. The electrodes were 2 cm long and 1 cm wide. The interelectrode distance was 2 cm. Hence, in this case A=C. The back sides of the electrodes were insulated. The upper edges of the electrode touch the free surface of the solution and the lower edge the bottom of the cell.

The dependence of the total resistance of a system with plane parallel electrodes on the distance between the electrode edges and the cell side walls is shown in Fig. 4.4.

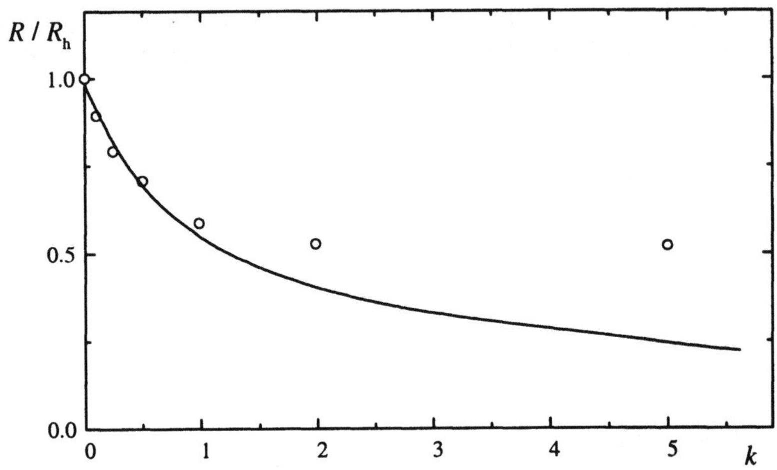

Figure 4.4. The dependence of the total resistance of the system on k: (o) experimentally determined values; (—) calculated values.[10] (Reprinted with permission from Elsevier Science).

The open circles represent the experimental values, and the curve is obtained using Eq. 4.6. The good agreement between the experimental results and the values predicted by Eq. 4.6. extends to $k \approx 1$. It can be concluded that for this system Eq. 4.6. is valid for $k < 1$. This means that the maximum penetration of the current lines occurs when $L = C = A$ in this case, and that the maximum length of the current line, l' is $l\sqrt{2}$

4.1.2 The very edge ohmic resistance

This consideration of the very edge current density can be elaborated mathematically in the following way[11]. Assuming total ohmic control, the voltage drop in the solution between the electrodes inside the homogenous field is given by:

$$U - E = \rho \, l \, j \tag{4.8}$$

and outside of the homogeneous field by:

$$U - E = \rho \, l_i \, j_i \tag{4.9}$$

where U is the cell voltage, E is the equilibrium potential difference, ρ is the specific resistivity of the electrolute, l the interelectrode distance, j is the current density, l_i is the length of the i-th current line and j_i is the current density corresponding to the i-th current line, as can be seen from Fig. 4.5.

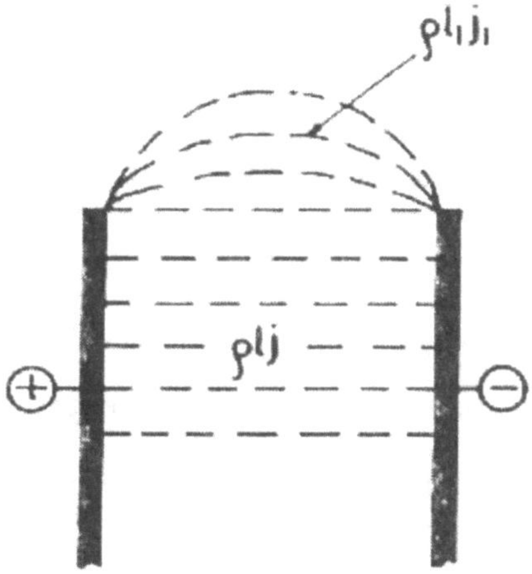

Figure 4.5. Current lines between the electrodes with edges not touching the side walls of the cell.[11] (Reprinted with permission from the Serbian Chemical Society, Belgrade, Yugoslavia).

The difference in the current lines outside of the homogeneous field is given by:

$$j_i + \Delta j_i - j_i = \frac{U-E}{\rho}\left(\frac{1}{l_i} - \frac{1}{l_i + \Delta l_i}\right) \tag{4.10}$$

or in the differential form

$$\frac{\mathrm{d}j_i}{\mathrm{d}l_i} = \frac{U-E}{\rho}\frac{1}{l_i^2} \tag{4.11}$$

When Eq. 4.11 is integrated from the inter-electrode distance l to the maximum length of the current line, l', j', the maximum contribution to the edge current density due to current line propagation between the electrode edges and the side walls of the cell, is obtained:

$$j' = \frac{U-E}{\rho}\left(\frac{1}{l} - \frac{1}{l'}\right) = \frac{U-E}{\rho l}\frac{l'-l}{l'} \tag{4.12}$$

Taking into accounts Eq. 4.8 one obtains

$$j' = j\frac{l'-l}{l'} \tag{4.13}$$

The edge current density j_e, can be written as

$$j_e = j + j' \tag{4.14}$$

The maximum value of j' is obtained from Eq. 4.13 as:

$$j'_{max} = j\ \frac{2-\sqrt{2}}{2} \tag{4.15}$$

Combining Eq. 4.15 and 4.14, the maximum edge current density can be given as[12]:

$$j_{e,\,max} = \left(2-\frac{\sqrt{2}}{2}\right)j \approx 1.3\,j \tag{4.16}$$

for $l' = l\sqrt{2}$, as follows from Figs. 4.2 and 4.4.

This means that the very edge resistance is lower than in the homogenous field and that the minimum effective interelectrode distance, $l_{eff,e,min}$ between the edges of the anode and cathode will be:

$$l_{eff,e,min} = \frac{2}{4-\sqrt{2}}\,l \tag{4.17}$$

because of

$$\rho l j = \rho\, l_{eff,e,min}\, j_{e,max} \tag{4.18}$$

4.1.3 The edge effect

In a cell with parallel plate electrodes, if the electrode edges do not touch the cell side walls, the potential difference between two points in the homogenous field symmetrically positioned on the electrodes is given by:

$$U = E + \eta_a + \eta_c + \rho l j \tag{4.19}$$

Analogously, the cell voltage at the edges can be expressed as:

$$U = E + \eta_{a,e} + \eta_{c,e} + \rho\, l_{eff,e}\, j_e \qquad (4.20)$$

where η_a and η_c are, respectively, the anodic and cathodic overpotentials corresponding to the homogenous field, and $\eta_{a,e}$ and $\eta_{c,e}$ are, respectively, the anodic and cathodic overpotentials corresponding to the edges.

Elimination of U from Eqn. 4.19 and 4.20 gives

$$\eta_{a,e} + \eta_{c,e} = \eta_a + \eta_c + \rho\, lj - \rho\, l_{eff,e}\, j_e \qquad (4.21)$$

In this case $\rho lj > \rho\, l_{eff,e}\, j_e$, because the increasing of the current density leads to the increasing of the cathodic and anodic overpotentials also.

In this way, a part of the ohmic potential drop in a homogenous field transforms into electrochemical overpotential for points at the plane electrode edges, or in a similar position, meaning the edge current density is larger than in the homogenous field. In this way it is possible to explain the change in the quality of the metal deposit near the edge and at the very edge of an electrode. It should be noted, however, that according to the proposed model the entire edge current is located at the very edge of the electrode. In other words, a homogeneous electric field and, consequently, a uniform current distribution is assumed over the entire electrode surface up to the very edge of the electrode, where the current density increases abruptly, which is quite close to the real state described by other authors[2-5]. The illustration of this effect will be given in Section 4.2.2.

4.1.4 The depth of the penetration of a current line between the electrode edges and the cell side walls

4.1.4.1 Mathematical model

Equation 4.12 can be rewritten in the form

$$l' = \frac{l}{1 - \dfrac{\rho l j'}{U - E}} \qquad (4.22)$$

and j' can be majorized by $j\,\dfrac{2-\sqrt{2}}{2}$ giving:

$$l' = \frac{l}{1 - \dfrac{\rho l j}{U - E}\,\dfrac{2-\sqrt{2}}{2}} \qquad (4.23)$$

as the maximum length of a current line. $U - E$ in Eqs. 4.22 and 4.23 is the ohmic potential drop, but it can be substituted by the cell potential due to the following facts.

The current along each line should be very low, and because of this the electrochemical overpotentials at the edges of electrodes due to one current line can be neglected relative to the ohmic potential drop. Hence, the cell potential transforms into the ohmic potential drop along each current line and $U - E$ in Eq. 4.23 can be substituted by the cell potential from Eq. 4.19.

Substitution of $U - E$ in Eq. 4.23 by the cell voltage from Eq. 4.19 gives:

$$l' = \frac{l}{1 - \frac{\rho l j}{\eta_a + \eta_c + \rho l j} \frac{2 - \sqrt{2}}{2}} \tag{4.24}$$

or after rearrangement:

$$l' = l \frac{\eta_a + \eta_c + \rho l j}{\eta_a + \eta_c + \frac{\sqrt{2}}{2} \rho l j} \tag{4.25}$$

Assuming a linear approximation of the propagation of a current line the relation between L', the maximum depth of the propagation of a current line penetration in the space between the edges of the electrodes and the cell side walls, l and l' is given by:

$$L' = \frac{1}{2} \sqrt{l'^2 - l^2} \tag{4.26}$$

Substituting l' from Eq. 4.25 into Eq. 4.26 and rearranging gives:

$$L' = \frac{l}{2} \left[\left(\frac{\eta_a + \eta_c + \rho l j}{\eta_a + \eta_c + \frac{\sqrt{2}}{2} \rho l j} \right)^2 - 1 \right]^{1/2} \tag{4.27}$$

It can be shown that if:

$$\eta_a + \eta_c >> \rho\, l j \tag{4.28}$$

then $l' \to l$ and $L' \to 0$, and in the opposite case $l' \to l\sqrt{2}$ and $L' \to l/2$. This shows that the ability of an electrolyte to distribute the current density uniformly increases with decreasing ρlj product, i.e., with decreasing ohmic polarization. Furthermore, is to be expected that the larger the spacing (the distance between the polarization j – U curves for $L = 0$ and $L > 0$), the worse current density distribution becomes.

4.1.4.2 Cell voltage-current density dependencies

Taking into account Eqn. 4.5 and 4.7 for $L > 0$ (assuming that the current density in the homogenous field is equal to the overall one, i.e. that current density distribution is uniform over all the electrode, except for the very edge), Eq 4.19 can be rewritten in the form

$$U = E + \eta_a + \eta_c + \frac{\rho\, ljA}{2L} \ln\left(1 + \frac{2L}{A}\right) \tag{4.29}$$

If $L' < L$, L in Eq. 4.29 should be substituted by L'.

$$U = E + \eta_a + \eta_c + \frac{\rho\, ljA}{2L'} \ln\left(1 + \frac{2L'}{A}\right) \tag{4.29a}$$

Hence, calculations of the current density-cell voltage dependencies must begin with the determination of the relation between L and L'.

It is obvious that if:

$$\eta_a + \eta_c >> \frac{\rho ljA}{2L} \ln\left(1 + \frac{2L}{A}\right) \tag{4.30}$$

electrolysis is predominantly under electrochemical control, and if:

$$\eta_a + \eta_c << \frac{\rho ljA}{2L} \ln\left(1 + \frac{2L}{A}\right) \tag{4.31}$$

the electrolysis is predominantly under ohmic control.

In the first case an S-shaped polarization curve can be expected, and a straight line in the second one. Hence, the shape of the polarization curve is a good indicator of the nature of the electrolysis process. On the other hand, the degree of current line penetration into the solution between the edges of the electrodes and the side walls of the cell is an indicator of the ability of an electrolyte to distribute the current density uniformly over the entire electrode.

The experiments were carried out in the cell presented in Fig. 4.6

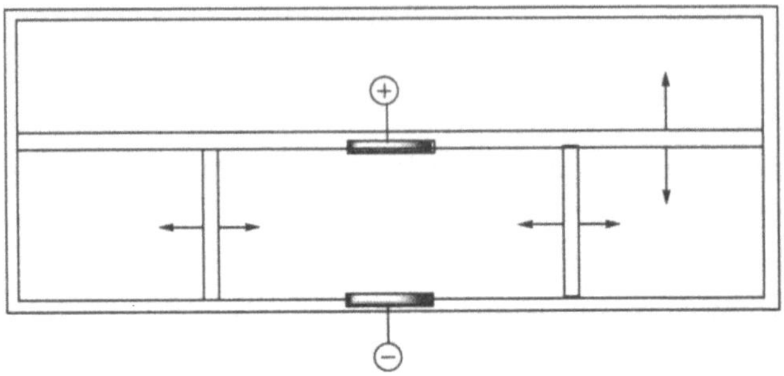

Figure 4.6. A cell with paralel electrodes.[9] (Reprinted with permission from the Serbian Chemical Society, Belgrade, Yugoslavia).

The dependences of the current density on the cell voltage for different interelectrode distances and different distances between the edge of the electrode and side wall, for the system (-)Cu | $CuSO_4$, H_2SO_4, H_2O | Cu(+) at a temperature of 20° C are shown in Figs. 4.7 – 4.10.

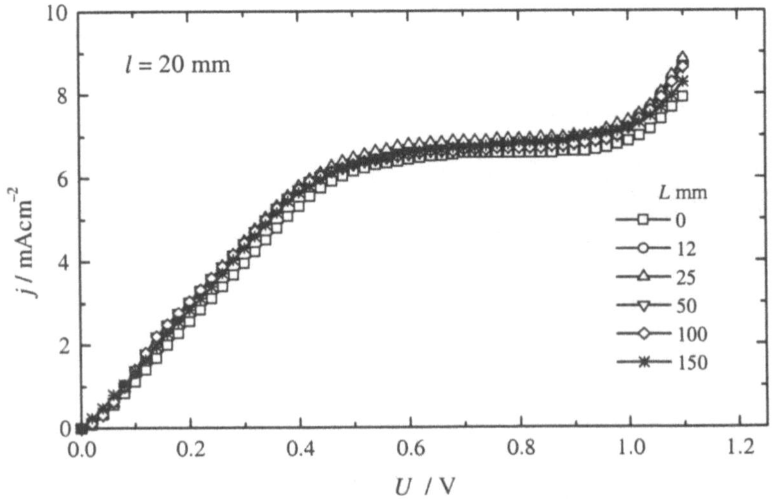

Figure 4.7. Current density-cell voltage dependencies of the system (-) Cu | 0.1 mol dm^{-3} $CuSO_4$, 0.1 mol dm^{-3} H_2SO_4 | Cu (+) with an interelectrode distance of 20 mm for different distances between the edge of the electrode and the side wall of the cell indicated in the diagram.[9] (Reprinted with permission from the Serbian Chemical Society, Belgrade, Yugoslavia).

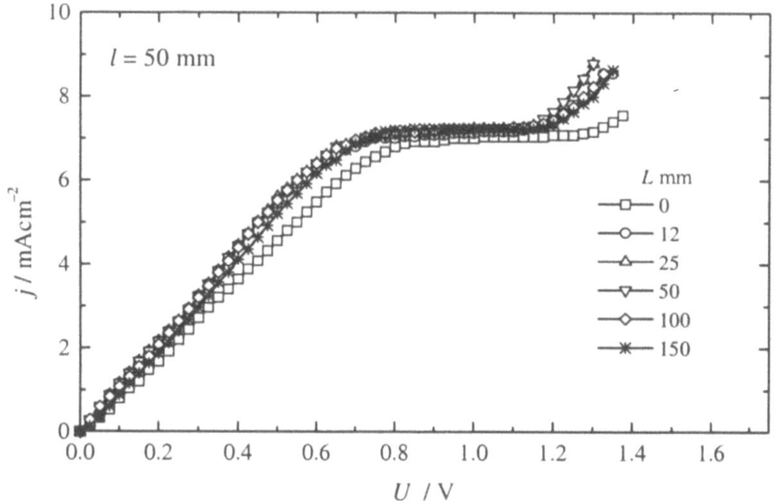

Figure 4.8. Current density-cell voltage dependencies of the system (-)Cu | 0.1 mol dm^{-3} $CuSO_4$, 0.1 mol dm^{-3} H_2SO_4 | Cu(+) with an interelectrode distance of 50 mm for different distances between the edge of the electrode and the side wall of the cell indicated in the diagram.[9] (Reprinted with permission from the Serbian Chemical Society, Belgrade, Yugoslavia).

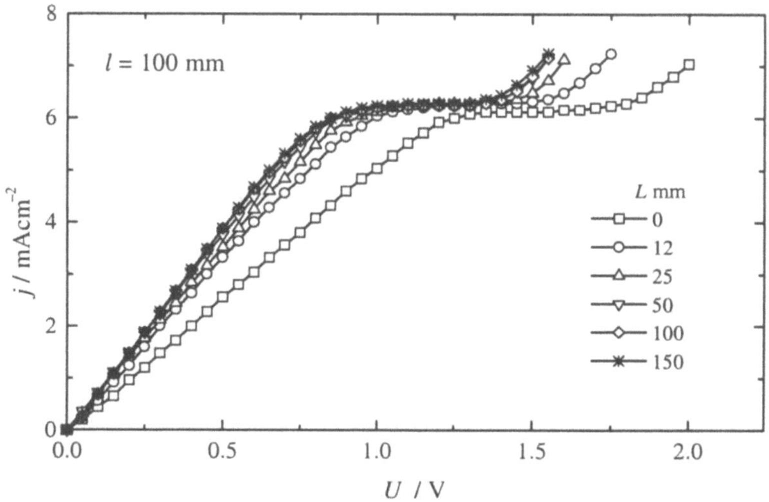

Figure 4.9. Current density-cell voltage dependencies of the system (-)Cu | 0.1 mol dm^{-3} $CuSO_4$, 0.1 mol dm^{-3} H_2SO_4 | Cu(+) with an interelectrode distance of 100 mm for different distances between the edge of the electrode and the side wall of the cell indicated in the diagram.[9] (Reprinted with permission from the Serbian Chemical Society, Belgrade, Yugoslavia).

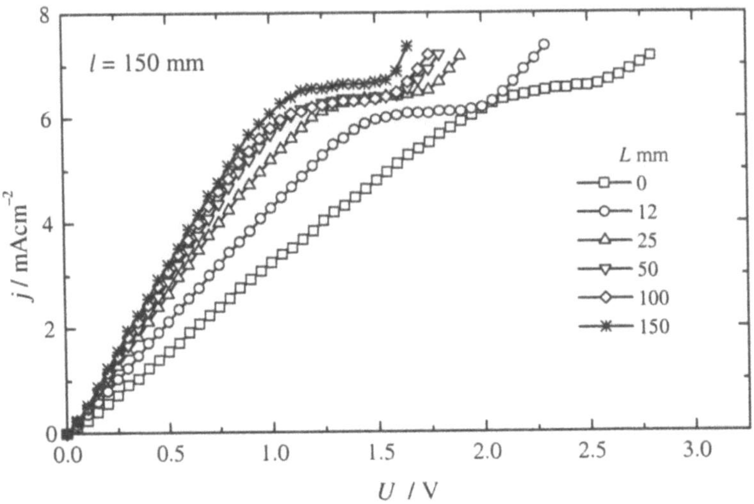

Figure 4.10. Current density-cell voltage dependencies of the system (-) Cu | 0.1 mol dm^{-3} $CuSO_4$, 0.1 mol dm^{-3} H_2SO_4 | Cu with an interelectrode distance of 150 mm for different distances between the edge of the electrode and the side wall of the cell indicated in the diagram.[9] (Reprinted with permission from the Serbian Chemical Society, Belgrade, Yugoslavia).

From Figs. 4.7-4.10, the change in the shape of the *j-U* dependencies as the inter-electrode distance increases can be seen. In the region of lower current densities, at the shortest interelectrode distance, the system is under mixed activation-ohmic control. At higher current densities, although a concentration overvolatge appears, the control becomes ohmic because the magnitude of the overvoltages is negligible in comparison to the ohmic voltage drop in the electrolyte. As the interelectrode distance increases, the electrolyte resistance in the cell increases too, giving rise to an increase in the contribution of ohmic control of the system which in turn causes a larger spreading out of the current lines between the edges of the electrodes and the side walls. Therefore, the difference between the *j-U* curves for different distances between the side wall and the edge of the electrode becomes larger, meaning a worse current density distribution. At very large current densities, approaching the limiting current, the contribution of the concentration overvoltage to the total cell voltage becomes significant and the system is under mixed diffusion-ohmic control. Finally, in the region of limiting current densities, the system is entirely under diffusion control. The polarisation curves for different distances between the edge of the electrode and the side wall are very similar indicating a minimal spreading of the current lines out of the homogeneous electric field.

4.1.4.3 Determination of the current density distribution

The effect of the geometry of a cell on the ability of the electrolyte to distribute the current density uniformly can be illustrated by plotting the ratios of the current density in cells with different interelectrode distance l (20, 50, 100 and 150 mm) and an electrode edge – cell side wall distance $L =$ 150 mm, and in cells with different electrode edge-side wall distance L (12.5, 25, 50, 100, 150 mm) and an interelectrode distance l=150 mm to the current density in a cell with the same l values and L = 0, as a function of current density in a cell with L = 0, normalized to the limiting diffusion current density, as shown in Figs. 4.11 and 4.12. It should be noted that the increase of the limiting diffusion current density in a cells with L>0 over the value in a cell with L=0 was not taken into account in the derivation of the curves in Figs. 4.11-4.13.

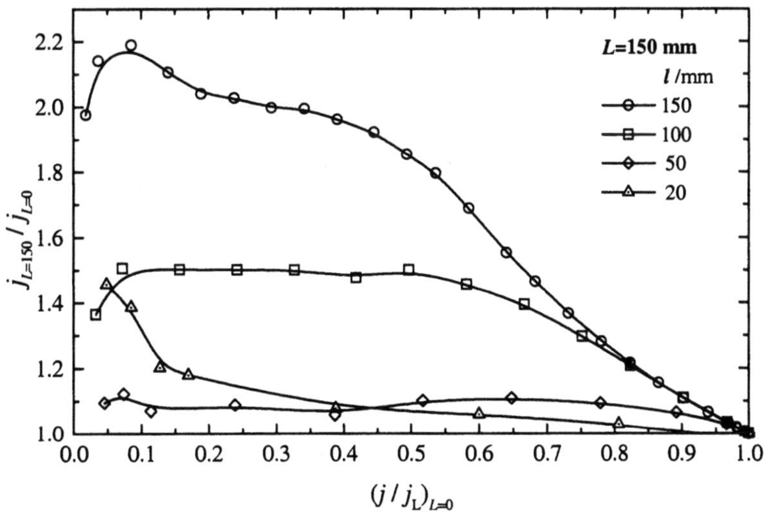

Figure 4.11. The $j_{L=150} / j_{L=0}$ ratio in the cell Cu|0.1 mol dm^{-3} $CuSO_4$, 0.1 mol dm^{-3} H_2SO_4|Cu with electrode edges–cell side wall distance L = 150 mm and different interelectrode distances a) l = 20 mm, b) l = 50 mm, c) l = 100 mm and d) l = 150 mm, as a function of the current density in a cell with L = 0, normalized relative the limiting diffusion current density. Data from Figs. 4.7-4.10.[13] (Reprinted with permission from the Serbian Chemical Society, Belgrade, Yugoslavia).

Obviously, the larger the current density ratio, the lower is the ability of the electrolyte to distribute homogeneously the current in the cell.

Hence, increasing the interelectrode distance leads to a worsening of the current density distribution, as does increasing the electrode edges – cell side wall distance.

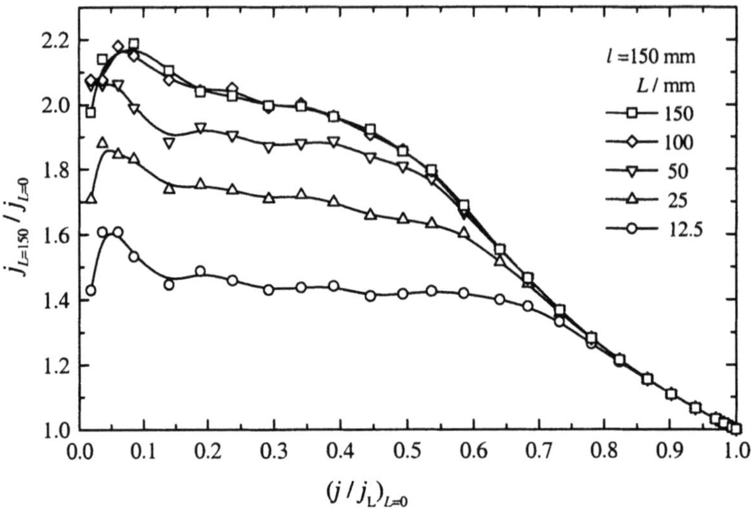

Figure 4.12. The j_L / $j_{L=0}$ ratio in cell Cu|0.1 mol dm^{-3} $CuSO_4$, 0.1 mol dm^{-3} H_2SO_4|Cu with electrode distance l = 150 mm and different electrode edges–cell side wall distances a) L = 12.5 mm, b) L = 25 mm, c) L = 50 mm and d) L = 100 mm, as a function of the current density in cell with L = 0, normalized relative to the limiting diffusion current density. Data from Fig. 4.10.[13] (Reprinted with permission from the Serbian Chemical Society, Belgrade, Yugoslavia).

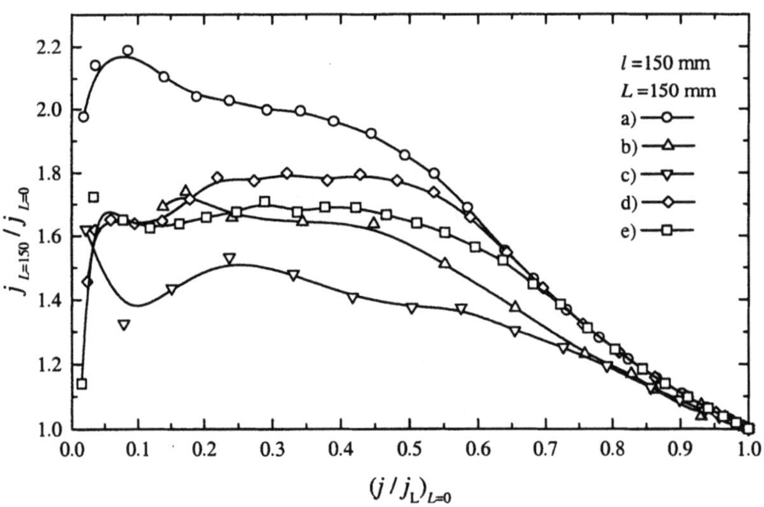

Figure 4.13. The $j_{L=150}$ / $j_{L=0}$ ratio in a cell with interelectrode distance l = 150 mm and electrode edges–cell side wall distance L = 150 mm for a) Cu|0.1 mol dm^{-3} $CuSO_4$, 0.1 mol dm^{-3} H_2SO_4|Cu, t = 20°C; b) Cu|0.03 mol dm^{-3} $CuSO_4$, 0.1 M H_2SO_4|Cu, t = 20°C; c) Cu|0.1 mol dm^{-3} $CuSO_4$, 0.5 M H_2SO_4|Cu, t = 20°C; d) Cu|0.1 mol dm^{-3} $CuSO_4$, H_2SO_4|Pb, t = 20°C; e) Cu|0.1 mol dm^{-3} $CuSO_4$, H_2SO_4|Cu, t = 40°C. Data from Figs. 3 and 7, ref. 9 and 2, 4 and 8 ref 11.[13] (Reprinted with permission from the Serbian Chemical Society, Belgrade, Yugoslavia).

The effects of the supporting electrolyte and the depositing ion concentration, insoluble anode and temperature on the ability of an electrolyte to distribute homogeneously the current density are illustrated in Fig. 4.13.

All the above facts can be explained by discussing L', the depth of the current line penetration between the electrode edges and cell side wall for $L'<L$. Previously it was shown that:

$$L' = \frac{l}{2}\left[\left(\frac{\eta_a + \eta_c + \rho l j}{\eta_a + \eta_c + \rho j l \frac{\sqrt{2}}{2}}\right)^2 - 1\right]^{1/2} \tag{4.27}$$

and

$$U = E + \eta_a + \eta_c + \frac{\rho l j A}{2L'}\ln\left(1 + \frac{2L'}{A}\right) \tag{4.29a}$$

For one and the same current density in a cell, the electrochemical part of the cell voltage does not depend on the interelectrode distance, but the ohmic drop in a homogeneous field is strongly dependant on it, which leads to an increase of L' with increasing l. This produces decrease of the cell voltage at a fixed current density, because of the decrease of the overall ohmic resistance. Hence, at a fixed cell voltage, increasing L' will results in an increase of the current density relative to a cell with $L = 0$ and, hence, a worse current density distribution. The same will happen with increasing L at $l = 150$ mm if the condition $L' < L$ is not satisfied.

In the similar way, in the cells with the same $l = 150$ mm and $L = 150$ mm, L' will depend on the fractional contribution of the ohmic drop $\rho l j$ to the cell voltage. At a fixed l, the value of the product $\rho l j$ increases faster with increasing j than the electrochemical part of cell voltage. Hence, increasing the depositing ion concentration and the stirring rate will cause the ability of the electrolyte to distribute the current density uniformly worsen, as well as a decreasing of supporting electrolyte concentration. Increasing value of the Tafel slope and, probably decreasing exchange current densities will improve the current distribution.

Increasing the temperature also has a significant effect, regardless of the simultaneous decrease in ρ and increase in j. This means that the decrease in ρ is more pronounced than the increase in j. In all cases, the increase in the degree of diffusion control leads to a better current density distribution on a macroprofile. On the other hand, in such a situation the current density distribution worsens on a microprofile (see section 3.2).

4.1.5 Quantitative treatment

Using data from Figs. 4.7-4.10, for $L = 0$ but with different value of l, the cell voltage – interelectrode distance dependencies for different current densities can be obtained. The results of these calculations are shown in Fig. 4.14.

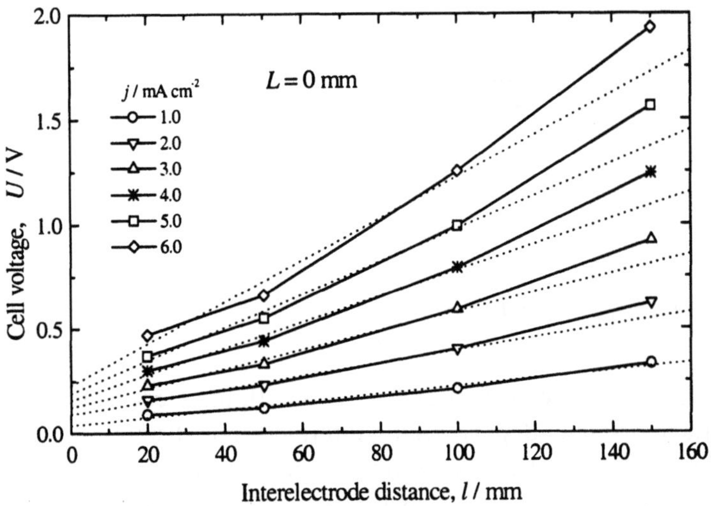

Figure 4.14. The cell voltage – interelectrode distance dependencies for $L = 0$ in Cu | 0.1 mol dm^{-3} $CuSO_4$, 0.1 mol dm^{-3} H_2SO_4 | Cu. Data from Figs. 4.7-4.10.[14] (Reprinted with permission from the Serbian Chemical Society, Belgrade, Yugoslavia).

Obviously, the intercepts represent the difference of the anodic and the cathodic overpotentials for selected values of the current density, and the resistivity of the electrolyte can be determined from corresponding slopes. The obtained values are given in Table 4.1.

Table 4.1. The and cathodic overpotentials and resistivity of the electrolyte for selected values of current density, obtained from Fig.4.14.[14] (Reprinted with permission from the Serbian Chemical Society, Belgrade, Yugoslavia).

j, $mAcm^{-2}$	$\eta_a + \eta_c$, V	ρ, Ω cm
1	0.06	14.0
2	0.09	14.5
3	0.13	14.7
4	0.17	14.8
5	0.20	15.2
6	0.23	16.2

4.1.5.1 Calculation of the cell voltage-current density distribution dependences

The diagrams from Figs. 4.7-4.10, can also be presented in the form shown in Figs. 4.15-4.18.

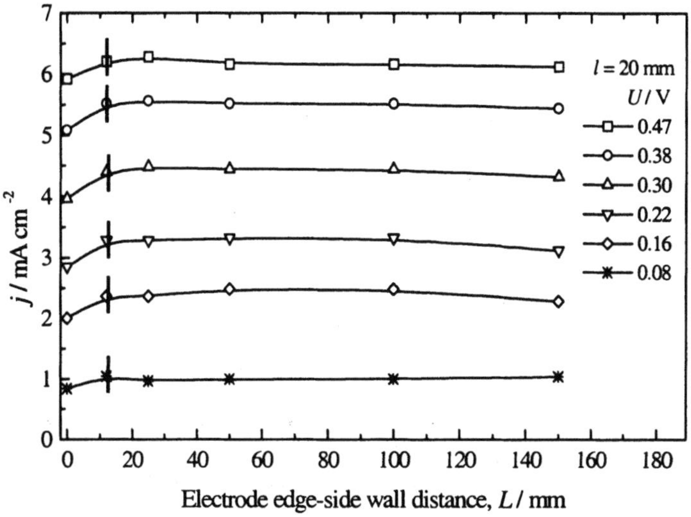

Figure 4.15. Current densities for different cell voltages as functions of the electrode edges – cell side wall distances in a cell with interelectrode distance l = 20 mm for the system Cu|0.1 mol dm^{-3} $CuSO_4$, 0.1 mol dm^{-3} H_2SO_4|Cu. Data from Fig. 4.7.[14] (Reprinted with permission from the Serbian Chemical Society, Belgrade, Yugoslavia).

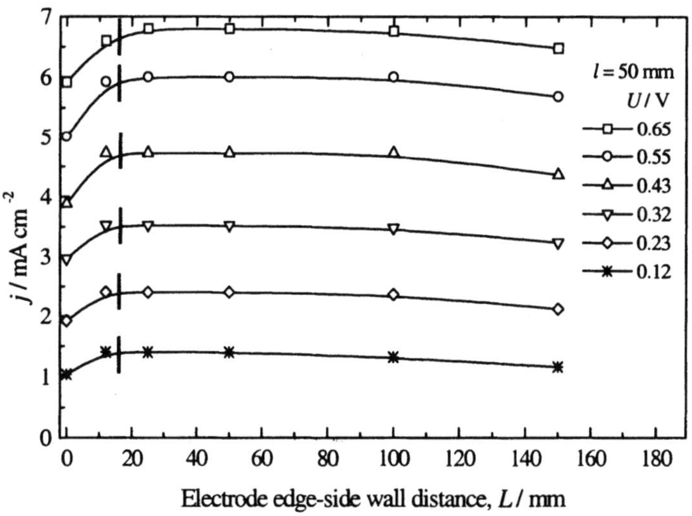

Figure 4.16. Current densities for different cell voltages as functions of the electrode edges – cell side wall distances in a cell with interelectrode distance l = 50 mm for the system Cu|0.1 mol dm^{-3} $CuSO_4$, 0.1 mol dm^{-3} H_2SO_4|Cu. Data from Fig. 4.8.[14] (Reprinted with permission from the Serbian Chemical Society, Belgrade, Yugoslavia).

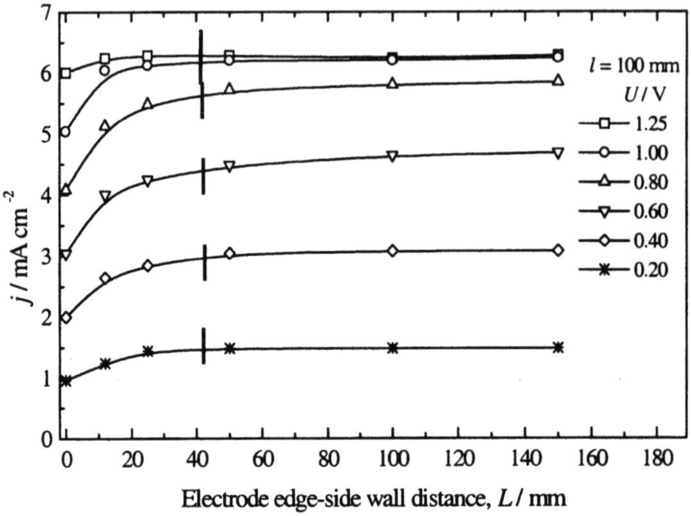

Figure 4.17. Current densities for different cell voltages as functions of the electrode edges – cell side wall distances in a cell with interelectrode distance l = 100 mm for the system Cu|0.1 mol dm^{-3} $CuSO_4$, 0.1 mol dm^{-3} H_2SO_4|Cu. Data from Fig. 4.9.[14] (Reprinted with permission from the Serbian Chemical Society, Belgrade, Yugoslavia).

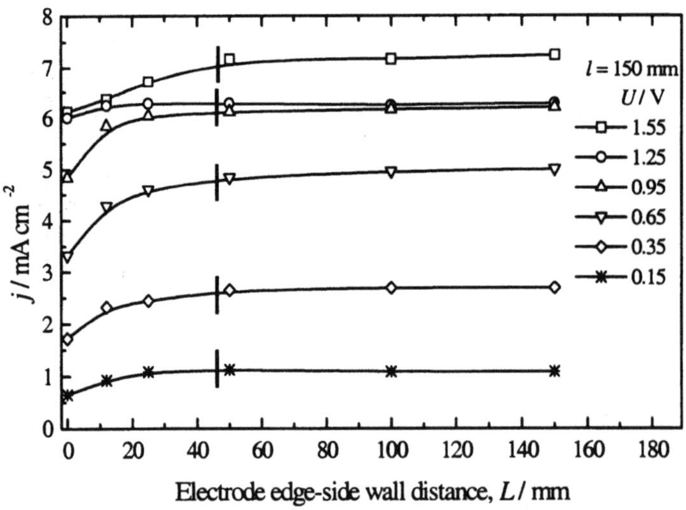

Figure 4.18. Current densities for different cell voltages as functions of the electrode edges – cell side wall distances in a cell with interelectrode distance l = 150 mm for the system Cu|0.1 mol dm^{-3} $CuSO_4$, 0.1 mol dm^{-3} H_2SO_4|Cu. Data from Fig. 4.10.[14] (Reprinted with permission from the Serbian Chemical Society, Belgrade, Yugoslavia).

It can be seen from Figs. 4.15-4.18 that the change of current with increasing L virtually ends at $L = l/2$ for $l < 2A$, at $L = A$ for $l > 2A$ and that L' can be successfully calculated using Eqn. 4.27 and the values from Table 4.1. Calculated L' is marked on the j-L dependences in Figs. 4.15-4.18 by a thin vertical line.

Hence, the Eq. 4.27 is valid for $l \leq 2A$ and $L > L'$. At larger interelectrode distances for $l > 2A$, L' remains constant and equal A.

Using equations 4.27 and 4.29, together with the $\eta_a+\eta_c$ and ρ values from Table 4.1, dependencies like those from Figs. 4.7-4.10. were calculated and the results are presented in Figs. 4.19 and 4.20.

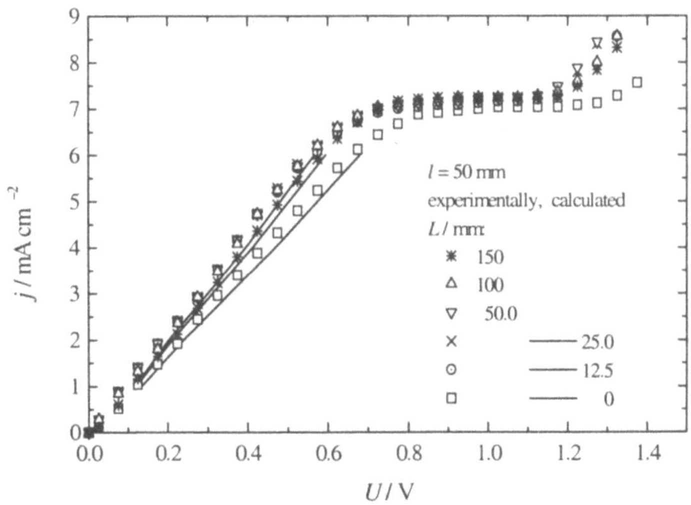

Figure 4.19. The cell voltage – current density dependencies (lines) for l = 50 mm, calculated using Eqs 4.27 and 4.29 and data from Fig. 4.14, the points are taken from Fig 4.8.[14] (Reprinted with permission from the Serbian Chemical Society, Belgrade, Yugoslavia).

It can be seen from Figs. 4.19 and 4.20 that the agreement between the calculated and the measured values is very good. In this way the method for the calculation of L' is verified, as is the possibility of the calculation of cell voltage – current density dependences.

Hence, the current density distribution in cells with plane parallel electrodes can be calculated without experimental measurements by using the data from simple polarization measurements, or literature data for kinetic parameters.

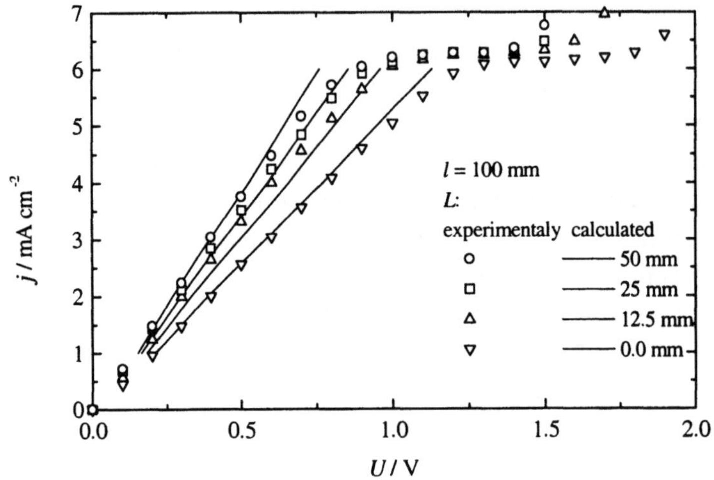

Figure 4.20. The cell voltage–current density dependencies (lines) l = 100 mm, calculated using equations 4.27 and 4.29 and data from Fig. 4.14, the points are taken from Fig 4.9.[14] (Reprinted with permission from the Serbian Chemical Society, Belgrade, Yugoslavia).

4.1.5.2 The critical current density for dendritic growth initiation at the edges

The polarization curve equation is given by:

$$j = \frac{j_0 f_c}{1 + \frac{j_0 f_c}{j_L}} \tag{2.28}$$

for $f_c \gg f_a$. The critical overpotential for dendritic growth initiation, η_i is given by

$$\eta_i = \frac{b_c}{2.3} \ln \frac{j_L}{j_0} \tag{3.60}$$

Substitution of η_i from Eq. 3.60 into Eq. 2.28 and further rearranging produces:

$$j_i = \frac{1}{2} j_L \tag{4.32}$$

where j_i is the critical current density for the dendritic growth initiation. On the other hand the edge current density j_e is given by:

$$j_e = j + j' \tag{4.14}$$

where

$$j' = j\frac{l'-l}{l'} \tag{4.13}$$

and

$$l' = l\frac{\eta_a + \eta_c + \rho l j}{\eta_a + \eta_c + \frac{\sqrt{2}}{2}\rho l j} \tag{4.25}$$

If $L > L'$, the edge current density could be obtained by combining Eqs. 4.13, 4.14 and 4.25 as:

$$j_e = j\frac{\eta_a + \eta_e + \left(2 - \frac{\sqrt{2}}{2}\right)\rho l j}{\eta_a + \eta_e + \rho l j} \tag{4.33}$$

Assuming that maximum edge current density is given by Eq. 4.32 the substitution of j_i in Eq. 4.33 instead of j_e and further rearranging produce:

$$j_{max} = \frac{1}{2}\frac{\eta_a + \eta_e + \rho l j}{\eta_a + \eta_e + 2 - \frac{\sqrt{2}}{2}\rho l j} j_L \tag{4.34}$$

from which the maximum current density, j_{max}, in the homogenous field at which dendrites at the edges do not grow can be calculated. It follows from Eq. 4.34 that for:

$$\eta_a + \eta_c >> \rho l j, \quad j_{max} \sim 0.5\, j_L \tag{4.35}$$

and if

$$\eta_a + \eta_c << \rho l j, \quad j_{max} \sim 0.4\, j_L \tag{4.36}$$

being in both cases larger than the current density corresponding to the end of the Tafel linearity, which is the optimum current density for the deposition of compact metal (see section 3.2.1.3.1). Hence, if deposition current density corresponds to the end of the Tafel linearity dendrites will not grow at the edges of the electrode. It should be noted that in metal electrorefinning working current density can be determined relative to the initial concentration of depositing ions, because it remains constant or increases during refining process. In electrowinning processes the working current density must be determined relative to the final concentration of depositing ions, because it is lower than initial one. The same reasoning is valid in the case of $L < L'$, meaning, in general, that if the current density in cell is lower than $0.4 j_L$ the dendrites and probably carrot like protrusion on the electrode edges can not grow.

4.2 CELLS WITH LOW ANODE POLARISATION

Cells with a small cathode and a large anode are often used in electroplating technology. In this case, a homogenous distribution of the deposit over the entire cathode is required.

4.2.1 The dependence of the current density at the tip of a stationary wire electrode on the current density in the middle of the electrode

It can be seen from Fig. 4.21 that the dissipation of current lines from the tip of a stationary wire electrode is more pronounced than in the case of the edges of plane parallel electrodes. This is because the dissipation in the former case occurs through the space, while in the latter case it takes place in one plane, normal to the electrodes, to which two symmetrically positioned points belong. Hence, it can be taken that the overall resistance between the tip of the cathode and anode will be equal to an infinitely large number of resistances, as in the case of the edges of two plane parallel electrodes connected in parallel, being equal to zero.

The cell voltage, U, for the part of the system where the current density is homogeneously distributed is given by:

$$U = E + \eta_a + \eta_c + RI \tag{4.37}$$

where R is the ohmic resistance of the electrolyte and I current in the cell, and for the tip of wire electrode:

$$U = E + \eta_{a,e} + \eta_{c,t} \tag{4.38}$$

where $\eta_{c,t}$ and $\eta_{a,e}$ are the overpotential at the tip of wire electrode and at the edge of the cylindrical anode, respectively, or after elimination of U-E from Eqs. 4.37 and 4.38

$$\eta_{a,e} + \eta_{c,t} = \eta_a + \eta_c + RI \tag{4.39}$$

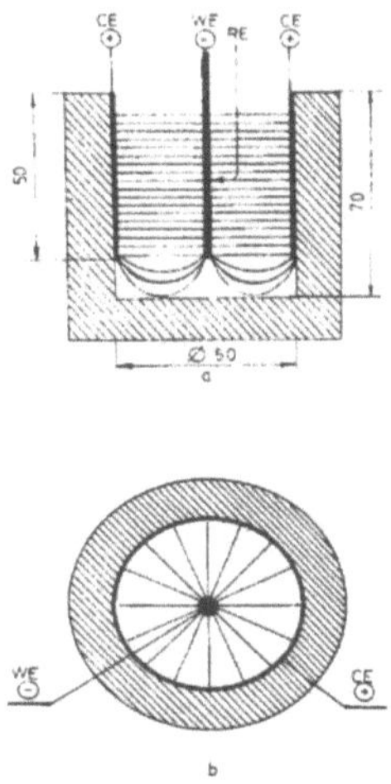

Figure 4.21. Current distribution in a cylindrical electrode geometry: a) front cross section, b) top cross section.[15] (Reprinted with permission of Elsevier Science).

In this way the ohmic potential drop in a homogenous field transforms into the electrochemical overpotential for points at the tip of a wire electrode or in a similar position. This means a larger tip current density than in a homogenous field. Finally, if the anode surface area is much larger than that of the cathode i.e., if:

$$\eta_c >> \eta_a \tag{4.40}$$

Eq 4.38 can be rewritten in the form:

$$\eta_{c,t} = \eta_c + RI \tag{4.41}$$

The ability of an electrode to distribute uniformly current density on a whole cathode can be easily estimated by comparing the cathodic polarization curve with the cathodic current density-cell voltage dependance. The lower is the difference between them, the better distribution of the current density should be expected.

On the other hand, Eq. 4.41 can be rewritten in the form:

$$j_t = \frac{j\, j_L \exp\left(\dfrac{2.3RI}{b_c}\right)}{j_L - j + j\exp\left(\dfrac{2.3RI}{b_c}\right)} \tag{4.42}$$

assuming that in mixed control deposition:

$$\eta = \frac{b_c}{2.3}\ln\frac{j}{j_0} + \frac{b_c}{2.3}\ln\frac{1}{1-\frac{j}{j_L}} \quad (2.39)$$

or in activation controlled deposition:

$$j_t = j\exp\left(\frac{2.3RI}{b_c}\right) \quad (4.43)$$

where

$$\eta = \frac{b_c}{2.3}\ln\frac{j}{j_0} \quad (2.37)$$

It can be taken to the first approximation that the same relation is valid also for the edge of a small, square stationary vertical cathode placed in the middle of a large cylindrical anode because the current line distribution is similar to the one from tip of wire electrode.

It follows that for $R \rightarrow 0$ Eq. 4.42 can be rewritten in the form:

$$j_t = j \quad (4.44)$$

This means that the lower the resistance of the electrolyte the is lower the difference in the current densities in the middle and at the tip of the electrode. The increase of b_c leads to a more uniform current distribution, also. This can happen in the presence of strongly adsorbed species or during deposition from some complex salt solution.

It is also seen that current density distribution on a macroprofile becomes uniform if $j \rightarrow j_L$, but in this case rough and dendritic deposit appears, being unuseful for plating purpose.

The difference, Δj, between the current density at the tip (edge) and in the middle of the electrode is obviously:

$$\Delta j = \frac{j\, j_L \exp\left(\frac{2.3RI}{b_c}\right)}{j_L - j + j\exp\left(\frac{2.3RI}{b_c}\right)} - j \quad (4.45)$$

and

$$\frac{\Delta j}{j}=\frac{j_{\mathrm{L}}\exp\left(\dfrac{2.3RI}{b_{\mathrm{c}}}\right)}{j_{\mathrm{L}}-j+j\exp\left(\dfrac{2.3RI}{b_{\mathrm{c}}}\right)}-1 \tag{4.46}$$

It follows from Eq. 4.42 that the exchange current density does not effect the current distribution, but there is an exception in the cases of electrodeposition processes characterized by very large value of the exchange current density. When the deposition can be under diffusion or the ohmic control and the deposit will be formed only at the edges or in a similar position, where ohmic resistance is low. Decreasing the exchange current density by complexing the depositing ion leads to a more homogenous distribution. The above discussion is illustrated by the following examples.

4.2.2 Experimental evidence

4.2.2.1 The effect of ohmic resistance

In order to illustrate the effect of the ohmic resistance of a cell, deposition of copper was performed at room temperature on to a stationary copper wire electrodes (length 40 mm and diameter 0.8 mm) placed in the middle of a cylindrical cell (length 5 cm and diameter 6 cm). The surface of the cell was covered by a high purity copper plate from electrolytes containing 0.1 mol dm^{-3} $CuSO_4$ and 0.1 mol dm^{-3} $CuSO_4$ in 0.5 mol dm^{-3} H_2SO_4. The plot of cell voltage and overpotential vs. current density for the copper deposition are shown in Figs. 4.22 and 4.23.

The relation between the current density at the electrode tip and in the homogenous field can be estimated by considering the cathodic overpotential – current density and cell voltage –current density dependencies. According to Eq. 4.41, the tip overpotential is equal to the cell voltage for a wire electrode.

It could be seen that the tip overpotential (cell potential) is larger than the overpotential in the middle of the electrode during deposition from 0.1 mol dm^{-3} $CuSO_4$, while during deposition from 0.1 mol dm^{-3} $CuSO_4$ in 0.5 mol dm^{-3} H_2SO_4 the overpotentials are practically the same. In the former case there is a large difference in the morphology of the deposit at the tip and the rest of the electrode (Fig. 4.24a), while in the latter case the quality of the deposit is the same over the whole surface (Fig. 4.24b).

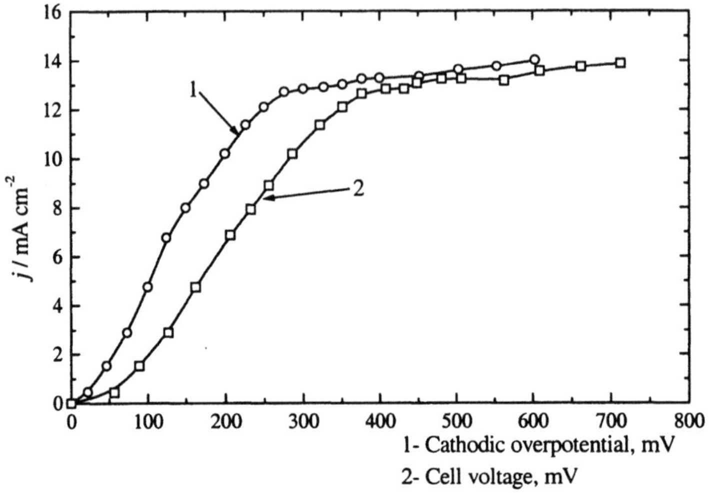

Figure 4.22. Plots of overpotential (1) and cell voltage (2) vs. current density for the copper deposition from 0.1 mol dm^{-3} $CuSO_4$.[15] (Reprinted with permission from Elsevier Science).

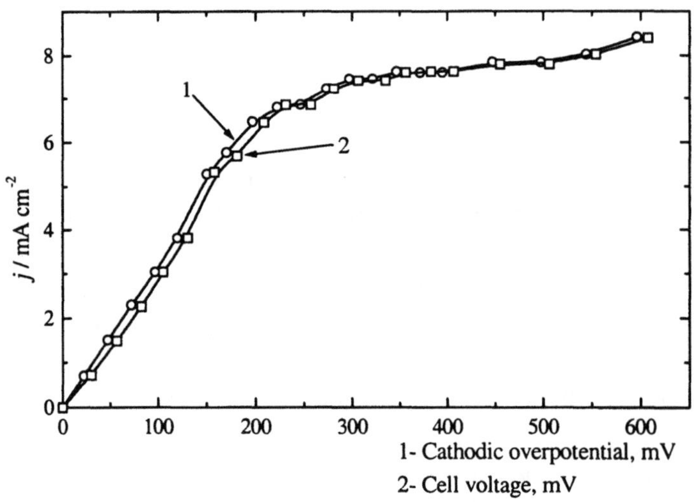

Figure 4.23. Plots of overpotential (1) and cell voltage (2) vs. current density for copper deposition from 0.1 mol dm^{-3} $CuSO_4$ in 0.5 mol dm^{-3} H_2SO_4.[15] (Reprinted with permission from Elsevier Science).

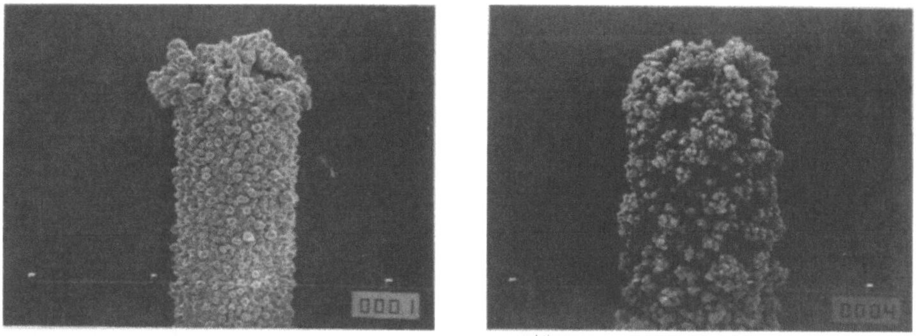

Figure 4.24. Copper deposit obtained on a stationary copper wire electrode from a) 0.1 mol dm^{-3} $CuSO_4$ b) 0.1 mol dm^{-3} $CuSO_4$ in 0.5 mol dm^{-3} H_2SO_4. Quantity of electricity 40 mA h cm^{-2}. Overpotential 250 mV.[15] (Reprinted with permission from Elsevier Science).

This is in perfect agreement with Eq. 4.44.

4.2.2.2 Deposition in the presence of strongly adsorbed organic additives (effect of increased cathodic Tafel slope)

In order to illustrate the effect of a strongly adsorbed organic additive[16], cadmium was deposited onto a stationary vertical flat copper electrode of area 1 cm x 1 cm placed in the middle of a cylindrical cell of diameter 6 cm, and height 5 cm. The cell surface was covered by the anode, which was made from high purity cadmium plate. The reference electrode was a high purity cadmium wire. The electrolyte used in all the experiments was a solution of 0.25 mol dm^{-3} $CdSO_4$ in 0.5 mol dm^{-3} H_2SO_4 to which 3.3 g dm^{-3} polyoxyethylene alkylphenol (9.5 mol ethylene oxide) was added.

The overpotential-current density and cell voltage-current density plots for cadmium deposition are presented in Fig. 4.25. The cathodic polarization curve obtained from potentiostatic polarization measurements has a similar shape to that found for an anode which can become passive; above a certain overpotential, increasing the cathode polarization leads to a decrease in the cathodic current followed by a range of potential in which the overpotential has little effect on the current. Current oscillations were observed at the beginning of this plateau in some cases (see also section 3.1.6.2).

It was shown in a section 3.2.1.3.1 that the optimum plating overpotential is determined by the upper limit of validity of the Tafel equation for the deposition process. In this case, as can be seen from Fig. 3.13 that the optimum deposition overpotentials for cadmium deposition are about 40 and 530 mV in the absence and in the presence of additive, respectively.

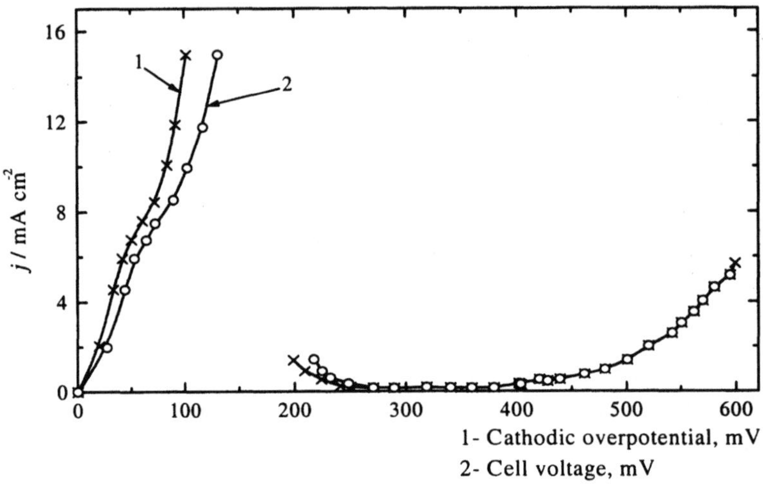

Figure 4.25. Plots of overpotential (1) vs. current density and cell voltage (2) for cadmium electrodeposition from sulfate solution with additive.[16] (Reprinted with permission from Elsevier Science).

Figure 4.25 shows that there is a large difference between the deposition overpotential and the cell voltage (tip overpotential) at low overpotentials which becomes negligible at high overpotentials, indicating a uniform current density distribution[16], due to the additive adsorption, as illustrated in Fig 4.26.

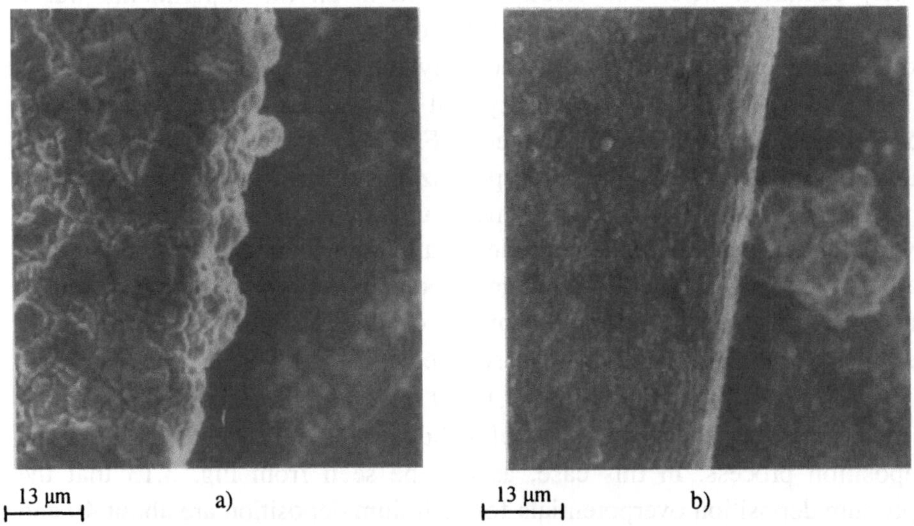

Figure 4.26. Cadmium deposits obtained from sulfate solution with additive, at the edge of a copper electrode. The thickness of deposit was 3μm. Deposition overpotential: a) 40 mV, b) 530 mV.[16] (Reprinted with permission from Elsevier Science).

4.2.2.3 Deposition from a complex salt solution (effect of exchange current density)

In order to illustrate the effect of deposition from complex salt solutions silver was deposited from simple and complex salt solutions[17]. The electrolytes used throughout the experiments were 0.1 mol dm^{-3} $AgNO_3$ in 0.2 mol dm^{-3} HNO_3 solution and 0.1 mol dm^{-3} $AgNO_3$ in 0.5 mol dm^{-3} $(NH_4)_2SO_4$ to which was added ammonium hydroxide to dissolve the precipitate of silver sulfate. The conductivities of the above solutions were almost the same. Silver was deposited onto a stationary vertical platinum cathode (1 cm x 1 cm) placed in the middle of a cylindrical cell (diameter, 6 cm, height 5 cm); the surface of the cell was covered by the anode, a high purity silver plate. Polarization curves were obtained for the platinum wire cathode, which had previously been plated with silver from the ammonium complex salt solution.

The overpotential-apparent current density and cell voltage (edge overpotential)-apparent current density plots for silver deposition from the nitrate solution and from the ammonium salt solution are presented in Figs. 4.27 and 4.28, respectively.

Silver deposition from the nitrate bath is under pure diffusion control at all overpotentials because $j_0 >> j_L$ and at $j << j_L$ the nucleation in the middle of the electrode does not occur because of ohmic control before nucleation. Hence, deposition from the nitrate bath is expected only at the edges where ohmic resistance is better, as predicted by Fig. 4.27. This is illustrated in Fig. 4.29a and b.

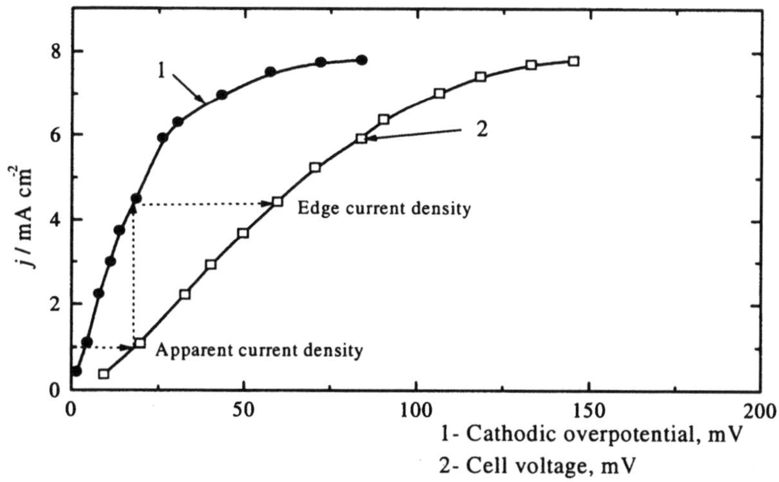

Figure 4.27. Overpotential-apparent current density (1) and cell voltage-apparent current density (2) dependencies for deposition from the nitrate bath.[17] (Reprinted with permission from Elsevier Science).

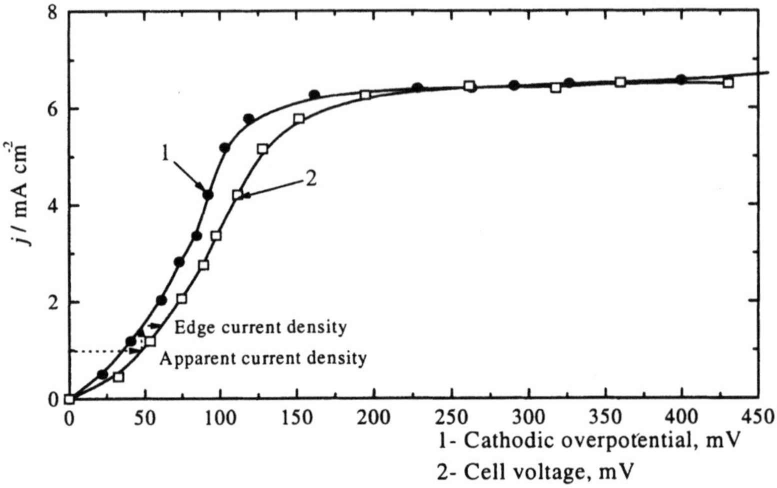

Figure 4.28. Overpotential-apparent current density (1) and cell voltage-apparent current density (2) dependencies for deposition from the ammonium complex bath.[17] (Reprinted with permission from Elsevier Science).

For the ammonium bath there is a region where deposition is under activation control because $j_0 < j < j_L$, $j_0 = 0.25$ mA cm^{-2} and a slope of 60 mV dec^{-1}. Hence, nucleation occurs over all surface. For deposition from the ammonium bath, as predicted by Fig. 4.28, a more homogenous distribution of the deposit is obtained, which is illustrated in Figs. 4.29c and 29d.

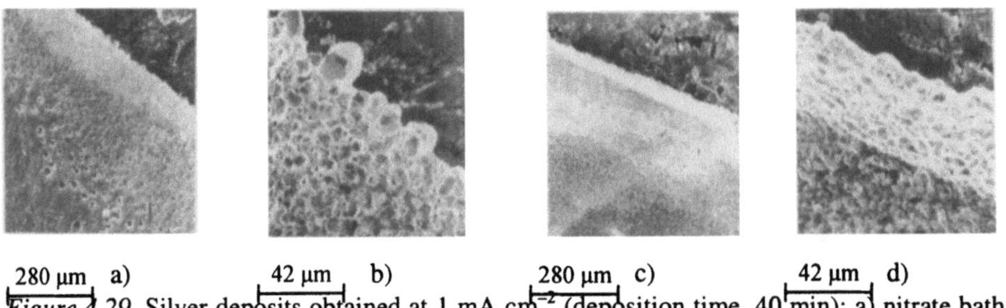

Figure 4.29. Silver deposits obtained at 1 mA cm^{-2} (deposition time, 40 min): a) nitrate bath; b) nitrate bath; c) ammonium complex bath; d) ammonium complex bath.[17] (Reprinted with permission from Elsevier Science).

In this way the effects of the deposition process parameters, ohmic resistivity on the current distribution are shown. It is obvious that decrease of the ohmic resistance and the increase of cathodic Tafel slope improve the current density distribution and in the case of $j_0 >> j_L$ the decrease of j_0 to the value $j_0 < j_c$. The effect of cell geometry and deposition conditions was treated in the previous section.

4.3 CORNER WEAKNESS PHENOMENA IN ELECTROFORMING

"Corner weakness" occurs in heavy deposits or electroforms at screened cathode parts *i.e.* corners. The deposit is thinner and at these areas,, in extreme cases, there is no deposition at all along the line of the corner bisector[18].

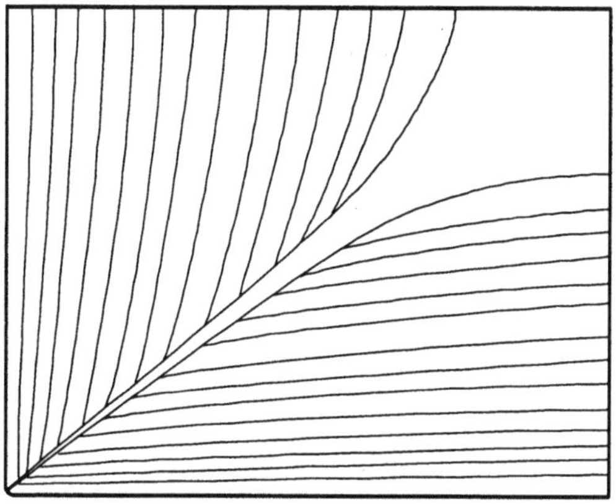

Figure 4.30. Schematic representation of microphotographs illustrating the "corner weakness" effect. (From Popov and Stevanović[19] with the kind permission of the Serbian Chemical Society, Belgrade, Yugoslavia).

The consequence is the emergence of fracture under negligible load along the line of the corner bisection, instead of fracture at much higher loads across the narrowest cross-section of an electroform normal to the line of pull.

To the best of our knowledge, a theoretical analysis of this phenomenon has not been reported so far. The purpose of this work was to undertake one, using the following assumptions[19]:

- the potential difference between each of two points on the anode and cathode is equal to the cell voltage,
- the current lines are normal to the electrode surface,
- along each current line a corresponding ohmic resistance exists and the current lines are independent of and insulated from each other,
- current lines in the vicinity of a protrusion divide into components which are normal to the electrode surface and
- the Kirchoff laws are valid for current lines branching.

4.3.1 Ohmic controlled deposition

According to the assumed model of current line division it follows that there is no deposition along the line of bisection if the division of the current lines occurs along the line indicated by the dashed line in Fig. 4.31. It can be seen that this configuration provides the same density of current lines at the cathode as at the anode.

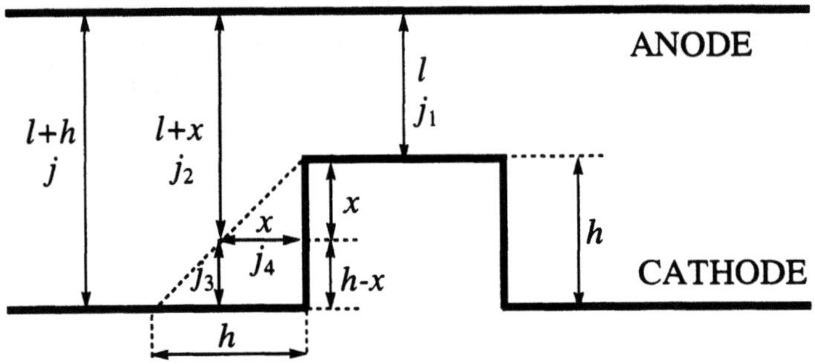

Figure 4.31. Evaluation of a current distribution in the cell using the concept of current line division.[19] (Reprinted with permission from the Serbian Chemical Society, Belgrade, Yugoslavia).

The ohmic potential drops along the current lines j and j_1 are given by Eqs. 4.47 and 4.48 respectively, asuming E=0.

$$U = \rho\,(l+h)j \tag{4.47}$$

$$U = \rho\, l j_1 \tag{4.48}$$

where U is the cell voltage and ρ is the resistivity of the solution. The ohmic resistance along the current line j_2 is somewhat different. It consists of the resistance between the anode and the dividing point (DP) and two resistances, proportional to x and $h-x$ connected in parallel between the DP and the cathode. Hence, the ohmic potential drop along current line j_2 can be written as:

$$U = \rho\, j_2\left[l + x + \frac{x(h-x)}{h}\right] \tag{4.49}$$

Elimination of U from Eqs. 4.47 and 4.49 and further rearrangement gives

$$j(l+h) = j_2\frac{hl + 2hx - x^2}{h} \tag{4.50}$$

The current densities j_3 and j_4 are given by

$$j_3 = \frac{\rho(l+h)j - \rho(l+x)j_2}{\rho(h-x)} \tag{4.51}$$

and

$$j_4 = \frac{\rho(l+h)j - \rho(l+x)j_2}{\rho x} \tag{4.52}$$

Substitution of j_2 from Eq. 4.50 into Eq. 4.51 and 4.52 and further rearrangement with use of Eqn. 4.47 gives

$$j_3 = j\frac{(l+h)x}{hl+2hx-x^2} \tag{4.53}$$

and

$$j_4 = j\frac{(l+h)(h-x)}{hl+2hx-x^2} \tag{4.54}$$

which enable the calculation of the deposit profiles at the cathode represented by Fig. 4.31

The proposed model implies that there is no current component in the direction of the corner vertex, and that the appearance of a crack along the corner bisector is to be expected.

A compact deposit cannot be obtained directly, but rather by the build up of the deposit in the x and y direction. An overlap of the x and y oriented deposits should occur when the current density virtually does not depend on the distance from the very corner. However, if the current density decreases upon approaching the corner vertex, the deposits would not overlap and a flaw would be created.

Equations 4.53 and 4.54 may be utilized for the calculation of the current density distribution at the beginning of deposition. The results of this calculation are shown in Fig. 4.32, as well as for different deposition times.

It is obvious that there should be no overlap of the deposit upon prolonged deposition. Moreover, it should be noted that the profiles were calculated assuming a constant current density, which is not the case in a real system where the space in the vicinity of the corner vertex is increasingly screened as the deposit grows. This implies that the real distribution of the metal deposit in the corners is worse than that calculated and shown in Fig. 4.32.

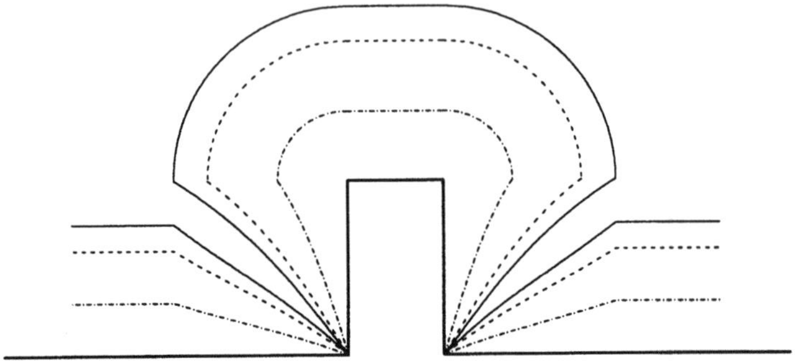

Figure 4.32. Simulation of a growth of the deposit from the model protrusion (h=5cm, l =15cm) calculated for pure ohmic control employing Eqs. 4.53 and 4.54.[19] (Reprinted with permission of the Serbian Chemical Society, Belgrade, Yugoslavia).

A number of microphotographs of deposit cross-sections illustrating the "corner weakness" effect can be found in the literature. They are schematically exemplified by Fig. 4.30. It can be seen that the calculated deposit profile (Fig. 4.32), with a crack appearing along the corner bisector, looks very similar to that typically obtained in plating practice (Fig. 4.30).

4.3.2 Mixed activation - diffusion - ohmic controlled deposition

In this case, it is not possible to perform an analysis of the current density distribution in the same manner as in the case of ohmic controlled deposition. Only a numerical solution can be obtained, regardless of the fact that the current line distribution is the same as shown in Fig. 4.31.

It is to be noted that in the model used anode and cathode are not isopotentional. Because of this the overpotentials in different points are calculated by direct using of Eqs.2.39 and 2.44 and corresponding current densities. It is assumed that cathode deposition is under mixed activation diffusion control and that anode dissolution is under activation control.

In this case, the first step is the calculation of the voltage current curves, for interelectrode distances l and $l + h$, using Eqs 4.55 and 4.56.

$$U = \frac{b_a}{2.3}\ln\frac{j_1}{j_0} + \frac{b_c}{2.3}\ln\frac{j_1}{j_0}\cdot\frac{j_L}{j_L - j_1} + \rho l j_1 \tag{4.55}$$

$$U = \frac{b_a}{2.3}\ln\frac{j}{j_0} + \frac{b_c}{2.3}\ln\frac{j}{j_0}\cdot\frac{j_L}{j_L - j} + \rho(l + h)j \tag{4.56}$$

where U is the cell voltage, b_a and b_c and j_0 are the anodic and cathodic Tafel slopes and exchange current density respectively and j_L is the limiting diffusion current density for the cathodic process, assuming the zero value of the reversible potential difference. The corresponding cell voltage-current curves shown in Fig. 4.33 enable the determination of the current densities on the frontal parts of the cathode (j and j_1) for any cell voltage.

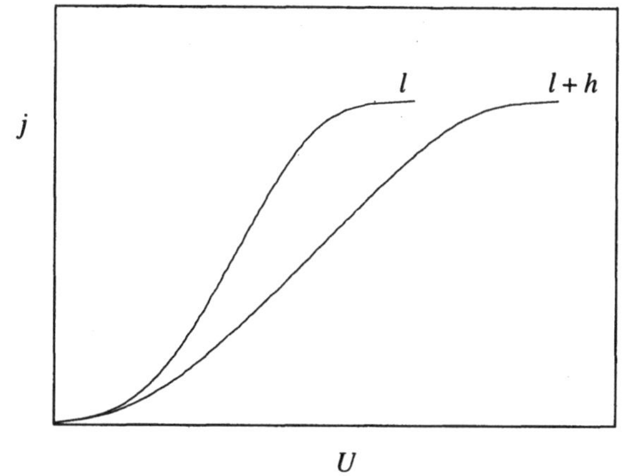

Figure 4.33. Cell voltage current curves for interelectrode distances l and $l + h$.[19] (Reprinted with permission from the Serbian Chemical Society, Belgrade, Yugoslavia).

The overall current density along the current line from the anode to the DP is obviously the sum of the partial ones branching at the DP, *i.e.* $j_3 + j_4$. Hence, Eqs. 4.57 and 4.58 are valid

$$U = \frac{b_a}{2.3}\ln\frac{j_3 + j_4}{j_0} + \rho(l + x)(j_3 + j_4) + \frac{b_c}{2.3}\ln\frac{j_3}{j_0}\frac{j_L}{j_L - j_3} + \rho j_3(h - x) \quad (4.57)$$

$$U = \frac{b_a}{2.3}\ln\frac{j_3 + j_4}{j_0} + \rho(l + x)(j_3 + j_4) + \frac{b_c}{2.3}\ln\frac{j_4}{j_0}\frac{j_L}{j_L - j_4} + \rho j_4 x \quad (4.58)$$

Elimination of U from Eqs. 4.57 and 4.58 gives

$$\frac{b_c}{2.3}\ln\frac{j_3}{j_0}\frac{j_L}{j_L - j_3} + \rho j_3(h - x) = \frac{b_c}{2.3}\ln\frac{j_4}{j_0}\frac{j_L}{j_L - j_4} + \rho j_4 x \quad (4.59)$$

Obviously, both sides of Eq. 4.59 represent the potential drop between the DP and the cathode, U_{DPC}, which can be plotted as a function of either j_3 or j_4, as shown in Fig. 4.34a and 4.34b, respectively.

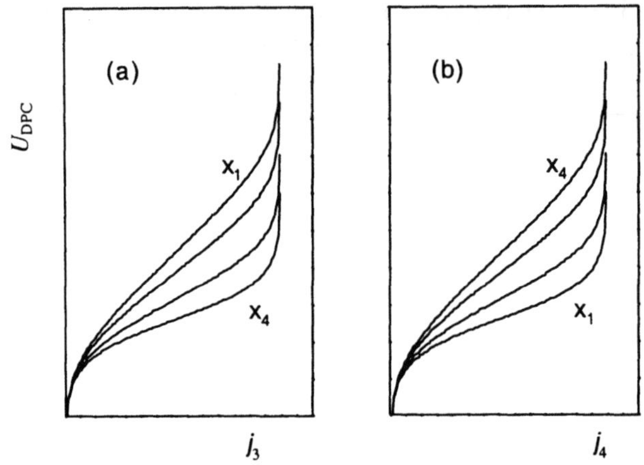

Figure 4.34. Potential drop between the DP and the cathode, U_{DPC}, as a function of (a) j_3 and (b) j_4, for different values of x.[19] (Reprinted with permission from the Serbian Chemical Society, Belgrade, Yugoslavia).

From the dependencies shown in Fig. 4.34, the corresponding values of j_3 and j_4 can be extracted by interpolation, for any U_{DCP} and x. These are required for the calculation of the potential drop between the anode and the DP, U_{ADP}, according to Eq. 4.60 and the overall cell voltage, U, according to Eq. 4.61.

$$U_{ADP} = \frac{b_a}{2.3} \ln \frac{j_3 + j_4}{j_0} + \rho(l + x)(j_3 + j_4) \tag{4.61}$$

$$U = U_{ADP} + U_{DPC} \tag{4.62}$$

Now, U, j_3 and j_4 can be tabulated or plotted as functions of the overall current density, $(j_3 + j_4)$, as exemplified in Fig. 4.35, for one value of x.

The situation is similar in the case of mixed diffusion-activation control. The profiles of the deposit at different times (Fig. 4.36) were calculated according to the procedures given above and the corresponding parameters: j_0 =0.05 mA cm^{-2}; j_L = 7 mA cm^{-2}; b_a =40 mV dec^{-1}; b_c =120 mV dec^{-1} and ρ =5 Ω cm, h=10 cm and l=5 cm.

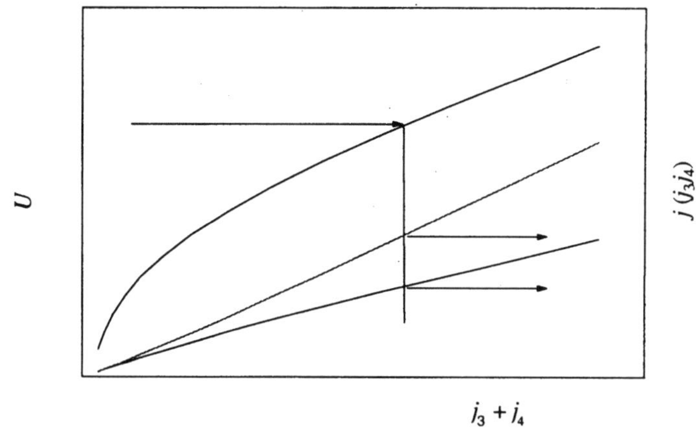

Figure 4.35. Cell voltage U, j_3 and j_4 as functions of the overall current ($j_3 + j_4$) for one chosen value of x.[19] (Reprinted with permission from the Serbian Chemical Society, Belgrade, Yugoslavia).

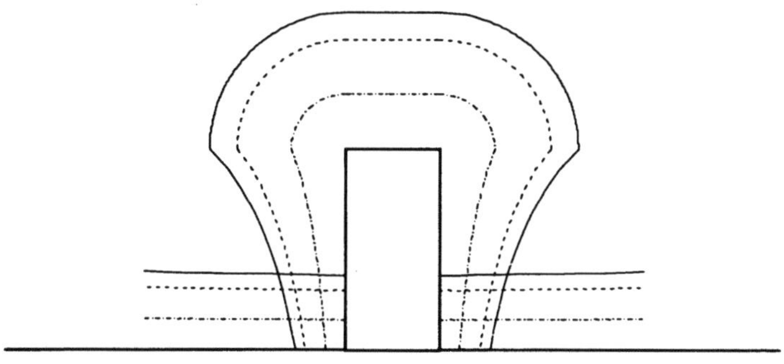

Figure 4.36. Simulation of the growth of the deposit obtained at U of 300 mV.[19] (Reprinted with permission from the Serbian Chemical Society, Belgrade, Yugoslavia).

Comparing the calculated deposit profiles for pure ohmic control (Fig. 4.32) with those pertaining to diffusion-activation control (Fig. 4.36), a significant difference in the cross section can be noticed. In the latter case, despite a much poorer system geometry (*cf.* h to l ratio) from the viewpoint of current distribution, no failure in the deposit appears. Even so, a flaw in the deposit along the corner bisector may be expected in both cases.

4.3.3 Activation - diffusion controlled deposition

In this case, the current line between the DP and the cathode splits into two equal parts. Hence, for $j_{3,4} = j_3 = j_4$, one can write:

$$U = \frac{b_a}{2.3} \ln \frac{2 j_{3,4}}{j_0} + \frac{b_c}{2.3} \ln \frac{j_{3,4}}{j_0} \frac{j_L}{j_L - j_{3,4}} \tag{4.62}$$

and simultaneously:

$$U = \frac{b_a}{2.3} \ln \frac{j_1}{j_0} + \frac{b_c}{2.3} \ln \frac{j_1}{j_0} \frac{j_L}{j_L - j_1} \tag{4.63}$$

Elimination of U from Eqs. 4.62 and 4.63 permits the correlation between j and j_1 in the form of Eq. 4.64, which cannot be solved explicitly if $b_a \neq b_c$.

$$b_a \ln \frac{2 j_{3,4}}{j_1} = b_c \ln \frac{j_1}{j_{3,4}} \frac{j_L - j_{3,4}}{j_L - j_1} \tag{4.64}$$

If $j_L >> j_{3,4}$ and $j_L >> j_1$ Eq. 4.64 can be rewritten in the form:

$$b_a \ln \frac{2 j_{3,4}}{j_1} = b_c \ln \frac{j_1}{j_{3,4}} \tag{4.65}$$

or

$$j_{3,4} = j_1 2^{-\frac{b_a}{b_a + b_c}} \tag{4.66}$$

which holds for pure activation control.

Evenly distributed deposits, without "corner weakness", may be obtained only by deposition under complete activation control at high Tafel slopes (Fig. 4.37). In practice, this is usually achieved by employing appropriate surface-active additives, as seen in Fig. 4.38.

In this way not only is the "corner weakness" effect fully explained, but also a new method of current density distribution evaluation in electrochemical cells is promoted.

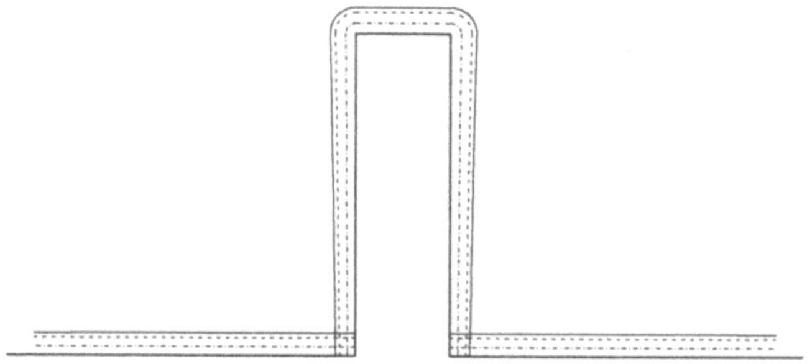

Figure 4.37. Simulation of the growth obtained at a cell voltage of 100 mV.[19] (Reprinted with permission from the Serbian Chemical Society, Belgrade, Yugoslavia).

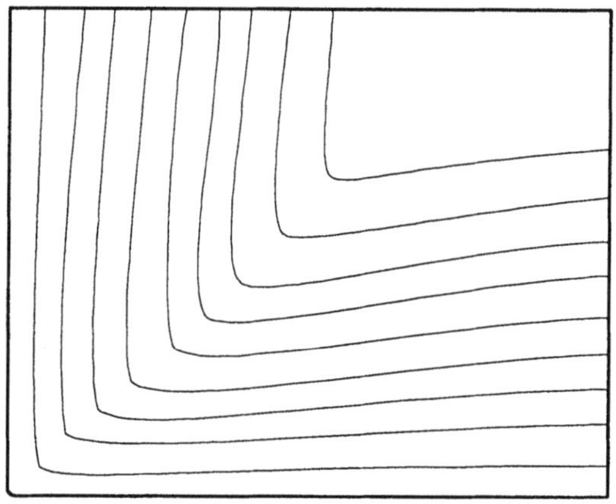

Figure 4.38. Schematic representation of microphotographs of cross-sections of deposits illustrating the effect of surface active agents on the elimination of "corner weakness".[19] (Reprinted with permission of the Serbian Chemical Society, Belgrade, Yugoslavia).

4.4 CONCLUSIONS

The current density distribution in electrochemical cell was considered in a new manner on the basis of equation of the electrode kinetics. In this way it was possible to illustrate the effect of deposition conditions, geometry of the system and kinetic parameters of deposition processes. Besides, three main problems of current density distribution in electrometallurgy were treated semiquantitatively. The edge effect in metal electrorefinning and electrowinning processes was discussed in details and it was shown that the dendritic growth at the edges of the electrodes can be avoided by keeping the deposition current density below some critical value, probably little larger than this corresponding to the end of the Tafel linearity. At this current density the less coarse deposit in the homogenous field without the dendrites at the edges of electrodes can be expected.

The current density distribution in electroplating and electroforming were also treated semiquantitatively and it was shown that decrease of ohmic resistance of electrolyte and increase of the Tafel slope for the cathodic process improves it. Obviously, all above discussion is valid if the local values of limiting diffusion current density do not varies along electrode surface, i.e. if the effect of the hydrodynamics can be neglected. As a metter of fact, the diffusion layer thickness may vary along the electrode interface due to hydrodynamic conditions and cause the different deposition conditions. This phenomenon is treated elsewhere[20,21]

4.5 FURTHER READINGS

1. Ibl, Norbert. "Current Distribution." In *Comprehensive Treatise of Electrochemistry,* Vol. 6, Ernest E. Yeager, John O'M. Bockris, Brian E. Conway and S. Sarangapani, eds. New York, NY: Plenum Press, 1983.
2. Ibl N., Current distribution in electrolysis (in French). Oberflähe-Surface 1975; 16:23-32
3. Wagner C. Theoretical analysis of the current density distribution in electrochemical cells. J. Electrochem. Soc. 1951; 98:116-28
4. Newman, John., *Electrochemical Systems.* N.J: Engelwood Clifts, Prentice Hall, 1973
5. Marathe V., Newman J. Current distribution on a rotating disc electrode. J. Electrochem. Soc. 1969; 116: 1704-19
6. Kasper C. The theory of the potential and the technical practice of electrodeposition. I. The general problem and the case of uniform flow, *Trans. Electrochem. Soc.*, 140; 77:131-42
7. Hoar T.D., Agar. Factors in throwing power illustrated by potential-current dependencies. Disc. Faraday. Soc. 1947; 1:162-68
8. Haring H.E., Blum W.M. Current distribution and throwing power in electrodeposition. Trans. Am. Electrochem. Soc. 1923; 43:365-97
9. Popov K.I., Zečević S.K., Pešić S.M. The current distribution in an electrochemical cell. Part I: The current voltage relationship for a cell with parallel plate electrodes. J. Serb. Chem. Soc. 1995; 60:307-16
10. Popov K.I., Maksimović M.D., Totovski D.Ć, Nakić V.N. Some aspects of current density distribution in electrolytic cells I: Dendritic growth of cadmium at the cathode edge in galvanostatic electrodeposition. Surf. Technol. 1983; 19:173-80
11. Popov K.I., Zečević S.K., Pešić S.M. J. The current distribution in an electrochemical cell. Part II:Qualitative considerations of the basis of polarization curve shape.J. Serb. Chem. Soc. 1996; 61:583-90
12. Popov K.I., Pešić S.M., Kostić T.M. The current distribution in an electrochemical cell. Part V: The determination of the depth of the current line penetration between the edges of the electrodes and the side walls at the cell. J. Serb.Chem. Soc. 1999; 64:795-800

13. Popov K.I., Pešić S.M., Živković P.M. The current distribution in an electrochemical cell. Part VII: The concluding remarks. J. Serb.Chem. Soc. in press
14. Popov K.I., Pešić S.M., Živković P.M. The current distribution in an electrochemical cell. Part VI: The quantitative treatment for cells with three plane parallele electrode arangments. J. Serb.Chem. Soc. 2001; 66: 491-98
15. Popov K.I., Pavlović M.G., Stojilković E.R., Stevanović Z.Ž. The current density distribution on stationary wire electrodes during copper and lead electrodeposition. Hydrometallurgy 1997; 46: 321-36
16. Popov K.I., Rodaljević Z.P., Krstajić N.V., Novaković S.D. Fundamental aspects of plating technology, V. The effect of strongly adsorbed species on the morphology of metal deposits. Surf. Technol. 1985; 25:217-22
17. Popov K.I., Krstajić N.V., Popov S.R. Fundamental aspects of plating technology, III. The effect of electrodeposition from complex salt solutions on metal distribution over macroprofiles. Surf.Technol. 1984; 22:245-50
18. Spiro, Peter, *Electroforming*. Teddington, Robert Draper Ltd, 1968.
19. Popov K.I., Stevanović R.M. A new line division concept for the determination of the current distribution in electrochemical cell. Part I. Theoretical background of the "corner weaknes" effect in electroforming. J.Serb.Chem.Soc. 2000; 65:905-14
20. Levich, Veniamin, *"Physicochemical Hydrodunamics"* N.J: Engelwood Clifts, Prentice Hall, 1962.
21. Ibl, Norbert. Dossenbach, O "Convective Mass Transport." In *Comprehensive Treatise of Electrochemistry,* Vol. 6, Ernest E. Yeager, John O'M. Bockris, Brian E. Conway and S. Sarangapani, eds. New York: Plenum Press, 1983.

Chapter 5

ELECTRODEPOSITION AT A PERIODICALLY CHANGING RATE

5.1 BASIC DEFINITIONS

It has been known for a relatively long time that the application of a periodically changing current in metal electrodeposition practice leads to improvements in the quality of electrodeposits. Three types of current variation have been found useful: reversing current (RC); pulsating current (PC); and sinusoidal, alternating current superimposed on a direct current (AC)[1-11]. In recent years, the beneficial effects of pulsating overpotential (PO) have also been discussed[3]. Even though this kind of electrodeposition at a periodically changing rate (EPCR) is important from a theoretical point of view and offers a variety of experimental possibilities, it is as yet not frequently used in metal electrodeposition practice.

5.1.1 Reversing current

Reversing current is represented schematically in Fig. 5.1. It is characterized by the cathodic current density, j_c, and the anodic current density, j_a, as well as by the duration of flow of the current in the cathodic and the anodic direction, t_c and t_a, respectively. Naturally,

$$t_c + t_a = T \tag{5.1}$$

where T is the full period of the RC wave.

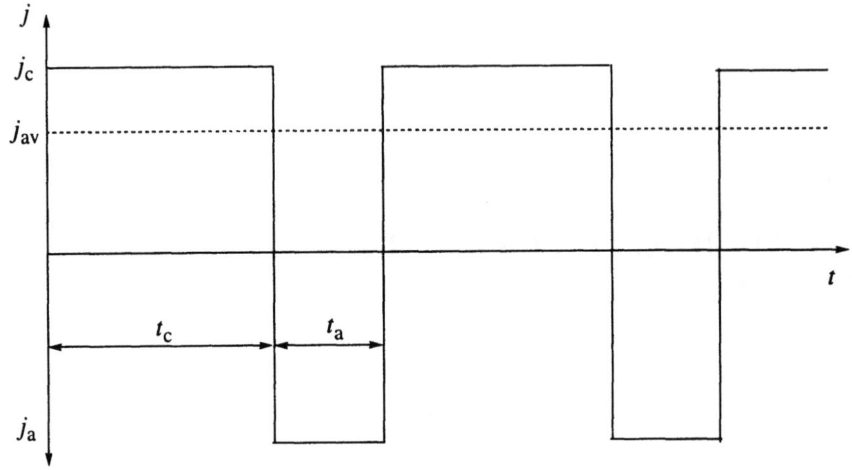

Figure 5.1. Waveform of a reversing current cycle.[7]

The average current density is then given by:

$$j_{av} = \frac{j_c t_c - j_a t_a}{t_c + t_a} \tag{5.2}$$

and for $j_c = j_a = j_A$

$$j_{av} = j_A \frac{1-r}{1+r} \tag{5.3}$$

where

$$r = \frac{t_a}{t_c} \tag{5.4}$$

RC is used in the second and millisecond range[7].

5.1.2 Pulsating current

Pulsating current consists of a periodic repetition of square pulses. It is similar in shape to RC except for the absence of the anodic component, as is shown in Fig. 5.2. PC is characterized by the amplitude of the cathodic current, j_c, the cathodic deposition time, t_c (on period), and the time interval t_p, in which the system relaxes (off period).

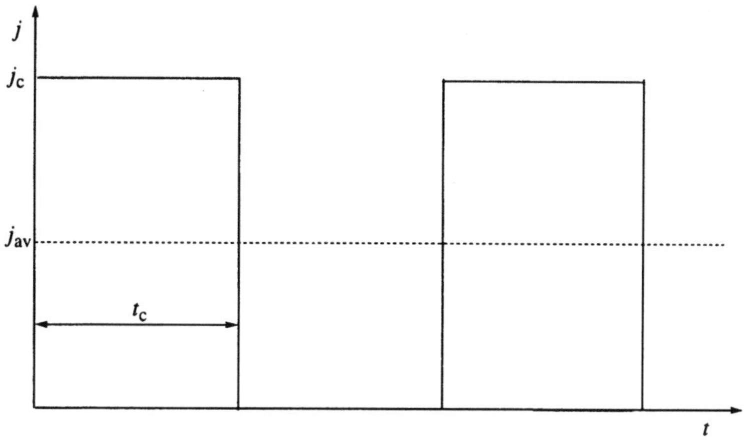

Figure 5.2. Waveform of a pulsating current cycle.[7]

The full period, T, is given by

$$t_c + t_p = T \tag{5.5}$$

and the average current density by

$$j_{av} = \frac{j_c t_c}{t_c + t_p} \tag{5.6}$$

or

$$j_{av} = \frac{j_c}{1+p} \tag{5.7}$$

where

$$p = \frac{t_p}{t_c} \tag{5.8}$$

It should be noted that rectified sinusoidal AC, especially half-rectified sinusoidal AC, often termed pulsating current in the literature, shows similar effects to those of PC[7].

5.1.3 Alternating current superimposed on direct current

Sinusoidal AC superimposed on a direct cathodic current (DC) is represented in Fig. 5.3. It is characterized by j_{dc}, j_p, and the frequency, which is usually 50 or 60 Hz. The resultant is termed an asymmetric sinusoidal current. The average current is equal to j_{dc}.

At a given DC value, three different types of current can be obtained, which can be denoted as follows: $j_p < j_{dc}$ "rippling current"; $j_p = j_{dc}$, "pulsating current"; $j_p > j_{dc}$, "current with an anodic component." The last type is mainly used in plating practice.

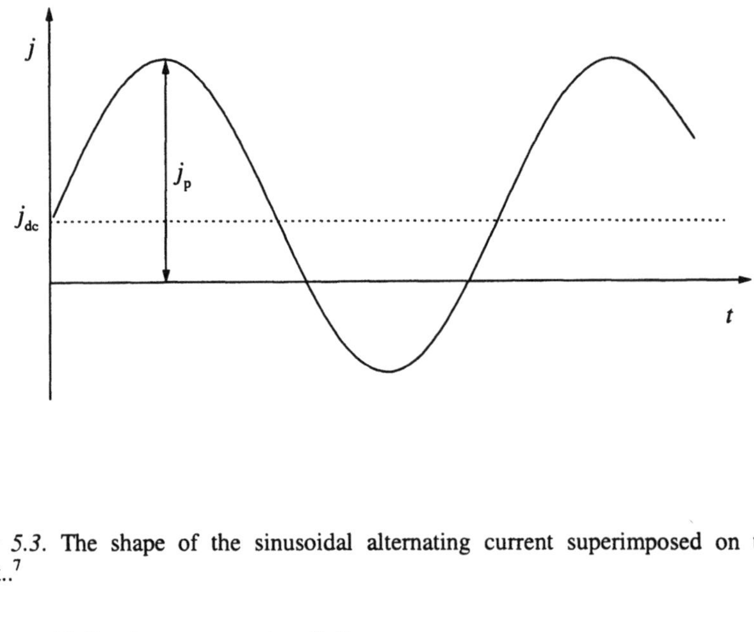

Figure 5.3. The shape of the sinusoidal alternating current superimposed on the direct current..[7]

5.1.4 Pulsating overpotential

Pulsating overpotential consists of a periodic repetition of overpotential pulses of different shapes. Square-wave PO is defined in the same way as PC except that the overpotential pulsates between the amplitude value η_A, and zero instead of current density. Non-rectangular pulsating overpotential is defined by the amplitude of the overpotential, η_A, frequency, and overpotential waveform[7].

There are a number of different current and overpotential waveforms used in EPCR[12,13], but the most important have been mentioned above.

5.2 SURFACE CONCENTRATION OF DEPOSITING IONS IN THE PERIODIC CONDITION

5.2.1 Electrodeposition with periodically changing range in the millisecond range

Electrodeposition with a periodically changing rate can be described in terms of time-and distance-dependent concentrations:

$$\frac{\partial C}{\partial t} = D\frac{\partial^2 C}{\partial x^2} \tag{5.9}$$

$$C(0, x) = C_0 \tag{5.10}$$

$$C(t, \delta) = C_0 \tag{5.11}$$

$$\frac{\partial C}{\partial x}\Big\|_{x=0} = \frac{j(t)}{nFD} \tag{5.12}$$

Equations 5.9 to 5.12 are solved for different $j(t)$ shapes and the solutions applied to different types of problems[7].

The current density $j(t)$ is the periodic function of time, which for periodic reverse currents[14] is given by:

$$j(t) = \begin{cases} j_c, & \text{for } mT < t \le [m + 1/(r+1)]T \\ -j_a, & \text{for } [m + 1/(r+1)]T < t \le (m+1)T \end{cases} \qquad m = 0, 1, 2..... \tag{5.13}$$

for pulsating currents[15] by:

$$j(t) = \begin{cases} j_c, & \text{for } mT < t \le [m + 1/(p+1)]\,T \\ 0, & \text{for } [m + 1/(p+1)]\,T < t \le (m+1)T \end{cases} \tag{5.14}$$

and for AC superimposed on DC[16] by:

$$j(f) = j_{dc} + j_p \sin(\omega t) \tag{5.15}$$

In the case of pulsating overpotential[3], $j(t)$ is given by:

$$j(t)= j_0\left[\frac{C(0,t)}{C_0}\exp\left(2.3\frac{\eta(t)}{b_c}\right)-\exp\left(-2.3\frac{\eta(t)}{b_a}\right)\right] \tag{5.16}$$

The surface concentration under periodic conditions can be evaluated as follows. For $j(t)$ given by Eq. 5.13, the solution of Eqs. 5.9 to 5.12 for $x = 0$, $t = [m + 1(r + 1)]\,T$, and $m\to\infty$, i.e., at the end of the cathodic pulses, under the periodic conditions is given by[14]:

$$C_c = C_0 - \frac{8\delta}{\pi^2 nFD}\sum_{k=0}^{\infty}\frac{1}{(2k+1)^2} x\left[j_c\frac{1-\exp[-\lambda_k T/(r+1)]}{1-\exp(-\lambda_k T)}\right. \left. - j_a\frac{\exp[-\lambda_k T/(r+1)]-\exp(-\lambda_k T)}{1-\exp(-\lambda_k T)}\right] \tag{5.17}$$

where $\lambda_k = \dfrac{(2k+1)^2\pi^2 D}{4\delta^2}$

The surface concentration, C_a, at the end of the anodic pulses under the same conditions, i.e., for $x = 0$, $t = (m + 1)\,T$, and $m\to\infty$ is given by:

$$C_a = C_0 - \frac{8\delta}{\pi^2 nFD}\sum_{k=0}^{\infty}\frac{1}{(2k+1)^2} x\left[j_c\frac{\exp[-\lambda_k rT/(r+1)]-\exp(-\lambda_k T)}{1-\exp(-\lambda_k T)}\right. \left. - j_a\frac{1-\exp(-\lambda_k rT/(r+1))}{1-\exp(-\lambda_k T)}\right] \tag{5.18}$$

It is easy to show that for a sufficiently long period T, ($T \gg t_0$), where $t_0 = \delta^2/(\pi^2 D)$, the system behaves as under DC conditions. For $T\to\infty$, and taking into account Eq. 2.29, Eqs. 5.17 and 5.18 become

$$\lim_{T\to\infty} C_c = C_0 - \frac{\delta\, j_c}{nFD} = C_0\left(1-\frac{j_c}{j_L}\right) \tag{5.19}$$

and

$$\lim_{T\to\infty} C_a = C_0 + \frac{\delta\, j_a}{n F D} \tag{5.20}$$

For a sufficiently small value of T, ($T << t_0$),

$$\lim_{T\to\infty} C_c = \lim_{T\to 0} C_a = C_s = C_0 - \frac{\delta}{n F D}\frac{j_c - r\, j_a}{r+1} \tag{5.21}$$

For $j(t)$ given by Eq. 5.14, solution of Eqs.5.9 to 5.12 for $x = 0$, $t=[m + 1/(p + 1)]$ T, and $m \to\infty$, i.e., at the end of the cathodic pulses under periodic conditions, is given by[15]:

$$C_{on} = C_0 - \frac{8\delta\, j_c}{\pi^2 n F D}\sum_{k=0}^{\infty}\frac{1}{(2k+1)^2}\frac{1-\exp(-\lambda_k T/(r+1))}{1-\exp(-\lambda_k T)} \tag{5.22}$$

The surface concentration at the end of pauses, C_{off}, under the same conditions [$x = 0$, $(m + 1)\ T$, $m \to \infty$] is given by:

$$C_{off} = C_0 - \frac{8\delta\, j_c}{\pi^2 n FD}\sum_{k=0}^{\infty}\frac{1}{(2k+1)^2}\frac{\exp(-\lambda_k pT/(p+1))-\exp(-\lambda_k T)}{1-\exp(-\lambda_k T)} \tag{5.23}$$

As in the previous case for $T >> t_0$, the system behaves as under DC conditions where

$$\lim_{T\to\infty} C_{on} = C_0 - \frac{\delta\, j_c}{nF D} \tag{5.24}$$

and

$$\lim_{T\to\infty} C_{off} = C_0 \tag{5.25}$$

For $T << t_0$, it follows from Eqs. 5.22 and 5.23 that

$$\lim_{T\to\infty} C_{on} = \lim_{T\to 0} C_{off} = C_s = C_0 - \frac{\delta\, j_c}{nFD(p+1)} \tag{5.26}$$

It is obvious from Eqs. 5.2, 5.4, and 5.7 and Eqs. 5.21 and 5.26 that, in both cases,

$$C_s = C_0 - \frac{\delta\, j_{av}}{nFD} = C_0\left(1 - \frac{j_{av}}{j_L}\right) \tag{5.27}$$

taking into account also Eq. 2.29

For *j(t)* given by Eq. 5.15, the surface concentration under periodic conditions is approximately given by[16]

$$C_s = C_0\left(1 - \frac{j_{dc}}{j_L}\right) - \frac{j_p}{nF(D\omega)^{1/2}}\sin\left(\omega t - \frac{\pi}{4}\right) \tag{5.28}$$

Hence, at sufficiently small value of T and not extremely high values of j_p, C_s, in AC will also be given by Eq. 5.27, implying that at sufficiently high frequencies, surface concentration is determined by the average current density regardless of the shape of the current wave.

In the case of a rectangular pulsating overpotential, $\eta(t)$ as a function of time, is given by[17]:

$$\eta(t) = \begin{cases} \eta_A, & \text{for } mT < t \le [m + 1/(p+1)]T \\ 0, & \text{for } [m + 1/(p+1)]T < t \le (m+1)T \end{cases} \tag{5.29}$$

Assuming that the surface concentration is determined by the average current density, Eq. 5.16 can be rewritten in the form

$$j = j_0\left(1 - \frac{j_{av}}{j_L}\right)\exp\left(2.3\frac{\eta(t)}{b_c}\right) - j_0\exp\left(-2.3\frac{\eta(t)}{b_a}\right) \tag{5.30}$$

For a sufficiently high value of η_A, Eq. 5.30 reduces during the on periods to:

$$j_{on} = j_0\left(1 - \frac{j_{av}}{j_L}\right)\exp\left(2.3\frac{\eta_A}{b_c}\right) \tag{5.31}$$

and during the off periods to:

$$j_{off} = -j_0\frac{j_{av}}{j_L} \tag{5.32}$$

The average current in PO deposition can easily be determined by

$$j_{av} = \frac{j_0}{p+1}\left[\left(1 - \frac{j_{av}}{j_L}\right)\exp\left(2.3\frac{\eta_A}{b_c}\right) - p\frac{j_{av}}{j_L}\right] \tag{5.33}$$

The overpotential amplitude is then given by

$$\eta_A = \eta_{dc} + \frac{b_c}{2.3}\ln\left(p+1+p\frac{j_0}{j_L}\right) \tag{5.34}$$

and

$$\eta_{av} = \frac{\eta_{dc}}{p+1} + \frac{b_c}{2.3p}\ln\left(p+1+p\frac{j_0}{j_L}\right) \tag{5.35}$$

where

$$\eta_{dc} = \frac{b_c}{2.3}\ln\left(\frac{j_{av}}{j_0}\right) - \frac{b_c}{2.3}\ln\left(1 - \frac{j_{av}}{j_L}\right) \tag{5.36}$$

and

$$\eta_{av} = \frac{\eta_A}{p+1} \tag{5.37}$$

Polarization curves for the average values for the copper deposition have been successfully calculated from the stationary polarization curve using Eq. 5.35[17] for $j_0 << j_L$. This is a good evidence that in PO deposition, the average current density also determines the surface concentration of the depositing ion.

The overpotential amplitude is larger than in the DC regime for one and the same average current density. At the same time, the diffusion overpotential remains constant, depending on the average current density only. Hence, the part of activation control in the overall amplitude overpotential increases with increasing pause to pulse ratio.

The situation is similar in pulsating or reversing current electrodeposition.

5.2.2 Capacitance effects

From the above discussion, it can be concluded that the useful range of frequencies is limited by mass-transfer effects at low frequencies. At high frequencies, the useful range is limited by the effect of the capacitance of the electrical double layer[15]. This is shown here for PC deposition. The time dependencies of the overpotential during the current pulses are shown in Fig. 5.4.

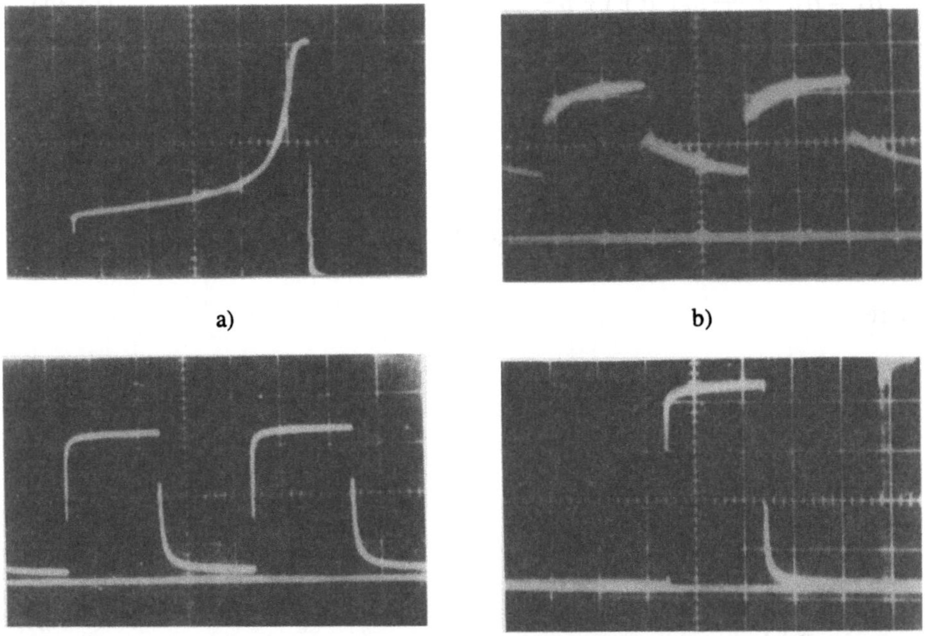

a) b) c) d)

Figure 5.4. The time dependence of the overpotential during current pulses in PC copper deposition from 0.5 mol dm^{-3} $CuSO_4$ in 1.0 mol dm^{-3} H_2SO_4: a) t_c = 10 s, p = 1, j_c = 1.2 j_L; x – axis, 2 s/div.; y-axis, 0.2 V/div.; b) t_c = 10^{-4} s, p = 1, j_c =1.2 j_L; x-axis, 5x10-5 s/div.; y-axis, 0.1 V/div.; c) t_c = 10^{-2} s, p = 1, j_c = 1.2 j_L; x-axis, 5x10^{-3} s/div.; y-axis, 0.1 V/div.; d) t_c = 10^{-2} s, p = 9, j_c = 6 j_L; x-axis, 5x10^{-3} s/div.; y-axis, 0.2 V/div.[15] (Reprinted with permission from Elsevier Science).

Mass-transfer limitations cause an increase of the overpotential at deposition times longer than the transition time as shown in Fig. 5.4a; the system enters full diffusion control at low frequencies if $j_c > j_L$. This is followed by an increase in the average overpotential[10,15]. At high frequencies, the PC is used both for double-layer charging and discharging, and for the deposition process, as illustrated by Fig. 5.5. The capacitance current during periodic charging and discharging of the double layer, at frequencies at which the effect of the double layer cannot be neglected, produces a smearing effect on the Faradic current wave, as illustrated by Fig. 5.4b.

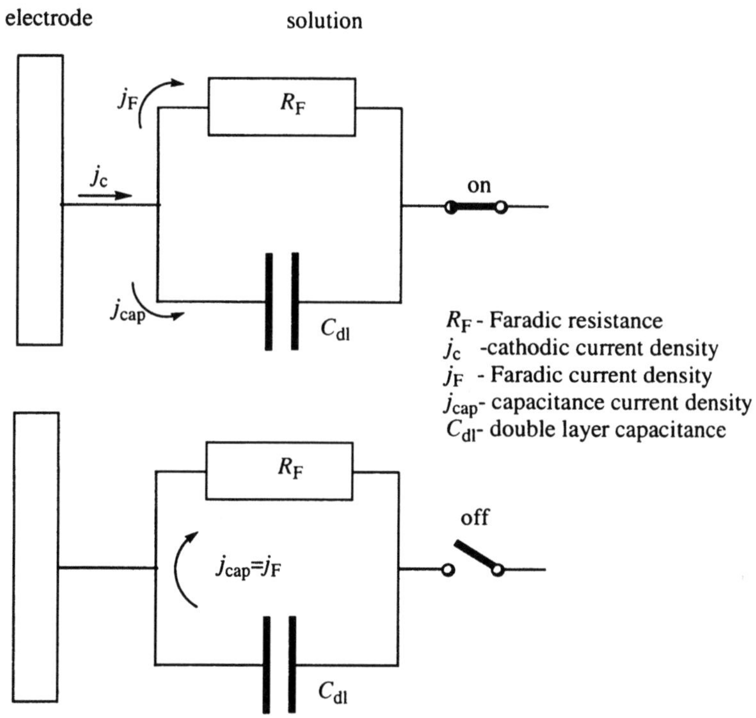

Figure 5.5. Schematic representation of the effect of the double layer capacitance on the faradic current during pulsating current electrodeposition.

Hence, as the frequency increases, the faradic current wave flattens, approaching a DC shape, and gives the same quality of deposit as DC even though the overall current appears to be pulsating. This is also followed by an increase in the average overpotential. Hence, the minimum average overpotential is a good indicator of the optimum frequency range of pulsation[3,10,15] in PC deposition. This range depends on the average current density and p, but, in general, the frequency lies in the range between 10 and 100 Hz, as illustrated in Figs. 5.4c and 5.4d.

In PO deposition, the effect of the double layer capacitance becomes less pronounced at higher frequencies compared to the other cases[10]. Also, at very high frequencies, the shape of the PO wave changes; for example, a square-wave PO becomes similar to a triangular one[10,17].

5.2.3 Reversing current in the second range

For T close to t_0, the behaviour of the system under RC conditions has to be analyzed using Eq. 5.18[7]. In this case, the concentration distribution inside the diffusion layer at the end of the anodic pulse is close to that given by Eq. 5.10. It follows from Eq. 5.18 that this will occur at:

$$C_a = C_0 \tag{5.38}$$

or

$$\sum_{k=0}^{x} \frac{1}{(2k+1)^2} \left[j_c \frac{\exp\left(\frac{-\lambda_k rT}{r+1}\right) - \exp(-\lambda_k T)}{1-\exp(-\lambda_k T)} - j_a \frac{1-\exp\left(\frac{-\lambda_k rT}{r+1}\right)}{1-\exp(-\lambda_k T)} \right] = 0 \tag{5.39}$$

It is known[14] that for $rT / (r+1) \geq 1.5\ t_0$, the series in Eq. 5.39 can be approximated using only the first term ($k = 0$). Hence, for $j_c = j_a$,

$$\exp\left(-\frac{r}{r+1} \cdot \frac{T}{4t_0}\right) - \exp\left(-\frac{T}{4t_0}\right) = 1 - \exp\left(-\frac{r}{r+1} \cdot \frac{T}{4t_0}\right) \tag{5.40}$$

or

$$r = \frac{\frac{4t_0}{T} \ln \frac{2}{1+\exp(-T/4t_0)}}{1 - \frac{4t_0}{T} \ln \frac{2}{1+\exp(-T/4t_0)}} \tag{5.41}$$

It is easy to show that for $T = 3\ t_0$, $r = 0.7$ and for $T = 16\ t_0$, $r = 0.2$ by assuming that Eq. 5.40 is valid for $T > 3\ t_0$ and that for $T > 16\ t_0$, the system behaves as under DC conditions. The optimum ratio t_c/t_a is given by

$$1.5 \leq t_c / t_a \leq 5 \tag{5.42}$$

for periods T such that

$$3\ \text{s} \leq T \leq 16\ \text{s} \tag{5.43}$$

if $t_0 = 1$ s for $\delta = 10^{-2}$ cm and $D = 10^{-5}$ cm^2 s^{-1}. A good agreement between the shape and the frequency of the RC calculated in this way and literature data is obtained, because in practically all cases, according to Bakhvalov[2],

$$1 \text{ s} \leq T \leq 30 \text{ s} \tag{5.44}$$

On the other hand, the solution of Eqs. 5.9 to 5.12 for

$$j(t) = j_c \tag{5.45}$$

is given by[18]

$$C_s = C_0 - \frac{j_c \delta}{nFD}\left\{1 - \frac{8}{\pi^2}\sum_{k=0}^{\infty}\frac{1}{(2k+1)^2}\exp\left[-\frac{(2k+1)^2 t}{4t_0}\right]\right\} \tag{5.46}$$

It follows from Eq. 5.46 that the surface concentration of depositing ions for $t \geq t_0$ can be given by:

$$C_s = C_0 - \frac{j_c \delta}{nFD}\left[1 - \frac{8}{\pi^2}\exp\left(-\frac{t}{4t_0}\right)\right] \tag{5.47}$$

The maximum amplitude of the current density variation, $j_{A,max}$, corresponding *to* $C_s = 0$ after a deposition time t_c, is given by

$$j_{A,max} = \frac{j_L}{1 - (8/\pi^2)\exp(-t_c/4t_0)} \tag{5.48}$$

It is obvious that using Eqs. 5.1 and 5.4 and $r = f(T)$ given by Eq. 5.41, Eq. 5.48 can be rewritten in the form

$$j_{A,max} = \frac{j_L}{1 - (16/\pi^2)\dfrac{\exp(-T/4t_0)}{1 + \exp(-T/4t_0)}} \tag{5.49}$$

In this way the complete RC wave can be estimated, or precisely calculated without approximations using a computer.

5.3 PREVENTION OF THE FORMATION OF SPONGY DEPOSITS AND THE EFFECT ON DENDRITIC PARTICLES

EPCR is used in the charging of silver-zinc storage batteries, to prevent, or to delay, the formation of spongy and dendritic deposits of zinc[19,20]. It is impossible to obtain smooth deposits of zinc from alkaline zincate solutions during prolonged deposition at a constant rate[20,21], because at lower overpotentials, spongy deposits arise while at higher overpotentials, dendritic zinc is formed.

It is well known that the reversible potential of a surface with radius of curvature r_{cur} would depart from that of a planar surface by the quantity[22]

$$\Delta E_r = \frac{2\gamma V}{F r_{cur}} \tag{5.50}$$

where γ is the interfacial energy. The filaments which form spongy deposits have extremely small tip radii. This makes the equilibrium potential of the spongy deposit 7 to 10 mV more cathodic than that of zinc foil[23,24].

Spongy deposit formation can, however, be completely prevented by PO deposition[20], as illustrated in Fig. 5.6.

Obviously, the more negative filaments dissolve faster during the off period than the flat surface, resulting in a smooth deposit. This is also valid for deposition using all current or overpotential waveforms that are characterized by some anodic current flow[3,7].

If spongy deposit formation is prevented or delayed the charging of the silver –zinc batteries is considerably improved[3].

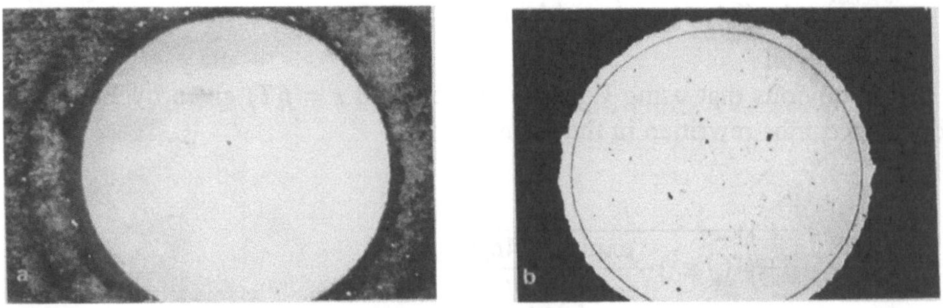

145 μm a) 145 μm b)

Figure 5.6. Cross-section photomicrograhs of zinc deposits plated out from a 50 g dm^{-3} ZnO in 10 mol dm^{-3} KOH solution onto a copper wire: a) constant overpotential, η = 22 mV, initial j_{av} = 4,0 mA cm^{-2}; deposition time 40 s; b) PO, p = 1, η_A=40 mV, ν = 100 Hz, initial j_{av} = 4,0 mA cm^{-3}; deposition time 4 h;[20]

The quantitative treatment of the selective dissolution during pauses can be performed as follows[25]. Equation 5.30 is valid for a flat electrode surface or protrusions with sufficiently large tip radii, where the surface energy term can be neglected. If it can not be neglected in zinc deposition, Eq. 5.30. can be rewritten for the tip of a protrusion inside the diffusion layer (h – height of protrusion and r_t – protrusion tip radius) in the form:

$$j_{\text{tip}} = j_0\left(1-\frac{j_{\text{av}}}{j_{\text{L}}}+\frac{j_{\text{av}}h}{j_{\text{L}}\delta}\right)\exp\left(\frac{2.3\eta(t)}{b_{\text{c}}}\right) - j_0\exp\left(\frac{2\gamma V}{RTr_{\text{t}}}\right)\exp\left(-\frac{2.3\eta(t)}{b_{\text{a}}}\right) \quad (5.51)$$

where j_{tip} is the tip current density and j_{av} is the average current density to the macroelectrode.

This is because:

$$C_{\text{tip}} = C_{\text{s}} + (C_0 - C_{\text{s}})\frac{h}{\delta} \quad (5.52)$$

if $C_s \neq 0$, or $j_{\text{av}}<j_{\text{L}}$ (C_s and C_{tip} are the concentration of the depositing ions on the electrode surface and on the tip of a protrusion, respectively), if $r_t/h \rightarrow 0$.

The output current during pauses ($\eta = 0$) become

$$j = j_0\left(1-\frac{j_{\text{av}}}{j_{\text{L}}}+\frac{j_{\text{av}}}{j_{\text{L}}}\frac{h}{\delta}\right) - j_0\exp\left(\frac{2\gamma V}{RTr_{\text{t}}}\right) \quad (5.53)$$

It is easy to show that the difference between the current density on the tip of a dendrite and the flat surface during the "off" period can be given by

$$\Delta j = j_0\frac{j_{\text{av}}h}{j_{\text{L}}\delta} - j_0\exp\left(\frac{2\gamma V}{RTr_{\text{t}}}\right) \quad (5.54)$$

This means that the dissolution of the a protrusion with tip radius r_t is faster relative to the flat surface or a protrusion with a sufficiently large value of r_t. It is obvious that spongy filaments can be completely (Fig. 5.6), and dendrites with low tip radii partially or completely dissolved during the pause, (Fig 5.7.). This means that both branching of dendrites and the formation of agglomerates can be prevented in square-wave pulsating overpotential deposition. In this way, even powder particle like that in Figs. 5.8 and 5.9 can be obtained.

Figure. 5.7. Lead deposit obtained a) at a constant overpotential from 0.1 mol dm^{-3} $Pb(CH_3C00)_2$ + 1.5 mol dm^{-3} $NaCH_3C00$ + 0.15 mol dm^{-3} CH_3COOH on a copper substrate, at 75 mV, for 50 s, b) The same as in a) but for. 350 s at 0 mV, c) The same as in a) but for 100 s at -10 mV (anodic) d) square-wave pulsating overpotential, overpotential amplitude 75 mV, pause to pulse ratio 3, pulse duration 0.5 s.[25] (Reprinted with permission of Elsevier Science).

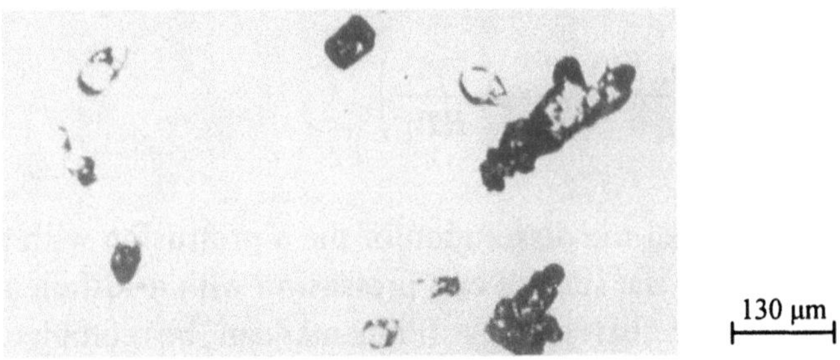

Figure.5.8. Silver powder particles obtained by pulsating overpotential. Pulse-to-pause ratio 1:5. Pulse duration 50 ms. Overpotential amplitude 160 mV. Deposition was carried out from an electrolyte containing 10 g dm^{-3} $AgNO_3$ and 100 g dm^{-3} $NaNO_3$ onto a graphite electrode.[26]

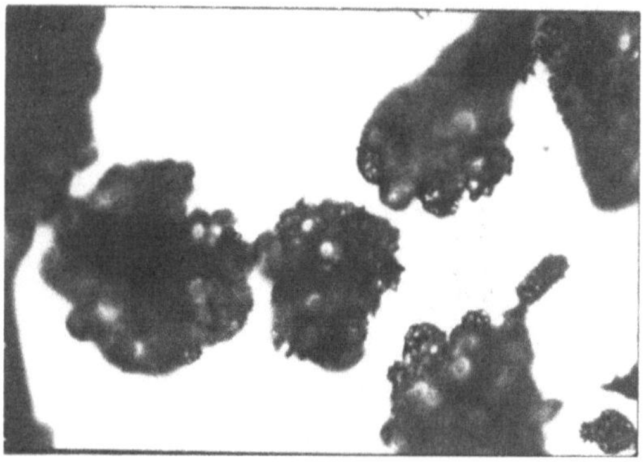

Figure. 5.9. Copper powder particles obtained using a square-wave pulsating overpotential of amplitude 600 mV, pulse duration 5 ms with a pulse to pause ratio of 1:5. The deposition was carried out from 0.1 mol dm^{-3} $CuSO_4$ in 0.5 mol dm^{-3} H_2SO_4 onto a platinum electrode painted with shellac. [27]

The monocrystal surfaces of silver powder particles can be explained by the assumption that during the off period of PO the adatoms in nonstable positions will dissolve faster than atoms in stable position in lattice. The similar effect on the morphology of powder particles can be seen in RC deposition[28,29], which leads to the strong effect on the apparent density of copper powders[29].

5.4 COMPACT DEPOSITS

5.4.1 Surface film

The first stage of metal film formation is nucleation on a foreign substrate. The nucleation rate, J, is given by Eq. 3.10 and depends strongly on the deposition overpotential. The nucleation overpotential is larger than the stationary one in galvanostatic deposition, and stationary values of the overpotential can be used in discussions of the effect of EPCR in galvanostatic as well as in potentiostatic deposition. It is obvious that in all cases of EPCR, the overpotential amplitude, η_A, is larger than in constant-current or constant overpotential deposition, η, for the same average current density.[30] Nucleation rate, J, in DC regime is given by:

$$J = K_1 \exp\left(-\frac{K_2}{\eta^2}\right) \tag{3.10}$$

Therefore,

$$\eta_A = \eta + \Delta\eta \qquad (5.55)$$

where $\Delta\eta > 0$ and

$$J_{EPCR} = K_1 \exp\left(-\frac{K_2}{(\eta + \Delta\eta)^2}\right) \qquad (5.56)$$

Hence, for the same current density:

$$J_{EPCR} > J \qquad (5.57)$$

as illustrated in Fig. 5.10.

On the other hand the increased amplitude on the current density leads to an increase of the ohmic potential drop during the pulses in EPCR relative to constant regimes and Eq. 3.5

$$r = f\frac{U_\Omega}{U_\Omega - \eta_{cr}} r_N \qquad (3.5)$$

can be rewritten in the form:

$$r_{EPCR} = f\frac{U_{\Omega(EPCR)}}{U_{\Omega(EPCR)} - \eta_{cr}} r_N \qquad (5.58)$$

It follows from Eq. 3.5 and Eq. 5.58 that:

$$r_{EPCR} < r \qquad (5.59)$$

because with increasing p

$$U_{\Omega(EPCR)} > U_{\Omega(CR)} \qquad (5.60)$$

Hence, the increasing nucleation density is also due to the decreasing zero nucleation zone radii. This effect leads to an increased coverage of the foreign substrate by the same quantity of deposited metal and to decreased porosity, surface resistance and increased density of deposit. Also, it can be expected that increase in compactness is associated with a decrease in internal stresses and increased ductility and hardness of metal deposits[7].

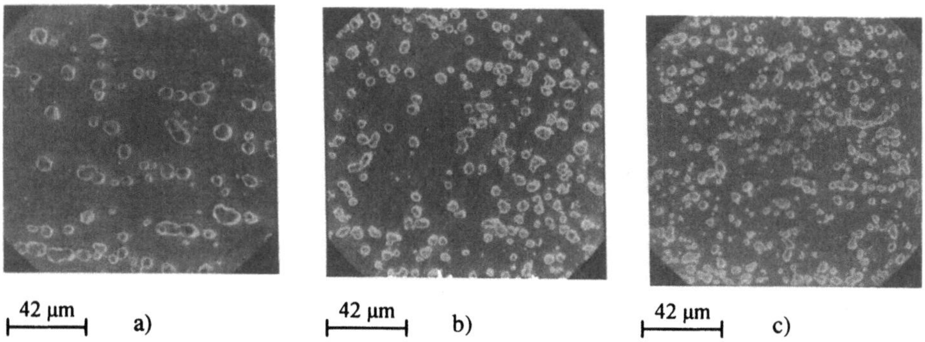

Figure. 5.10. SEM photomicrographs of cadmium deposits obtained from 1.0 mol dm^{-3} $CdSO_4$ in 0.5 mol dm^{-3} H_2SO_4 on a plane copper cathode by DC and PC deposition at different pause-to-pulse ratios. j_{av} = 10 mA cm^{-2}, t_c = 10 ms, t = 120 *s*. a) DC; b) PC p = 4, t_c = 10 ms, c) PC p = 9, t_c = 10 ms,.[30] (Reprinted with permission of Elsevier Science).

5.4.2 Electrode surface coarsening

The general equation of the polarization curve is given by:

$$j = \frac{j_0 f_c - j_0 f_a}{1 + \dfrac{j_0 f_c}{j_L}} \qquad (2.28)$$

for the flat part of an electrode and by

$$j_t = \frac{j_0 f_c - j_0 f_a}{1 + \left(\dfrac{j_0 f_c}{j_L}\right)\left(\dfrac{\delta - h}{\delta}\right)} \qquad (3.26)$$

for the tip of a protrusion (δ>>h), around which the lateral diffusion flux can be neglected.

Using Eq. 5.19 in the form

$$C_0 - C_s = C_0 \frac{j}{j_L} \qquad (5.62)$$

it is easy to show that

$$C_0 - C_s = C_0 \frac{j_0 f_c}{j_L + j_0 f_c} \tag{5.63}$$

and

$$C_0 - C_{s,t} = C_0 \frac{j_0 f_c}{\frac{j_L \delta}{\delta - h} + j_0 f_c} \tag{5.64}$$

if $f_c >> f_a$
where C_s and $C_{s,t}$ are the surface concentration of depositing ions on the flat electrode surface and on the tip of a protrusion respectively.

The rate of increase at the tip of a protrusion relative to the flat surface is given by[31]

$$\frac{dh}{dt} = VD\left(\frac{C_0 - C_{s,t}}{\delta - h} - \frac{C_0 - C_s}{\delta}\right) \tag{5.65a}$$

or

$$\frac{dh}{dt} = \frac{j^2}{j_L} \frac{V}{\delta n F} h \tag{5.65b}$$

after substitution of C_0 - $C_{s,t}$ and C_0 - C_s from Eqs.5.62 and 5.63 into Eq. 5.65a and further rearranging, assuming $\delta >> h$.

It was shown earlier that at sufficiently high frequencies, the average current density in electrodeposition at a periodically changing rate produces the same concentration distribution inside the diffusion layer as a constant current density of the same intensity. Hence, Eq. 5.65b is valid for all cases of electrodeposition at a constant and periodically changing rate at sufficiently high frequencies.

However, an increase in surface coarseness in deposition using a rectangular pulsating overpotential or pulsating current is only possible during the pulses of current or overpotential[31] and the integral form of Eq. 5.65 can be written as

$$h = h_0 \exp\left(\frac{j}{j_L} \frac{V}{\delta n F} \frac{Q}{p+1}\right) \tag{5.66}$$

if $j = j_{av}$

$$Q = jt \tag{3.17}$$

Equation 5.66 is valid for a pulsating current, square wave pulsating overpotential and reversing current in the millisecond range under the assumption that the entire surface dissolves uniformly during the pauses. The deposits obtained by constant and pulsating overpotential in the mixed control under other conditions are the same are shown in Fig 5.11. The deposit obtained by pulsating overpotential is considerably less rough.

The copper deposits obtained under activation and mixed control as those from Fig. 3.21 are shown in Fig 5.12. A considerable decrease in the grain size of deposit obtained at low current densities (in the activation controlled region Fig. 3.21a and Fig. 5.12a) due to the increase of the amplitude of the overpotential relative to the corresponding value in constant overpotential deposition, can be seen. There is no qualitative change, however, in the structure of the deposit.

A qualitative change in the structure of the deposit appears in mixed controlled deposition (Fig 3.21c and Fig. 5.12b). It is seen that the protrusions caused by mass transport limitations are strongly reduced relative to the deposits shown in Fig. 3.21c, but the grain size in enlarged. It is obvious that the grains obtained by pulsating overpotential with current densities belonging to the region of mixed control, Fig. 5.12b, are almost as regular as those deposited under activation control (Fig. 5.12a). This is obviously due to the increased degree of activation control during the overpotential pulses and the increased grain size relative to those in Fig. 3.21b and c is due to the selective dissolution during the "off" periods. The smaller nuclei formed during the overpotential pulse will be completely or partially dissolved during the overpotential pause and the current density and the current density on the partially dissolved ones during the next overpotential pulse will be considerably lower than on larger ones because of their more negative reversible potentials, and the growth of larger grains will be favorized.

In this way the appearance of the deposit shown in Fig. 3.21c changes and becomes that shown in Fig. 5.12b which was formed using PO deposition of the same quantity of deposited metal and average current density. It can be also seen from Fig. 5.12 that a good deposit can be obtained by PO deposition over a wide range of current densities. This means that in EPCR deposition current density can be considerably increased relative to DC case.

On the other hand, it is known that the orientation of nuclei strongly depends on the depositing overpotential and that the electrode reaction parameters can be different for different crystal planes. It is therefore not surprising that the effect of structure on EPCR has been reported for many cases. In some cases, deposits which behave as monocrystals[7] and deposits with improved crystal perfection can be obtained.

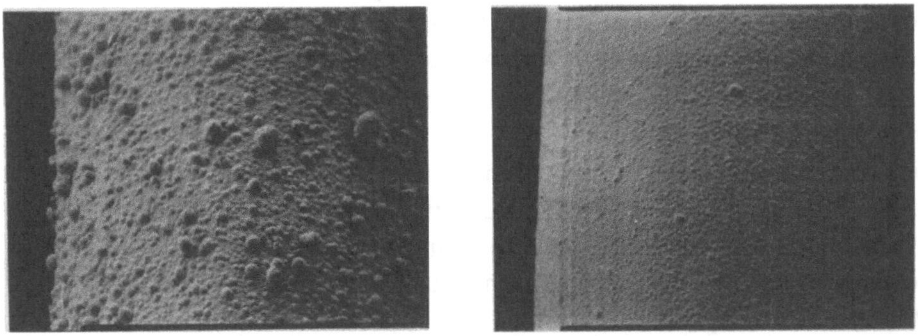

125 µm a) 125 µm b)

Figure. 5.11. Copper deposits obtained by constant and pulsating overpotential deposition from 0.1 M $CuSO_4$ in 0.5 M H_2SO_4 on a copper wire electrode. Quantity of electricity 20 mAh cm^{-2} a) constant overpotential 210 mV, initial current density, 6.5 mA cm^{-2}. b) pulsating overpotential, initial average current density 6.5 mA cm^{-2}, overpotential amplitude 322 mV, pulse to pause ratio 3.[31] (Reprinted with permission from Elsevier Science).

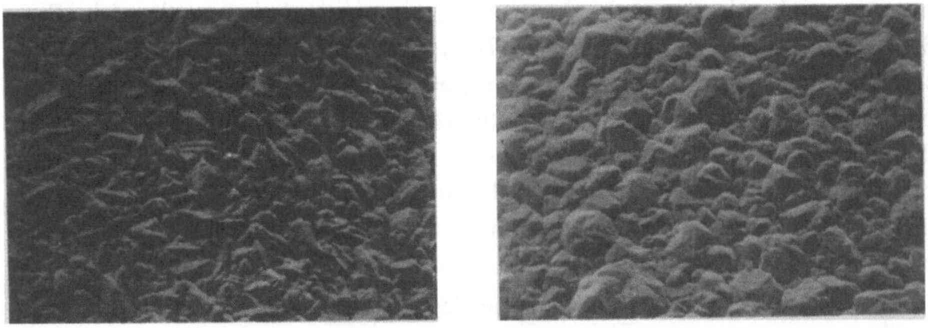

17 µm a) 17 µm b)

Figure. 5.12. Copper deposits obtained by pulsating overpotential deposition: a) initial average current 2.9 mA cm^{-2} overpotential amplitude 210 mV, pause-to-pulse ratio 3, b) initial average current density 6.5 mA cm^{-2}, overpotential amplitude 322 mV, pause-to-pulse ratio 3. Other conditions as in Fig. 5.11.[31] (Reprinted with permission from Elsevier Science).

It is obvious that the same reasoning is valid for RC in the millisecond range and PC. Some different situations appear in the case of RC in the second range[32].

The surface concentration changes during the cathodic pulse in RC deposition in the second range according to Eq. 5.46.

$$\frac{C_s}{C_0} = 1 - \frac{j_c}{j_L} f(t) \tag{5.67}$$

where

$$f(t) = 1 - \frac{8}{\pi^2} \sum_{k=1}^{\infty} \frac{1}{(2k+1)^2} \exp\left[-\frac{(2k+1)^2 t}{4t_0} \right] \text{ and } t_0 = \frac{\delta^2}{\pi^2} D \tag{5.68}$$

In this case

$$j_c = j_0 \frac{C_s}{C_0} f_c \tag{5.69}$$

is also valid and substitution of $\frac{C_s}{C_0}$ from Eq. 5.67 in Eq. 5.69 gives

$$j_t = j_0 f_c \left[1 - \frac{j_t}{j_L} \frac{\delta - h}{\delta} f(t) \right] \tag{5.70}$$

for the tip of a protrusion and

$$j_c = j_0 f_c \left[1 - \frac{j_c}{j_L} f(t) \right] \tag{5.71}$$

for a position a flat surface. If all the surface is isopotential, elimination of f_c from Eqs. 5.70 and 5.71 and further rearranging produces

$$j_t = \frac{j_c}{1 - \frac{j_c}{j_L} \frac{h}{\delta} f(t)} \tag{5.72}$$

The difference in the current densities at the tip of a protrusion and the flat portion of the electrode is then given by:

$$\Delta j = j_t - j_c = \frac{j_c^2}{j_L}\frac{h}{\delta} f(t) \tag{5.73}$$

for $\delta >> h$.

Now, according to Eqs. 5.72 and 5.73 it can be written

$$\frac{dh}{dt} = \frac{j_c^2}{j_L}\frac{h}{\delta}\frac{V}{nF} f(t) \tag{5.74}$$

or in the integral form:

$$h = h_0 \exp\left[\frac{j_c}{j_L}\frac{V}{nF} F(t)\, Q\right] \tag{5.75}$$

where

$$F(t) = \int_0^{t_c} f(t)\, dt \tag{5.76}$$

for the first phase.

Assuming that the surface will dissolute uniformly during the anodic period, it is obvious that because $f(0) = 0$ and $f(\infty) = 1$, the increase in the surface coarseness in the RC regime will be lower than in the DC regime until the condition

$$F(t) < \frac{1-r}{1+r} \tag{5.77}$$

is satisfied. The above reasoning is valid if a polycrystalline deposit is obtained and the same derivation as in case of CO deposition can be used.

It can be seen from Fig. 5.13 that the structure of the deposit obtained by RC in the second range is more similar to that obtained in the DC than in the PC regime, but the surface coarseness of this deposit is considerably lower than in DC case, being close to this in PC deposition.

This is because in RC deposition there is a considerable concentration polarisation, producing polycrystalline deposit.

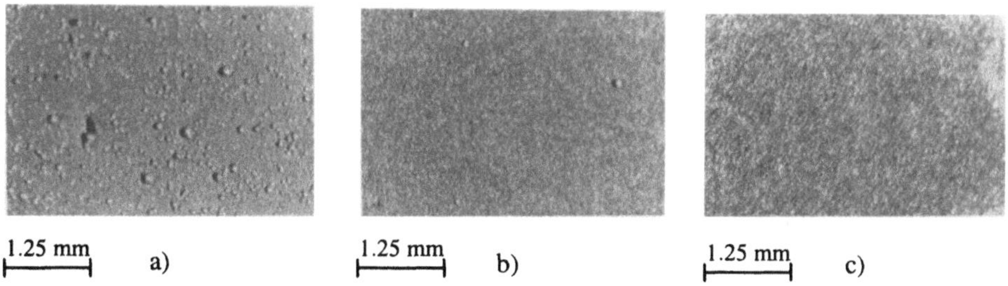

Figure 5.13. Photomicrographs of copper deposited on a flat platinum electrode from 0.5 M $CuSO_4$ in 0.5 M H_2SO_4; j_{av}, = 0.9 j_Lt, t = 60 min: a) DC; b) RC, r = 1/7, T = 16 s; c) PC, p = 1, ν = 10 Hz.[33] (Reprinted with permission of Elsevier Science).

5.5 Current density and morphology distribution on a macroprofile

The current density distribution on a macroprofile in EPCR has been treated in several papers[7]. It seems that this distribution improves the deposition by current waves with anodic flow but that without this flow it is worse than in DC deposition. This behaviour can be successfully explained[34].

Assuming that Eq. 4.43 is valid for the flat electrode in a cell with low anode polarization also, the current densities in the middle, j, and at the edge, j_e of a flat electrode in a cell with low anode polarization can be related by

$$j_e = j\exp\left(\frac{2.3RI}{b_c}\right) \tag{4.43}$$

if the deposition in both cases is under activation control in the Tafel region and that the limiting diffusion current density is the same in the middle and at the edge of the electrode, I is the cell current corresponding to j. During the current pulses the amplitude values of current densities and current should be substituted in Eq. 4.43 producing

$$j_{e,A} = j_c\exp\left(\frac{2.3RI_A}{b_c}\right) \tag{5.78}$$

where $j_{e,A}$ and j_c are the amplitude values of current densities in the middle and at the edge of electrode and I_A is the amplitude of the current in the cell. On the other hand, the amplitude in pulsating current deposition and average current density are related by Eq. 5.7

$$j_{av} = \frac{j_c}{p+1} \tag{5.7}$$

so Eq. 5.78 can be rewritten in the form

$$j_{e,av} = j_{av} \exp\left(\frac{2.3RI(p+1)}{b_c} \right) \tag{5.79}$$

assuming that $j = j_{av}$, if

$$\exp\left(\frac{2.3RI(p+1)}{b_c} \right) > \exp\left(\frac{2.3RI}{b_c} \right) \tag{5.80}$$

then $j_{av,e} > j_{av}$, meaning worse current density distribution in PC than in DC conditions

The effect of reversing current on the current distribution at the macroprofile level can easily be discussed for the case of activation-controlled deposition if the Tafel slopes of the anodic and cathodic processes are different, as they are for copper deposition and dissolution in sulphate solutions. With the assumption that the current density in RC deposition is sufficiently high so that the effect of the opposing processes can be neglected, the limiting diffusion current density is the same over all electrode surface. The difference between the current density at the edge and that at the middle of the electrode in cathodic deposition is

$$\Delta j_c = j_c \left[\exp\left(\frac{2.3RI_c}{b_c} \right) - 1 \right] \tag{5.81}$$

and for anodic case

$$\Delta j_a = j_a \left[\exp\left(\frac{2.3 \times 3RI_a}{b_c} \right) - 1 \right] \tag{5.82}$$

since $b_a = b_c/3$ for copper deposition, where I_c and I_a are the cell currents corresponding to j_c and j_a. It is obvious that for

$$\Delta j_c t_c = \Delta j_a t_a \tag{5.83}$$

which occurs when

$$\frac{t_c}{t_a} = \frac{\Delta j_a}{\Delta j_c} = \frac{\exp\left(\frac{2.3 \times 3RI}{b_c}\right) - 1}{\exp\left(\frac{2.3RI}{b_c}\right) - 1} \tag{5.84}$$

(if $j_c = j_a = j$ and $I_c = I_a = I$), deposits of equal thickness can be obtained at the edge and at the middle of the electrode. In this way, a completely uniform average current density distribution at the macroprofile level can be obtained in RC deposition. The diffusion limitations of the cathodic processes will improve the distribution in RC, but this approach is sufficient to explain the essence of the effect, as illustrated in Fig. 5.14[34].

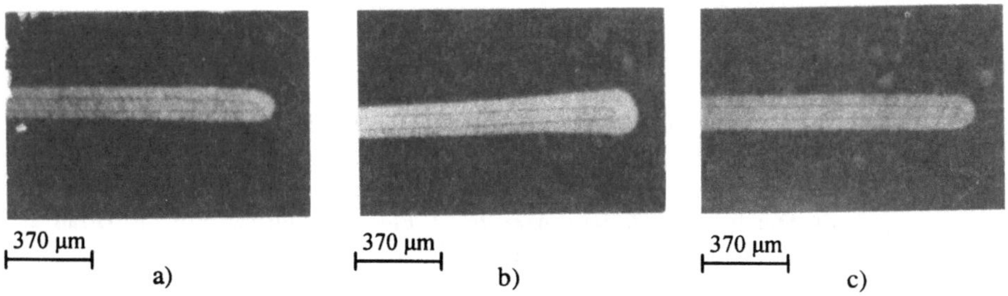

Figure. 5.14. Cross section of copper deposits obtained from 1.0 mol dm^{-3} $CuSO_4$ in 1.0 mol dm^{-3} H_2SO_4 at the edge of plane copper electrodes, previously plated with bright nickel. j_{av}= 10 mA cm^{-2}, t = 4 h. a) DC; b) PC, p = 1, 50 Hz, c) RC, r = 1/7, T = 8 s.[34] (Reprinted with permission of Elsevier Science).

The best current distribution is expected in the case of PO if all the electrode surface can be taken as an isopotential. Under the assumption that the limiting diffusion current density does not vary over the electrode surface area, the same current density can be expected over all points of the electrode. A good approximation of PO deposition can be the RC deposition by the current wave optimized relative both current density distribution on micro and macro profile as recently shown[35]. Hence, it seams that RC should be the optimum regime of EPCR. Besides, the crack-free chromium deposits with improved current density distribution on both micro and macroprofile and with practically no reduced hardness were obtained recently[36,37], by RC deposition. This means that the formation of unstable chromium hydride can also be prevented by RC, but this phenomenon has been not treated semiquantitatively so far.

5.6 CONCLUSIONS

The beneficial effect of EPCR in the electrodeposition of metals are clearly demonstrated and explained in a semiquantitative way. Therefore, for the same effective current density EPCR requires a somewhat larger energy consumption[7], than in DC deposition as well as the more expensive current generators. This should not be considered as an important drawback of such electrolysis, bearing in mind all the considerable improvements of the process of metal deposition that result.

It appears that in each particular case one should compare all the characteristic of both deposition with constant rate and EPCR, and then decide which one to apply, considering all the requirements in relation to the quality of the deposits, the productivity of the electrolyser, and other parameters of the electrolytic production operation.

5.7 FURTHER READINGS

1. Bibikov, Nikolai, *Electrodeposition of Metals by AC (in Russian).* Moscow, Leningrad: Mashgiz, 1961.
2. Bakhvalov, Grigorii, *New Technology of Metal Electrodeposition (in Russian).* Moscow: Metallurgiya, 1966).
3. Despić, Aleksandar; Popov, Konstantin. "Transport Controlled Deposition and Dissolution of Metals." In *Modern Aspects of Electrochemistry,* Vol. 7, Brian E. Conway, John O'M. Bockris, eds. New York, NY: Plenum Press, 1972.
4. Ibl N., Puippe J. A., Angerer H. Electrocrystallization in pulse electrolysis. Surf. Technol 1978; 6: 287-30
5. Polukarov Yu. M., Grinina V. V., "Electrodeposition of Metals by Periodically Changing Rate and Single Pulses" (in Russian) *Itogi nauki i tekhniki, Seriya Elektrokhimiya* Vol. 22, Yu. M. Polukarov, ed, Moscow: Acad. Sci. Moscow: 1985.
6. Theory and Practice of Pulse Plating, Jean-Claude Puippe, Frank Leaman, eds. American Electroplate and Surface Finishing Society, Orlando, Florida: 1986.
7. Popov, Konstantin; Maksimović, Miodrag. "Theory of the Effect of Electrodeposition at a Periodically Changing Rate on the Morphology of Metal deposits." In *Modern Aspects of Electrochemistry,* Vol. 19, Brian E. Conway, John O'M. Bockris, Ralph E. White, eds. New York, NY: Plenum Press, 1989.
8. Pesco, Anthony; Cheh, Huk. "Theory and Application of Periodic Electrolysis." In *Modern Aspects of Electrochemistry,* Vol. 19, Brian E. Conway, John O'M. Bockris, Ralph E. White, eds. New York, NY: Plenum Press, 1989.
9. Popov, Konstantin; Pavlović, Miomir.. "Electrodeposition of Metal Powders with Controlled Grain Size and Morphology." In *Modern Aspects of Electrochemistry,* Vol. 24, Ralph E. White, John O'M. Bockris, Brian E. Conway, eds. New York, NY: Plenum Press, 1993.
10. Maksimović M.D. The influence of the charge and discharge of the electrical double layer in electrodeposition by periodically changing rate. J. Serb. Chem. Soc. 1965; 60: 449-58

11. Maksimović M.D., Popov K.I. Mass transfer during electrodeposition of metals at a periodically changing rate. J. Serb. Chem. Soc. 1999; 64: 317-40
12. Viswanathan K., Cheh H.Y. Mass transfer aspect of electrolysis by periodic currents. J. Electrochem. Soc. 1979; 126: 398-401
13. Chin D.T. Mass transfer and current-potential relation in pulse electrolysis. J. Electrochem. Soc. 1983; 130: 1657-67
14. Popov K.I., Maksimović M.D., Simić M.S. The effect of periodic reverse current on the surface rougneses of metal deposits and maximal deposition rate, Surf. Technol. 1982; 16: 209-18
15. Popov K.I., Maksimović M.D., Ocokoljić B. M., Lazarević B.J. Fundamental aspects of pulsating current metal electrodeposition I: The effect of the pulsating current on the surface roughness and the porosity of metal deposits, Surf. Technol. 1980; 11: 99-109
16. Maksimović M.D., Totovski D.Ć., Ivić A.P. The effect of AC superimposed on DC in the electrodeposition of metals. Surf. Technol. 1983; 18: 233-241
17. Popov K.I., Maksimović M.D., Zečević S.K., Stojić M.R. Surface roughening and dendritic growth in pulsating overpotential copper electrodeposition. Surf. Coat. Technol. 1986; 27: 117-129
18. Popov K.I., Maksimovic M.D. Maximum deposition rate in metal electrodeposition by reversing current in the second range J. Serb. Chem. Soc. 1991; 56: 25-31
19. Romanov V.V. The Zinc spongy electrodeposits formation during deposition from zincate alkaline solution (in Russian). Zh. Prikl. Khim. 1963; 36: 1057-63
20. Popov K.I., Keča D.N., Anđelić M.D. Electrodeposition of zinc on copper from alkaline zinkate solutions. J. Appl. Electrochem. 1978; 8: 19-23
21. Diggle J.W., Despić A.R., Bockris J.O'M. The mechanism of the dendritic electrocrystallization of zinc. J. Electrochem. Soc. 1969; 116:1503-14
22. Barton J.L., Bockris J.O'M. The electrolytic growth of dendrites from ionic solutions. Proc. Roy.Soc. London 1962; A268: 485-505
23. Bek R.Yu., Kudryavtsev N.T. The effect of a periodically changing rate on the zinc electrodeposition from zincate alkaline solutions (in Russian). Zh. Prikl. Khim. 1961; 34: 2020-27
24. Arouete S., Blurton K.F., Oswin H.G., Controlled current deposition of zinc from alkaline solution, J. Electrochem. Soc. 1969; 116: 166-169
25. Popov K.I., Stojilković E.R., Radmilović V., Pavlović M.G. Morphology of lead dendrites electrodeposited by square-wave pulsating overpotential. Powder Technology 1997; 93: 55-61
26. Pavlović M.G., Maksimović M.D., Popov K.I., Kršul M.B. The effect of pulsating overpotential on the morphology of electrodeposited silver powder particles. J. Appl. Electrochem. 1978; 8: 61-65
27. Popov K.I., Maksimović M.D., Pavlović M.G., Ostojić G.R. Formation of powdered copper deposits by square-wave pulsating overpotential, J. Appl. Electrochem. 1977; 7: 331-37
28. Pavlović M.G., Pavlović Lj.J., Stoilković E.R., Radmilović V., Popov K.I. The effect of particle structure on apparent density of electrolytic copper powder. J. Serb. Chem. Soc. 2001; 66:913-933
29. Popov K.I., Pavlović Lj.J., Stoilković E.R., Radmilović V. The effect of reversing current deposition on apparent density of electrolytic copper powder. J. Serb. Chem. Soc. 2002; 67:61-67

30. Popov K.I., Maksimović M.D., Stevanović R.M. Krstajić N.V. Fundamental aspect of pulsating current metal electrodeposition. VIII: The effect pulse to pause ratio on microthrowing power of metal deposits. Surf. Technol. 1984; 22: 155-58
31. Popov K.I., Pavlović M.G., Pavlović Lj. J. Čekerevac M.I. Remović G.Ž. Electrode surface coarsening in pulsating overpotential copper electrodeposition. Surf. Coat. Technol. 1988; 34: 355-63
32. Popov, Konstantin "The Effect of Periodically Changing Rate in Electrodeposition on the Morphology of Metal Deposits." (in Serbian) In *Application of Pulsating Regimes in Electroplating*, Miomir Pavlović, Aleksandar Dekanski, eds. ICTM, Department of Electrochemistry, Belgrade, 1992.
33. Popov K.I., Maksimović M.D., Totovski D.Ć. Fundamental aspects of pulsating current metal electrodeposition. VI: The comparison of electrode surface roughening in pulsating current and periodic reverse current electrodeposition of metals. Surf. Technol. 1982; 17: 125-29
34. Popov K.I., Totovski D.Ć., Maksimović M.D. Fundamental aspects of pulsating current metal electrodeposition. VII: The comparison of current density distribution in pulsating current and periodic reverse current electrodeposition of metals. Surf. Technol. 1983; 19: 191-85
35. Popov K.I., Kostić T.M., Stoilković E.R., Nikolić N.D., Pavlović M.G. The determination of the optimum current wave in reversing current metal electrodeposition. J. Serb. Chem. Soc. 1998; 63: 537-44
36. Petrović B.M., Kostić T.M., The properties of chromium electrodeposited with programed currents. Part I: Direct current. J. Serb. Chem. Soc. 2000; 65: 55-63
37. Petrović B.M., Kostić T.M., The properties of chromium electrodeposited with programed currents. Part II: Reversing current. J. Serb. Chem. Soc. 2000; 65: 65-72

Chapter 6

ELECTROWINNING

6.1 THEORETICAL ASPECTS OF ELECTROWINNING

Electrowinning of a metal is based on the electrolysis of aqueous solutions or melts of metal salts with insoluble anode[1,2].

The basic reactions during electrowinning from aqueous solutions are[2]:
- cathodic deposition of the desired metal:

$$Me^{n+} + ne^- \rightarrow Me \qquad \text{(cathode)} \qquad (6.1)$$

and
- oxygen evolution on the anode:

$$2H_2O \rightarrow O_2 + 4H^+ + 4e^- \quad \text{(anode)} \qquad (6.2)$$

Electrowinning of a metal by the electrolysis is usually performed in an aqueous solution of the sulfates of a metal in dilute sulfuric acid, with the aim to avoid problems related with the deposition of the hydroxide of the metal. Sulfuric acid is chosen because of its relatively low price[2].

Several processes of preparations and enrichment of the metal ore proceed the electrolysis, which is the final phase in the electrowinning process.

Bearing in mind the fact that metals are found in nature usually in the form of their sulfide and oxide ores, the preparation usually consists in[2]:

I *Enrichment of the ore*, serves for the removal of the wastes by flotation which leads to a concentrate of metal sulfides and/or metal oxides.

II *Roasting – oxidation of the concentrate*, if the dominate form of the metal in the concentrate is the metal sulfide which is insoluble in sulfuric acid, then the roasting is performed in the air with the aim of transforming the sulfide into the oxide which is soluble in sulfuric acid.

III *Leaching of the concentrate*, is performed with dilute sulfuric acid, impoverished electrolyte from the electrolysis process, which gives a solution of the metal sulfate with some impurities.

IV *Refining of the leach* means the elimination of impurities with more positive potentials, which could be deposited together with the metal, thus decreasing its purity, or result in the hydrogen evolution reaction. Elimination of the impurities is performed by increasing the pH of the solution, by precipitation with hydrogen sulfide or by cementation.

V *Electrolysis* serves for the final extraction of the metal by cathodic deposition in the electrochemical reactor. Since oxygen is evolved at the anode according to Eq. 6.2, the concentration of metal ions decreases and the acidity of the electrolyte increases, which can result in hydrogen evolution together with metal deposition at the cathode. The compensation of the metal ions concentration is performed in phase III.

Although practically all metals can be obtained by the electrochemical procedure, the most wide spread technologies involve the electrowinning of zinc from a solution of its sulfate[1].

6.2 THEORETICAL ASPECTS OF ZINC ELECTROWINNING

Apart from the conventional metallurgical procedure of zinc production, today the most common procedure is the electrowinning of zinc by electrolysis of an aqueous solution of zinc sulfate. This electrochemical procedure gives zinc of 99.99 % purity and less loss of cadmium than the classical pyrometallurgical treatment.

During the electrolysis of an aqueous zinc sulfate solution, the basic electrode reactions are:
cathodic deposition of zinc

$$Zn^{2+} + 2e^- \longrightarrow Zn \tag{6.3}$$

and oxygen evolution at the insoluble anode

$$2H_2O \rightarrow O_2 + 4H^+ + 4e^- \quad (6.2)$$

Although the electrode potential of the reduction of zinc ions is much more negative than the electrode potential of the reduction of protons (about 0.8 V), it is possible to deposit zinc from the acid solution with a high current efficiency thanks to the extremely high value of the hydrogen overpotential on zinc.

However, the presence of impurities (*e.g* metal ions with reduction potentials more positive than that of protons) will cause these metals to be deposited together with zinc. The results is the formation of cathodic surfaces having a low value of the hydrogen overpotentials. Intensive hydrogen evolution will occur on such surfaces together with the dissolution (corrosion) of the already deposited zinc, following the mechanism of local galvanic cells. Impurities such as copper, bismuth, germanium and antimony do not just decrease the current efficiency, but completely prevent the deposition of zinc. It can be concluded that the electrolyte for zinc electrowinning has to be completely freed from such impurities before the electrolysis. Hence, the basic aim of zinc ore treatment is the production of pure zinc sulfate without impurities which would exhibit a bad influence on the cathodic reaction[2].

6.2.1 Technological scheme of zinc electrowinning

The composition of the zinc ore influences the manner of treatment of the zinc sulfate solution. Zinc sulfate is extracted from an oxide ore after removing the wastes, with the spent solution of zinc sulfate and sulfuric acid from the electrolysis, according to the reaction[2]:

$$ZnO + H_2SO_4 \rightarrow ZnSO_4 + H_2O \quad (6.4)$$

Extraction from a sulfide ore necessitates roasting in an air atmosphere, during which the main reaction is the formation of zinc oxide according to:

$$2ZnS + 3O_2 \rightarrow 2ZnO + 2SO_2 \quad (6.5)$$

Impurities in a zinc ore can be classified into three main groups[3]:

1. Fe, Si, Al, Sb, Bi, Ge, Sn
2. Cu, Cd, Ni, Co
3. Pb, Ag, SiO_2

During the dissolution of zinc sulfate, according to Eq. 6.4, the extent of the free acid decreases, which leads to an increase of pH of the solution. This results in the precipitation of Fe(III), Al and partly Cu in the form of hydroxides. Owing to the high specific surface of Fe(III) hydroxide, Sb, Bi and Ge are adsorbed on its surface and in that manner removed from the solution, while As forms the insoluble $Fe_4O_5(OH)_5As$. Fe(II) hydroxide precipitates at pH 6.5, at which pH zinc(II) hydroxide precipitates too. For this reason it is necessary to oxidize Fe^{2+} to Fe^{3+}. This is achieved by the addition of MaO_2 which reacts with Fe^{2+} according to the reaction:

$$2Fe^{2+} + 4H^+ + MnO_2 \rightarrow 2Fe^{3+} + Mn^{2+} + 2H_2O \quad (6.6)$$

Ni^{2+} and Co^{2+} also react with MnO_2 at lower values of pH, giving the corresponding three-valent ions which are deposited in the form of hydroxides, $Co(OH)_3$ and NiOOH, and so in this way partly removed.

During the electrolysis MnO_2 is formed by the anodic oxidation of Mn^{2+} ions according to the reaction:

$$Mn^{2+} + 2H_2O \rightarrow MnO_2 + 4H^+ + 4e^- \quad (6.7)$$

The impurities of the third group are insoluble in the leach.

The precipitate, containing impurities of the I, III, and partly of the II group, are removed from the leach by filtration. The removal of group II impurities is based on cementation of the metal ions from the solution with zinc powder, according to the reaction:

$$Me^{2+} + Zn \rightarrow Me\downarrow + Zn^{2+} \quad (6.8)$$

After these treatments, the concentration of zinc ions varies between 120 and 170 g dm^{-3}, and the concentration of the free acid between 150 and 200 g dm^{-3}. Typical contents of metal ions in the electrolyte for the electrowinning of zinc is given in Table 6.1[3].

Table 6.1. Typical contents of metal ions in the electrolyte for the electrowinning of zinc.[3] (Reprinted with permission from the Faculty of Technology and Metallurgy, Belgrade)

Contents of metal ions / g dm^{-3}				
Zn^{2+}	Mn^{2+}	Fe^{2+}	Ca^{2+}, Mg^{2+}	Cd^{2+}, Pb^{2+}, Cu^{2+}
120-170	<3	0.02-0.04	1-10	$1\text{-}10\times10^{-3}$

It is common to add between 10 and 30 mg dm^{-3} of glue which serves to enhance the quality of the deposit and to avoid the formation of dendrites which could lead to short circuits between the anode and cathode during electrolysis, and also to a decrease in the current efficiency[2].

The described procedures are schematically given in Fig. 6.1.

The cathode material used in the electrochemical cell is usually aluminium sheet 4 to 8 mm in thickness[3]. Aluminium is used owing to the poor adhesion of the deposited zinc, which can later be easily removed from the cathode[2]. Lead sheet 5 to 10 mm in thickness, and in average, having the dimensions of 100 × 600 mm, serves as the anode[2,3]. The life time of these lead anodes is 1.5 – 2 years, but if the lead anode is alloyed with 1 – 2 % of silver, the life time is increased to 3 – 4 years[2].

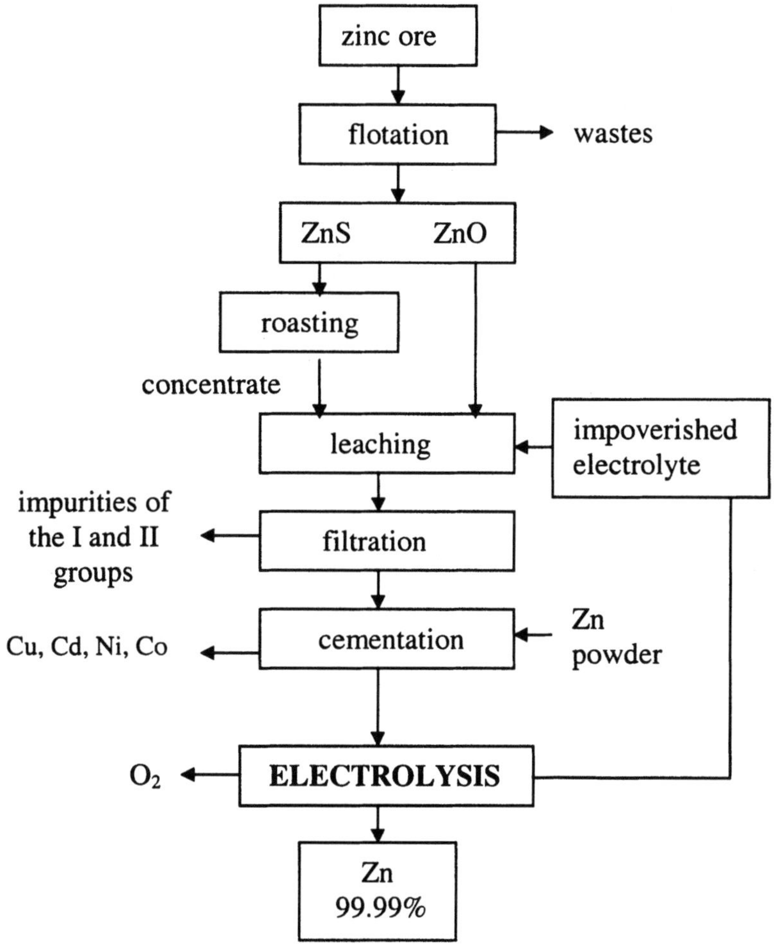

Figure 6.3. Technological scheme of zinc electrowinning.

The average electrolysis time is between 10 and 24 h. during which time the concentration of zinc should drop to 45 – 55 g dm^{-3}. After the electrolysis the cathodes are taken from the cell and the zinc is mechanically removed.

Typical conditions for the electrowinning of zinc are as follows[2,3,4]:

the cathodic current density:	400 – 700 A m^{-2}
current efficiency:	88 – 92 %
voltage:	3.3 – 3.6 V
temperature of the electrolyte:	30 – 40 °C
specific electrical energy consumption:	3000 – 3500 kW h t^{-1}

For the electrowinning of 1 t of zinc in average about: 2.5 t of the concentrate, 4 kg of MnO_2, 16 kg of Zn powder and 62 kg of sulfuric acid are consumed[2].

The zinc concentrate usually contains 0.5 – 1 wt. % of Cd, which is why in most zinc electrowinning plants, Cd is obtained parallel by a similar process[3,4].

6.3 FURTHER READINGS

1. Pletcher, Derek, *Industrial Electrochemistry,* New York: Chapman and Hall, 1984.
2. Zečević, Strahinja; Gojković, Snežana; Nikolić, Branislav, *Electrochemical Engineering (In Serbian),* Belgrade: Faculty of Technology and Metallurgy, 2001.
3. Đorđević, Spasoje, *Electrometallurgy (In Serbian),* Belgrade: Faculty of Technology and Metallurgy, 1972.
4. Kubasov, Vladimir; Bannikov, Vladimir, *Electrochemical Technology of Inorganic Compounds (In Russian),* Moscow: Khimia, 1989.

Chapter 7

ELECTROREFINING

7.1 THEORETICAL ASPECTS OF ELECTROREFINING

Electrorefining of a metal by electrolysis is a way of purification of a metal previously extracted by classical metallurgical or electrochemical processes, with the aim of removing impurities, which could exhibit in negative effects on the physico–chemical and mechanical properties of the metal. In principle, almost all metals can be electrorefined, but judging by the amount of 8 million tones per year, copper electrorefining outweighs all others[1,2].

The procedure of a metal electrorefining is schematically given in Fig. 7.1.

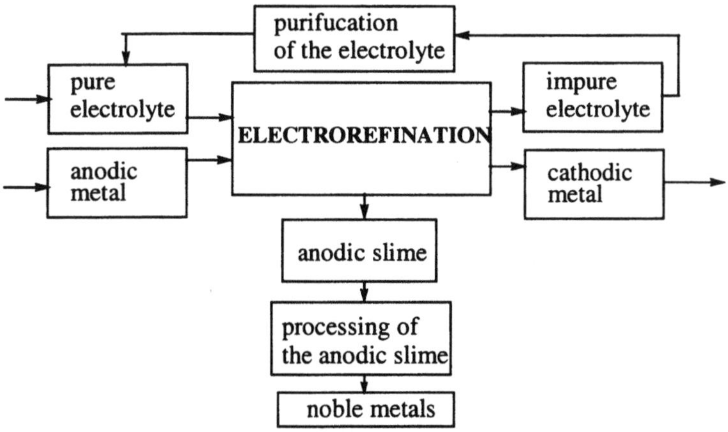

Figure 7.1. Schematic representation of electrorefining processes.

7.1.1 Selectivity of metal dissolution and deposition

The main principle of electrorefining is based on the anodic dissolution of a metal containing impurities and its cathodic deposition with an as much as possible reduced amount of impurities. The selectivity principle of metal dissolution and deposition can be explained starting from a metal M placed in a suitable solution containing its ions M^{n+}, which leads to the establishment of the equilibrium electrode potential $E_e(M^{n+}|M)$. Under some value of current density, j, the base metal is anodically dissolved

$$M \longrightarrow M^{n+} + ne^- \quad (7.1)$$

having an anodic potential $E_a(j)$, and cathodically deposited

$$M^{n+} + ne^- \longrightarrow M \quad (7.2)$$

having a cathodic potential $E_c(j)$, as shown in Fig. 7.2[2].

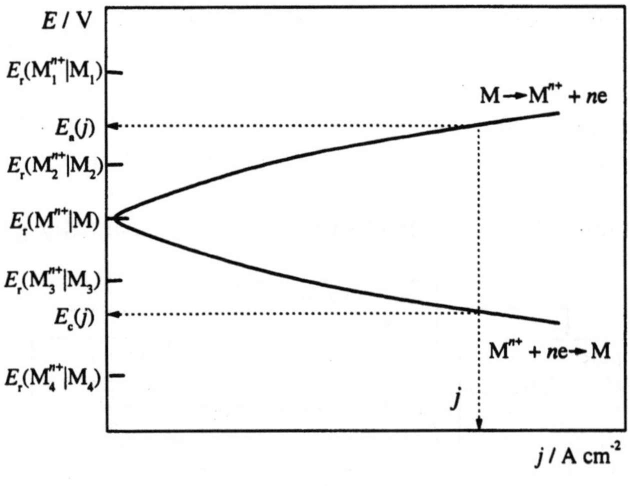

Figure 7.2. Polarization curves of metal dissolution and deposition together with the equilibrium electrode potentials of dissolution and deposition of impurities.[2] (Reprinted with permission from the Faculty of Technology and Metallurgy, Belgrade).

All metal-impurities having an equilibrium potential more positive than $E_a(j)$ (such is metal M_1) will not be anodically dissolved, but owing to the dissolution of the base metal (erosion) they will be transferred into the electrolyte and fall down to the bottom of the reactor forming the anodic slime.

Metal-impurities which have an equilibrium potential more negative than $E_a(j)$ (such as: M_2, M_3 and M_4) will be dissolved together with the base metal M, and transferred into the electrolyte in the form of metal ions. Exceptions are those metals whose cations can react with the anions in the electrolyte forming sparingly soluble salts which sediment as anodic slime[2].

During cathodic deposition of a metal, all the metals whose equilibrium potentials are more positive than the potential of the cathode $E_c(j)$ (metals M_2 and M_3), could be deposited, depending on their concentrations and deposition overpotentials. In the example given in Fig. 7.2. metal M_4 will be anodically dissolved, but not cathodically deposited, while metal M_1 will not be anodically dissolved, but transformed into the anodic slime[2].

It can be concluded that the selectivity principle of electrorefining is based on such electrolysis conditions which ensure that metal impurities that could be cathodically deposited will not be anodically dissolved (anodic slime) and impurities that could be anodically dissolved will not be cathodically deposited, but remain in the electrolyte.

7.2 THEORETICAL ASPECT OF COPPER ELECTROREFINING

Copper produced by metallurgical processes contains significant amount of impurities, between 0.5 and 2 %, which have a bed influence on the mechanical and electrical properties of copper[3]. Amounts of 0.15 % of phosphorous or 0.5 % of arsenic, dramatically decrease the electrical conductivity of copper, as shown in Table 7.1.[3] Hence, the improvement of the electrical properties of copper is the main reason for electrorefining.

The second reason for electrorefining is the extraction of some noble impurities, normally present in non-refined copper (anodic metal), such as: gold, silver, platinum and palladium etc[2].

Table 7.1. The influence of admixtures on the electrical conductivity (ρ) of copper. Electrical conductivity of pure copper is taken to be 100 %. [3] (Reprinted with permission from Faculty of Technology and Metallurgy, Belgrade).

Sb %	ρ / %	As %	ρ / %	Bi %	ρ / %	P %	ρ / %
0.02	97.5	0.005	98.5	0.01	99.5	0.02	96.0
0.05	95.0	0.01	95.5	0.05	98.5	0.07	60.0
0.10	88.0	0.05	85.0			0.15	42.0
0.25	75.0	0.5	40.0				

As stated previously, anodic copper contains between 0.5 and 2 % of admixtures that can be divided into four groups[2]:

1. Ni, Co, Fe, Sn, Zn, Pb
2. Cu_2O, Cu_2S, Cu_2Se, Cu_2Te
3. Au, Ag, Pt, Pd, Se, Te
4. As, Sb, Bi

The main reactions occurring at the electrodes, which determines the potentials of anode and cathode, during electrorefining of copper are as follows:

$$Cu \rightarrow Cu^{2+} + 2e^- \quad \text{(anode)} \tag{7.3}$$

$$Cu^{2+} + 2e^- \rightarrow Cu \quad \text{(cathode)} \tag{7.4}$$

while the impurities can react in the following manners[2,3].

– Metals of the first group, whose equilibrium potentials are more negative (less noble) than copper (see Table 2.1) will during anode dissolution be transferred together with copper into solution. Due to the presence of sulfate ions in the electrolyte, lead ions will form lead sulfate and fall into the anodic slime. Other metals, such as Ni, Co, Sn and Zn, will be concentrated in the electrolyte, since their deposition potential is more negative than the potential of the cathode, whose value is around 0.3 V. The presence of these metals in the electrolyte has no bad influence, but it necessitates periodical purification of the electrolyte. The presence of iron ions is completely undesirable. Iron can be anodically dissolved in the form of Fe^{2+}, but can not be deposited since its equilibrium potential is much more negative than the potential of the cathode. Fe^{2+} ions can be oxidized to Fe^{3+} at the anode, but the value of the standard electrode potential of this reaction (+0.77 V) is more positive than the potential of the anode ($E_a = 0.4$ V).

Bearing in mind the Nerst relation (Eq. 2.2) for the equilibrium electrode potential:

$$E_{Fe^{3+}/Fe^{2+}} = E^{\theta}{}_{Fe^{3+}/Fe^{2+}} + \frac{RT}{F} \ln \frac{a_{Fe^{3+}}}{a_{Fe^{2+}}} \tag{7.5}$$

it could be concluded that the potential is dependent on the ratio of the Fe^{2+} and Fe^{3+}concentration. If the concentration of Fe^{3+} is much lower than that of Fe^{2+}, the equilibrium potential will become more negative than the actual potential of the anode which could result in the oxidation of Fe^{2+} to Fe^{3+}. This would decrease the concentration of Fe^{2+} and increase the concentration of Fe^{3+}, which leads to a possibility for the reduction of Fe^{3+} to Fe^{2+} at the cathode. This unwanted closed circle would consume a part of the current during the electrorefining process.

All the metals with which the electrolyte is enriched lead to a consumption of free acid.

Insoluble impurities from the second group, such as Cu_2S, Cu_2Se and Cu_2Te, can not be anodically dissolved, but sediment and form anodic slime. Only Cu_2O can be dissolved, according to:

$$Cu_2O + 2H^+ \rightarrow Cu^{2+} + Cu + H_2O \tag{7.5}$$

Hence, half of the copper is dissolved into the electrolyte in the form of Cu^{2+} ions, while the other half forms a fine powder which partly reacts with the air oxygen and sulfuric acid in the electrolyte, and partly sediments to form anodic slime.

Metals of the third group as Au, Ag, Pt, Se and Te also sediment as the anodic slime. They are more noble than copper and according the rule that reactions with lower electrode potentials occur first (see section 2.3.1) at the anode, these metals act as insoluble inert inclusions and fall down as the anodic slime.

The most undesirable impurities are the metals of the fourth group: As, Sb, and Bi. Their presence drastically decreases the quality of the cathode copper deposit, as shown in Table 7.1. These metals are the most difficult to be removed becouse of the values of their reduction potential which are near the potential of copper, precisely between the equilibrium potential of copper and hydrogen. However, their concentration in the electrolyte is usually very small, so their deposition potential is more negative compared with the potential of the cathode, and under normal conditions they are not deposited. If for some reason the concentration of copper ions decreases, their deposition could be expected. On the other hand, Sb and Bi could react with As forming sparingly soluble arsenates, which floats in the form of fine foam on the electrolyte and make electrolysis process difficult.[4]

The effects of electrorefining could best be illustrated by Table 7.2,[2] which gives the distribution of impurities in the anodic slime, electrolyte and cathode metal[2].

Table 7.2. Typical distribution of the impurities in the anodic slime, electrolyte and cathode metal after copper electrorefining.[2] (Reprinted with permission from the Faculty of Technology and Metallurgy, Belgrade).

	slime, %	electrolyte, %	cathode, %
Cu	0.1	3	97
Au	99	0	1
Ag	98	0	2
Se, Te	98	0	2
Pb	98	0	2
Zn	4	95	1

	slime, %	electrolyte, %	cathode, %
Ni	4	95	1
Fe	5	94	1
As	25	40	2
Sb	58	40.4	2
Bi	80	18	2
Sn	99	0	1
S	96	0	4
SiO_2	100	0	0

7.2.1 Technological scheme of copper electrorefining

The process of copper electrorefining is schematically illustrated in Fig 7.3.

Anodes of 99 – 99.5 % pure copper obtained, by pyrometallurgical treatment, are prepared by casting after melting in the anode furnace. The surface area of the anode depending on the capacity of the electrolytic cell, is between 1 and 1.5 m^2, their width is between 40 and 50 mm, while their weight is between 300 and 350 kg[2,4]. A part of the anodes is prepared to cathodes by rolling into 1 mm thick copper sheets 5 kg in weight[2,4]. During the manufacture of the anodes and cathodes, hooks for hanging the electrodes onto electrode tracks are also produced.

After manufacture, the electrodes are placed into electrochemical cells, the number of which can, depending on the capacity of the plant, be a few tens connected in series or parallel.

Electrorefining of copper takes place in an electrolyte, a typical composition of which is given in Table 7.3[3].

Table 7.3. Typical composition (kg m^{-3}) of the electrolyte for copper electrorefining.[3] (Reprinted with permission from Faculty of Technology and Metallurgy, Belgrade).

Electrolyte composition, kg m^{-3}				
Cu^{2+}	H_2SO_4	Ni^{2+}	Sb^{3+}	As^{3+}
35-50	125-225	<20	<0.8	<12

Thiourea and gelatine (10-100 g per 1 t of copper) are used as additives for enhancing the quality of the deposit and preventing the formation of dendrites which can produce short circuits during the electrolysis process[3].

The conditions of the copper electrorefining process are[2,4]:

- cathodic current density: 200 – 300 A m^{-2}
- current efficiency 94 – 96 %
- cell voltage: 0.25 – 0.35 V
- temperature of the electrolyte: 55 – 60 ^{0}C
- specific electrical energy consumption: 250 – 360 kW h t^{-1}

After the electrorefining process, cathodic copper of 99.97 – 99.99 % purity is dried, and further processed by melting, rolling into thin metal sheets, wires or similar commercial products.

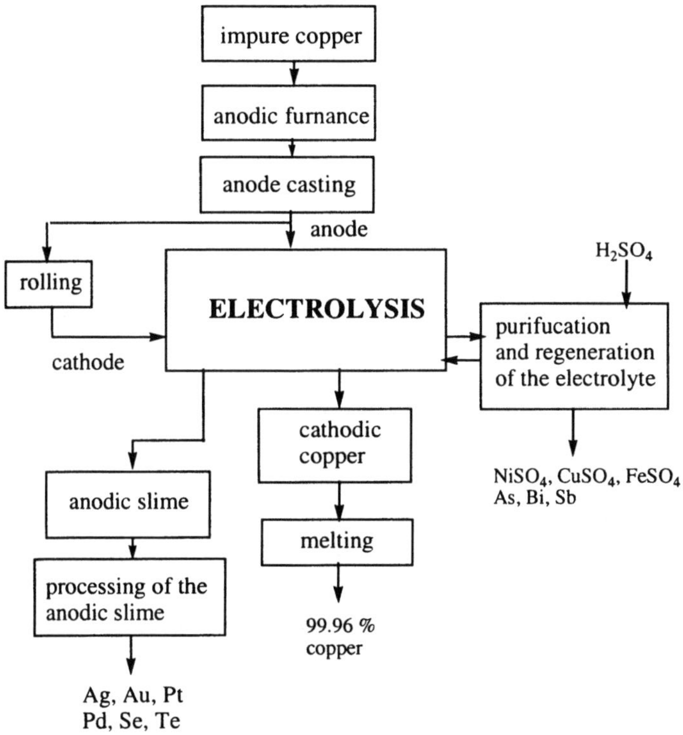

Figure 7.3. Technological scheme of copper electrorefining

During electrolysis process the electrolyte is enriched in Ni, Sb, and As, which are removed by regeneration, while the anodic slime is removed from the electrochemical cell and transported for further processing. It is important to mention that the process of the anode slime can bring in such high incomes that the entire cost of electrorefining is covered.

As previously stated, copper electrorefining is an electrolysis process with a soluble anode and theoretically the composition of the electrolyte should not change during the process. However, in a real system some composition changes do occur. First by the acid electrolyte dissolves some amount of anodic copper in the presence of air oxygen, which results in an increase in the concentration of copper ions, and a concurrent decrease in the free acid concentration. Secondly, impurities from the anodic copper, which did not sediment into the anodic slime, remain in the electrolyte and additionally decrease the concentration of free acid.

7.2.1.1 Refining of the electrolyte

In practice, two methods of correcting the electrolyte composition and refining are in common use[2].

The first way consists of removing copper, As and Sb by cathodic deposition in separate electrochemical cells with insoluble anodes. This procedure increases the concentration of the acid after which the solution is concentrated by vaporization whereby crystallization of nickel, iron and copper sulfates occurs.

The second procedure consists of the neutralization of the free sulfuric acid by dissolving copper. Copper sulfate crystallizes from the saturated solution. The rest of the electrolyte is freed from Cu, As and Sb by electrolysis in a separate cell and from Ni, Zn and Fe by evaporation.

After this refining, the electrolyte, practically contains only sulfuric acid, which is further, corrected by dissolving copper.

7.2.1.2 Processing of the anodic slime

The anodic slime is a precious crude from which noble and rare metals and metalloids such as the platinum group metals, gold, silver, selenium, tellurium etc are extracted[1,2].

Copper is removed from the anodic slime by dissolution in hot dilute sulfuric acid with air circulation, after which selenium and tellurium are removed. Subsequently, by melting after the addition of quartz sand, sodium carbonate and sodium nitrate, the anodic slime is transformed into an alloy composed of 93 % silver, 3% gold, 1% copper, 0.05% palladium, 0.03% platinum, and traces of other metals[3].

Silver of 99.99 % purity is extracted from the alloy by the process of anodic dissolution in a solution of silver nitrate. Owing to the high value of the exchange current density which leads to the poor adhesion of the silver deposit, some specific constructions of the refining cell are required which enable the separation of the anodic slime collected in polypropylene bags containing the anodes from the crystals of silver that drop from the cathode to the bottom of the cell[2].

After the process of silver electrorefining, the anodic slime is composed of 95% gold, 5% silver and about 1% copper, palladium, platinum and some other metals. After melting and casting, the anodes are subjected to a electrorefination in the electrolyte containing $HAuCl_4$ and free hydrochloric acid. Owing to the high value of the electrode potential of gold deposition of about 1.4 V, all of impurities, except silver which forms insoluble silver(I) chloride, are dissolved and remain in the electrolyte[2].

When the amount of platinum and palladium reaches a value of about 75 g dm^{-3}, the electrolyte is replaced. These metals are subsequently recovered in the form of ammonium salts: $(NH_4)_2PdCl_6$ and $(NH_4)_2PtCl_6$. By the reduction of these salts, palladium and platinum are obtained[2].

It is important to mention that the price of noble metals salts are a few times higher than the price of the noble metals themselves.

7.3 FURTHER READINGS

1. Pletcher, Derek, *Industrial Electrochemistry,* New York: Chapman and Hall, 1984.
2. Zečević, Strahinja; Gojković, Snežana; Nikolić, Branislav, *Electrochemical Engineering (In Serbian),* Belgrade: Faculty of Technology and Metallurgy, 2001.
3. • orđević, Spasoje, *Electrometallrgy (In Serbian),* Belgrade: Faculty of Technology and Metallurgy, 1972.
4. Kubasov, Vladimir; Bannikov, Vladimir, *Electrochemical Technology of Inorganic Compounds (In Russian),* Moscow: Khimia, 1989.

Chapter 8

OPTIMUM CONDITIONS FOR ELECTROPLATING

Metal coating represents a metal electrodeposit which changes the surface properties from those of the basic metal to those of electrodeposited one. It should be adherent, nonporous and without internal stresses. A good adhesion depends mainly on the preparation of the substrate for electrodeposition, resulting in a clean surface amenable to accepting a satisfactory deposit. Poor surface preparation can cause peeling of the coating making it useless. Crashing and peeling of a deposit can also be caused by internal stresses, which arise mainly from the incorporation of foreign materials in the lattice. If lattice of an electrodeposited metal is free of such inclusions the mechanical properties of the electrodeposit are practically the same as those of thermally prepared metal. Finally, the surface properties of the basic metal can not be completely transformed into the properties of the electrodeposited one if the deposit is porous.[1]

The preparation of a substrate for electrodeposition is not connected with metal electrodeposition and is not treated in this book. It is treated in details elsewhere[1,2]. The appearance of stresses, which is partially connected with the electrodeposition process, but has not yet been clarified is also not treated here.

On the other hand the effect of cementation on the adhesion of a coating and the effect of deposition condition on the porosity of the metal deposits can be treated in a semi-quantitative way.

8.1 CEMENTATION AND DEPOSITION FROM THE COMPLEX SALT SOLUTIONS

If zinc is immersed in a copper sulfate solution the reaction

$$Cu^{2+} + Zn \rightarrow Zn^{2+} + Cu \qquad (2.8)$$

occurs, whereby copper ions are converted to the metal, while the zinc dissolves. Such deposits are spongy and dendritic in a number of cases, and non-adherent as well. Hence, a good copper deposit on a zinc substrate can not be formed in this manner. (It should be noted that under some circumstances good deposits can be obtained by immersion deposition).[1]

In a cyanide-containing, bath the copper potential is sufficiently negative so cementation does not occur and copper can be successfully deposited onto zinc. This is due to the fact that the cyanide complexes of copper are very strong so the potential of copper in such a solution is much more negative than in simple salt solutions. On the other hand, the zinc cyanide complex is relatively weak and the potentials of two metals become comparable so an external power supply is required to deposit copper on the zinc from cyanide.

Analogously, from a copper sulfate solution, cementation of copper on immersed iron occurs according to the reaction

$$Cu^{2+} + Fe \rightarrow Fe^{2+} + Cu \tag{8.1}$$

but from a cyanide solution, copper can be successfully deposited using an external power supply. In this case the complexes of both copper and iron are very strong and cementation is theoretically possible if only the reversible potentials only are taken into the consideration. The fact that copper can be successfully deposited onto steal from cyanide containing solutions is explained by the fact that the reaction between iron and cyanide ions is very slow, and so the reversible potential is never reached.

It should be noted, concerning the electrodeposition from complex salt solutions that the best coatings are obtained from cyanide solutions. However, satisfactory deposits can also be obtained from some other solutions, but many complex-containing baths produce unsatisfactory deposits[1]. A list of suitable complex-containing baths and their working conditions can be found in the literature[3].

The quality of metal deposits obtained from complex salt solutions depends on the deposition process parameters and of the adsorption of anions onto the cathode. In the case of cyanide solutions, the deposition of good deposits is ascribed to the decrease of the exchange current density and increase of the value of the cathode Tafel slope due to the strong adsorption of cyanide anions onto the cathode[1,4].

8.2 THE POROSITY OF METAL ELECTRODEPOSITS

The porosity is the most important property of good adherent and stressless metal coating.

The coverage of an electrode surface with deposited metal increases with deposition time according to[5]:

$$\frac{d\theta}{dt} = kj(1-\theta) \tag{8.2}$$

where θ is the coverage, t is the time, k is a constant and j is the current density of metal electrodepositon onto an inert substrate at a given overpotential. Obviously, for systems with sufficiently low exchange current densities the deposition current densities, on the inert substrate and on the metal surface are the same (see section 3.1). The integral form of Eq. 8.2 is given by:

$$\theta = 1 - \exp(-kjt) \tag{8.3}$$

Assuming that porosity, π, corresponds to the uncovered electrode surface, it can be written:

$$\pi = 1 - \theta \tag{8.4}$$

and

$$\pi = \exp(-kjt) \tag{8.5}$$

Equation 8.5 is confirmed by an experiment as illustrated in Fig. 8.1.

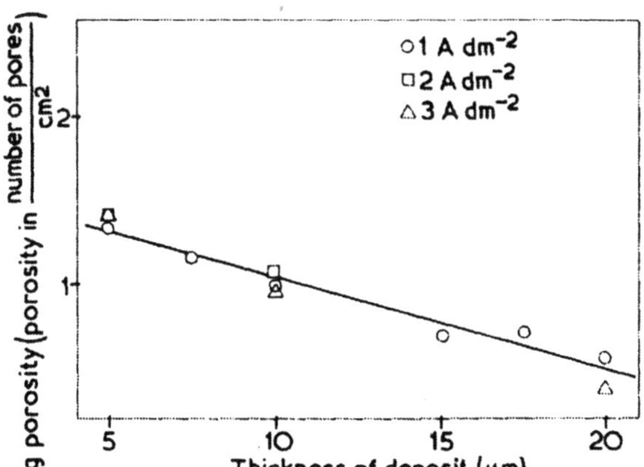

Figure 8.1. Relationship between the porosity and thickness of metal deposit for different values of current.[5]

The constant k in Eqs. 8.2, 8.3 and 8.5 can be evaluated in the following way. At $t = 0$

$$\frac{d\theta}{dt} = kj \tag{8.6}$$

and if one assumes a linear increases of θ with time and that the covering starts without overlapping, the equation

$$\theta = ktj = kQ \tag{8.7}$$

is valid. Obviously,

$$kQ_0 = 1 \tag{8.8}$$

where Q_0 is the quantity of electricity which corresponds to one monolayer of electrodeposited metal. Hence, k can be defined as the reciprocal of Q_0

$$k = \frac{1}{Q_0} \tag{8.9}$$

Equation 8.3 can be now rewritten in the form:

$$\pi_n = \exp(-n) \tag{8.10}$$

where n

$$n = \frac{Q}{Q_0} \tag{8.11}$$

is the average number of electrodeposited monolayers.

Although Eq. 8.10 is only qualitative, it can be successfully applied for discussing the dependence of the metal deposit porosity on the surface coarseness. Let the local thickness distribution of an n-monolayer thick (average thickness) deposit be:

fraction of surface	$1-2f$	f	f
average thickness of deposit	n	$n+1$	$n-1$

The porosity of such a deposit will be given, according to Eq. 8.10, by:

$$\pi = (1-2f)\exp(-n) + f\exp[-(n+1)] + f\exp[-(n-1)] \qquad (8.12)$$

Dividing Eq. 8.12 by Eq. 8.10 one obtains:

$$\pi = \pi_n[1-2f+f(e+e^{-1})] \qquad (8.13)$$

It is obvious that because

$$e + e^{-1} > 2 \qquad (8.14)$$

for $f > 0$, a smoother deposit will be less porous than a coarse one. This means that the better the current density distribution on the microprofile and the macroprofile, the lower is the porosity of a deposit.

8.3 THE CONDITION FOR THE DEPOSITION OF A COATING WITH A MINIMUM POROSITY

The porosity of metal deposits depends on the thickness of the thin surface film formation and the current density distribution on the microprofile and the macroprofile.

Small and large values of the exchange current density and cathodic Tafel slope enhance the formation of a surface film. This is realized by deposition from complex salt solutions or in the presence of strongly adsorbed additives. The best microprofile is obtained using deposition current densities a little larger than the current density which corresponds to the upper limit to the Tafel linearity in the presence of the some leveling and brightening agents. Finally, the current density distribution is improved also by increasing the Tafel slope of the deposition process and by decreasing the ohmic resistance of solution.

Hence, the basic characteristics of plating baths which produce good deposits should be:

– a simple or complex salt solution from which the metal is deposited at a sufficiently negative cathode potential relative to the substrate using a deposition process characterized by a low exchange current density and a large value of the cathodic Tafel slopes,

– a conducting electrolyte which makes a low as possible ohmic resistance of solution,

– the use of different kinds of additives, the synergetic effects of which improve the deposition conditions in the way described in the previous section.

There are also some other additives which are used occasionally for different purposes or activators of anode dissolution.

The current regimes are also very important in electroplating processes. The simplest current regime consist a short pulses of large current density followed by prolonged deposition at a many times lower current density. In this way nucleation takes place under more suitable conditions than exist during deposition of a low current density, which allows less coarse deposits to be obtained during prolonged electrodeposition. Regimes consisting of the periodic repetition of different current or overpotential waves can also improve the plating processes. It should be noted that the effect of electrodeposition at a periodically changing rate in the presence of organic additives is not completely understood yet.

In spite of the fundamentals of electroplating being very simple, (as shown in Chapters 3 to 5) the overall process is complicated and consists of a large number of sequences, the principles of which, as are given from the theory to the industrial practice in Ref. 1-3, are required for complete understanding of electroplating.

8.4 FURTHER READINGS

1. Lowenheim, Frederick, *Electroplating (Fundamentals of Surface Finishing)*. New York: McGrow-Hill book company, 1978.
2. Graham, Keuneth, *Electroplating Engineering Handbook,* New York: Von Nostrand Reinhold Company, 1971.
3. Lowenheim, Frederick, *Modern Electroplating*. New York: John Wiley & Sons, 1974.
4. Kabanov, Boris, *Electrochemistry of Metals and Adsorption (in Russian)*. Moscow: Nauka, 1966.
5. Popov K.I., Keča D.N., Drašković D.A., Vuksanović B.I. The effect of pulsating potential electrolysis on the porosity of metal deposits. J. Appl. Electrochem. 1976; 6:155-57

Chapter 9

ELECTROPLATING AND SURFACE FINISHING

The aspects of the electrodeposition of individual metals are discussed in other chapters. In this chapter, aspects of electrodeposition of alloys and composite materials, electroforming as well as electrodeposition of metals and alloys from nonaqueous solutions or room temperature molten salts are discussed. In addition, anodic processes i.e. electropolishing, electromachining and electrochemical oxidation of metals is presented. These processes are used in the production and developments of various materials, which include electronics, automotive, aerospace, biomedical, corrosion-protection and energy conversion applications.

9.1 ELECTRODEPOSITION OF ALLOYS

Electrodeposited alloys have attracted significant attention, due to their diverse applications in different industrial fields. Although different alloys can be electrodeposited from molten salts or from organic solutions, in this section only electrodeposition from aqueous solutions will be discussed.

Electrodeposition of over one hundred binary and ternary alloys has been investigated, however, only several alloy-plating systems have attained practical importance. Developments in the electronics, automotive and aerospace industries have driven research in the field of electrodeposition of alloys. Among these systems, alloys such as Fe-Ni, Ni-W, Ni-Mo, Pb-Sn, Cu-Ni, Fe-Zn etc. should be mentioned.

Simultaneous reduction of two metal ions is possible when their discharge potentials are equal, as presented by the equation:

$$E_1^\circ + \frac{RT}{n_1F}\ln a_1 + \eta_1 = E_2^\circ + \frac{RT}{n_2F}\ln a_2 + \eta_2 \qquad (9.1)$$

where E_1^o and E_2^o are standard electrode potentials of respective metals, a_1 and a_2 are metal ions activities, η_1 and η_2 are cathodic overvoltages, n_1 and n_2 are numbers of electrons, R is the gas constant, T is the absolute temperature and F is the Faraday's constant.

In the simple salt solutions, if the standard potentials of two metals are close, and if overvoltages are negligible, changing the activities can bring the discharge potential together. An example of this type is electrodeposition of Sn-Pb alloys from fluoroborate solutions.

If the standard potentials are significantly different, changing activities of metal ions cannot bring the discharge potential together. The most effective way of bringing close together discharging potentials of two metals, which are deposited simultaneously, is the formation of strong complexes with metal ions. In this case, not only activities of metal ions, but also mechanisms of deposition are changed. The complexing agents are chosen in a way to reduce the activity of ions of more positive metal to a greater extent then the activity of ions of less noble metal. It is important that the overvoltage of more noble metal is higher than the overvoltage of less noble metal. Among complexing agents, used in the electrodeposition of alloys, cyanides, pyrophosphates, ammonia, fluorides, citrates, tartars etc. should be mentioned. Sometimes, the addition of surface-active compounds to the solution decreases the rate of reduction of more noble metals.

For the same composition, properties of electrodeposited alloys differ from metallurgical (thermally) prepared alloys, which is a consequence of differences in the crystallisation process. The electrodeposited alloys, depending on the system, composition and electrolysis conditions may represent true solid solutions, and as well they may contain different phases consisting of various intermetallic compounds and of the mixture crystals of pure components (eutectic-type of alloys).

The change in the phase structure for the alloys with the same composition is often observed with a change in the conditions of electrodeposition. It is obvious, then, that the properties, including composition of electrodeposited alloys are determined by the electrodeposition conditions. The main conditions determining properties of electrodeposited alloys are classified in the following groups:

- Composition of the plating solution, which includes concentration of metals being deposited, concentration of complexing agents, concentration of conducting electrolyte, pH, concentration of additives etc,
- Operating conditions, which include current density, temperature and bath agitation, type of current (i.e. constant, pulsating, reversing etc.),

- Other parameters, such as cell geometry, shape of the cathode, thickness of the deposit, nature of the substrate, etc..

All of these parameters influence the composition and properties of electrodeposited alloys in different ways, which usually depends on the system being deposited. Therefore, a generalization of the effect of different variables on the properties of electrodeposited alloys would be difficult, unless impossible.

According to Brenner[1], electrodeposited alloy systems are classified into following five groups:

(i) Regular codeposition,
(ii) Irregular codeposition,
(iii) Equilibrium codeposition,
(iv) Anomalous codeposition and
(v) Induced codeposition.

The first three types, i.e. regular, irregular and equilibrium codepositions are referred to as a normal codeposition. In the normal type of codeposition, relative ratios of metals in the electrodeposited alloys are expected or predictable on the basis of the equilibrium potentials of these metals in the solution. While the equilibrium codeposition is a true normal codeposition, in regular and irregular codepositions the more positive metals deposit preferentially.

In the preferential deposition, the metal ratio in the deposited alloy is greater than the ratio of metal ion concentration in the solution, as presented by the equation:

$$\frac{[A]}{[B]} > \frac{C_A}{C_B} \tag{9.2}$$

where *[A]* and *[B]* are contents of metals *A* and *B* in the deposit, and C_A and C_B are metal ion concentrations in the solution.

In the regular codeposition, the alloy is deposited under diffusion control conditions. In this case, the amount of more noble metal in the alloy increases with an increase in the total metal ion concentration in the solution, a decrease in the current density, a raise in temperature and with stirring. The regular codeposition usually occurs when the potentials of the metals are far apart, and when metals do not form solid solutions. Typical examples of regular codeposition include Bi-Cu, Mn-Ni, Cd-Zn and Ag-Cu.

Irregular codeposition is related to systems in which deposition is not under diffusion-control. Deposition in these systems is controlled by irregularities of the potentials of metals in solution. This type of deposition occurs with metals which form solid solutions, and when potentials of metals being deposited are close together. Examples of irregular codeposition are Cd-Cu, Cu-Zn, Sn-Zn etc.

Equilibrium codeposition is characterized by the deposition, which is in equilibrium with the both parent metals. This is the only type of deposition in which the ratio of metal content in the deposit (plated at low current density) is equal to their ratio in the solution. The equilibrium deposition is rare and only a few plating systems, such as Cu-Bi and Pb-Sn (electrodeposited from acidic solutions) were investigated. The alloys, which do not have equilibrium with the both parent metals, belong to regular or irregular plating systems.

Anomalous and induced codepositions are classified as abnormal electrodeposition of alloys.

A very important factor that affects not only the composition, but also the properties of electrodeposited alloys based on metals of the iron group is simultaneous hydrogen evolution during electrodeposition. Most of the hydrogen produced during electrodeposition forms molecular H_2 which, as bubbles, is removed from the cathode surface. Another small fraction of the hydrogen becomes adsorbed in the crystal lattice of the electrodeposited metals. The quantity of hydrogen included in deposits of ferromagnetic alloys is approximately 0.45 at.%. According to Frumkin, hydrogen can be incorporated in solid solutions or as hydride phases. [2] Thermal treatment of these electrodeposits leads to removal of the incorporated hydrogen, causing deformation of the crystal lattice and ultimately changing the properties of the electrodeposited metal or alloy. In terms of electrocatalytic effects in the H_2 evolution reaction changing the surface composition of an electrodeposited alloy can lead to a time dependent current efficiency for metal deposition relative to the H_2 evolution rate.

The hydrogen evolution reaction on metals occurs in two main steps:

(i) discharge step

$$H^+ + e^- \rightarrow H_{ads.} \tag{9.3}$$

(ii) recombination-desorption step

$$H_{ads.} + H_{ads.} \rightarrow H_2 \tag{9.4}$$

or electrochemical desorption step

$$H_{ads.} + H^+ + e^- \rightarrow H_2 \tag{9.5}$$

Sometimes, a third step involving sorption into the metal arises, especially at transition metals

$$H_{ads.} \rightarrow H_{incl.} \qquad (9.6)$$

In the electrodeposition of alloys of the iron group of metals (Fe, Co and Ni) from simple- or complex- salt solutions, the *less* noble metal is reduced preferentially, and its relative content in the deposit is higher than that in the bath. Such phenomenon is known as *anomalous* codeposition. Anomalous codeposition is one of the most studied phenomena, especially on the Fe-Ni alloys, due to their applications in the electronics industry as magnetic materials. In the electrodeposition of Fe-Ni alloys the less noble metal, iron, deposits preferentially and its relative content in the alloy deposit is higher than that in the solution. The anomalous codeposition is observed in the electrodeposition of all alloys based on the iron group of metals (Fe, Co, Ni). Besides Fe-Ni, typical examples include Co-Ni, Fe-Zn, Ni-Zn etc.

There are elements in the periodic table, which cannot be deposited alone from the aqueous solutions (e.g. Mo, W, P and Ge). These elements can readily be deposited with the iron group of metals. The phenomenon is known as induced codeposition. In the induced codeposition, the iron group of metals is referred as inducing metals, while Mo, W and P are reluctant elements.

One of the more widely studied examples of anomalous codeposition is electrodeposition of Ni-Fe alloys. Processes occurring during electrodeposition of Ni-Fe alloys can be summarized in terms of reactions:

$$Fe^{2+} + 2e^{-} \rightarrow Fe \qquad (9.7)$$

and

$$Ni^{2+} + 2e^{-} \rightarrow Ni \qquad (9.8)$$

with

$$2H^{+} + 2e^{-} \rightarrow H_2 \qquad (9.9)$$

or

$$2H_2O + 2e^{-} \rightarrow H_2 + 2OH^{-} \qquad (9.10)$$

The experimental studies of electrodeposition of Ni-Fe alloys have shown that:

(i) The ratio of Fe to Ni is higher in the alloy than in the electrolyte, and
(ii) The presence of Fe(II) in the solution inhibits the discharge of Ni.

Several models appeared in the literature with an attempt to explain the anomalous nature of electrodeposition of Ni-Fe alloys. The early models did not receive significant attention. Dahms and Croll postulated one of the most cited models for the electrodeposition of Ni-Fe alloys in 1965. [3] This model is based on the assumption that due to simultaneous evolution of hydrogen during electrodeposition an increase in pH near the electrode surface occurs. This increase in pH, as the authors assumed, causes the precipitation and adsorption of hydroxides at the electrode surface. The inhibition of discharge of Ni(II) ions was considered as a consequence of a blockage of Ni(II) ions by Fe(II) hydroxide, which is formed to a greater extent than Ni(II) hydroxide. In this way, an increase in the amount of iron content was attributed to the adsorption and incorporation of $Fe(OH)_2$ into the alloy deposit. According to the Dahms and Croll hypothesis, the inhibition of nickel discharge occurs at potentials where hydrogen reduction exceeds its mass transport limit.

Several modifications of the hydroxide-based model appeared in the literature. These models are often in agreement with experimental data, however, their validity is determined by their ability to simulate the measurable macroscopic aspects of deposition.

Although simultaneous hydrogen evolution is an important factor influencing properties of electrodeposited alloys, according to later experimental studies it appears unlikely that this reaction causes a significant increase in pH near the cathode surface. On the other hand, the anomalous codeposition occurs even at low current densities, where the hydrogen evolution reaction is negligible[4], and therefore cannot cause a high increase in pH which may lead to the precipitation of Fe(II)-hydroxides.

To explain the anomalous codeposition, Matlosz proposed a two-step reaction mechanism in which electrochemical (rather than chemical) kinetics is emphasized.[5] By a combination of a two-step reaction mechanism for the electrochemical reduction of the single metals depositing alone, a competitive adsorption model was developed. This mechanism is described by the following reactions:

$$M(II) + e^- \rightarrow M(I)_{ads.} \tag{9.11}$$

$$M(I)_{ads.} + e^- \rightarrow M \tag{9.12}$$

where the symbol M denotes either Fe or Ni.

The experimental results showed that nickel deposits normally at low overpotential. Deposition of nickel is suddenly inhibited at a specific cathodic polarization and, the codeposition process does not influence deposition of iron. This behavior is attributed to a higher surface coverage of the adsorption sites on the electrode by the iron intermediate species, which is in a good agreement with the proposed model. The mechanism suggests that the pH near the cathode surface plays only a secondary role and is not the necessary condition for an occurrence of the anomalous codeposition. The hydroxide concentration, at the electrode surface does not change the mechanism. The electrodeposition of iron-reach alloys at higher potentials could be avoided with a decrease in the Fe(II) concentration in the solution.

Based on the theoretical calculations and experimental observations Vaes et al. concluded that the metal hydrolysis does not play a determining role in anomalous codeposition of Ni-Fe alloys.[6]

The composition of Ni-Fe alloys depends on plating conditions including the composition of the electrolyte. Generalization of these dependencies is difficult due to different experimental conditions, and the various electrolytes used in production of these alloys.

Based on the experimental evidence that the current efficiency for nickel deposition is practically constant a simple equation for the dependence of alloy composition on the parameters of periodically changing current is derived.[7] This equation is presented as follows:

$$\frac{100}{100-\%\text{Fe}} = \frac{1}{CE(\text{Ni})} - \frac{k}{CE(\text{Ni})}\frac{Q_c}{Q} \quad (9.13)$$

where %Fe is the percent of iron in the deposit, CE(Ni) is the current efficiency of nickel deposition, Q_c is the cathodic charge, Q is the total charge and k is a constant. The function Q_c/Q depends on parameters of periodically changing current. As this equation shows, there is a linear relationship between the composition of Ni-Fe alloys and Qc/Q function. The experimental results confirm the validity of the equation (9.13), as presented in Figure 9.1.

Similarly, for the electrodeposition of Ni-Fe alloys under the potentiostatic conditions, the following equation is applicable:

$$\log\left(\frac{\%\text{Ni}}{\%\text{Fe}}\right) = \log k - k'E \quad (9.14)$$

where k and k' are constants and E is depositing potential. For a certain range of potential, a linear relationship between log (%Ni/%Fe) and E is observed as it is confirmed by the experimental results (Figure 9.2).

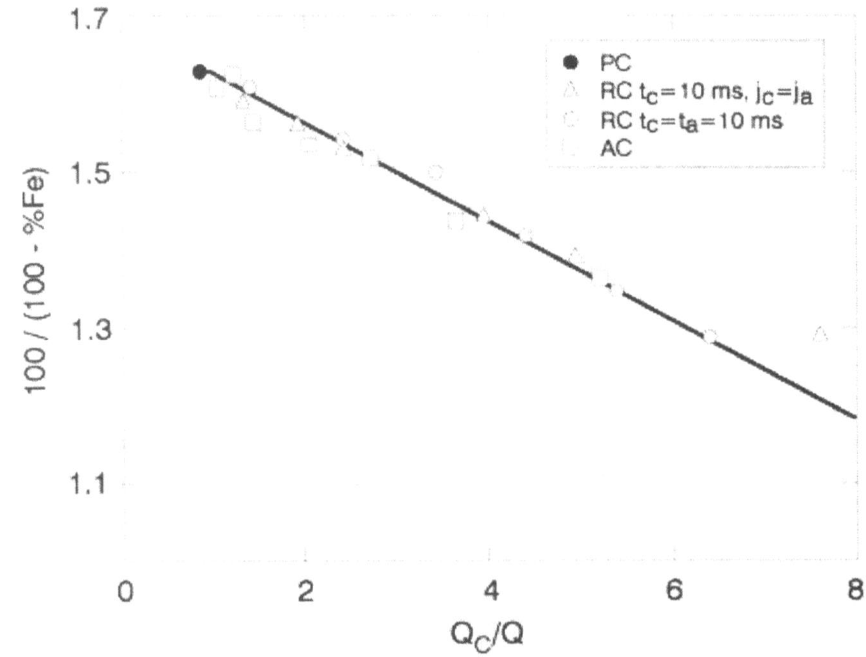

Figure 9.1. The effect of Q_c/Q on the Fe content in the electrodeposition of Ni-Fe alloys under pulsating current (●), alternating sinusoidal current superimposed on direct current (□, ν = 50 Hz), reversing current (Δ, t_c = 10 ms, $i_c = i_a$; $t_c = t_a$ = 10 ms)[7]

One of the most studied systems of induced codeposition is electrodeposition of Ni-Mo alloys because of their corrosion and wear- resistance properties as well as for their electrocatalytic effect on the hydrogen evolution reaction in alkaline solutions. Ni-Mo alloys are electrodeposited from citrate or pyrophosphate solutions. The composition of Ni-Mo alloys depends on concentration and mass transport of Ni (II) and molybdate species. [8] If the relative concentration of Ni(II) in the solution is significantly higher than the relative concentration of molybdate ions, the molybdenum content decreases with current density, but increases with the rotation rate. Contrary, if the concentration of molybdate ions is larger than the concentration of Ni(II), an increase in the current density leads to an increase in molybdenum content in the alloy. However, the alloy composition for this case is independent on the rotation rate. This suggests that that the mass transport does not influence the composition of Ni-Mo alloy when the concentration of molybdate is significantly larger than the concentration of Ni(II) in the plating solution.

The researchers proposed several mechanisms for the electrodeposition of Ni-Mo alloys. It seems that the most feasible is the mechanism proposed by Chassaing *et al.* [9]. They reported a two step mechanism for the electrodeposition of Ni-Mo alloys, which can be presented as follows:

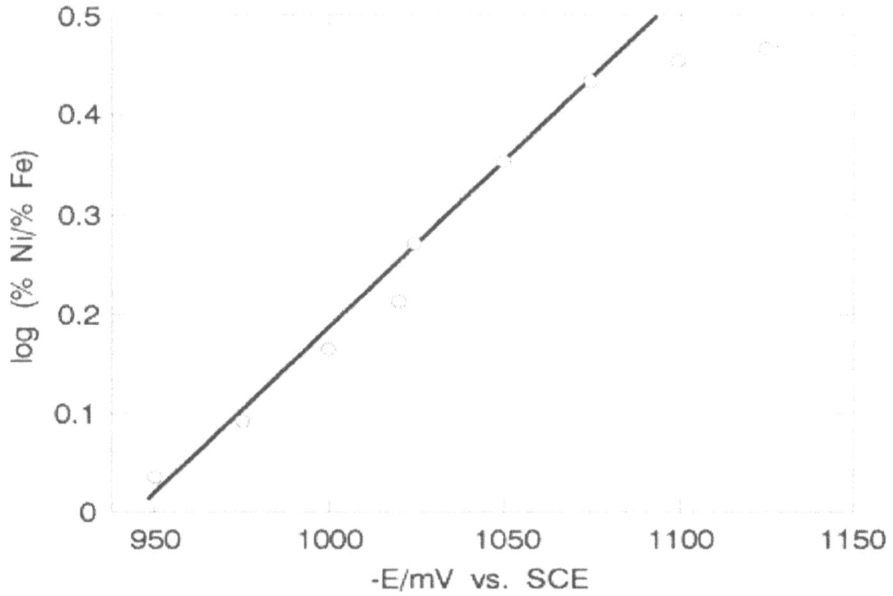

Figure 9.2 The effect of potential on composition of electrodeposited Ni-Fe alloys [7]

(i) Formation of MoO_2 which, in the presence of Ni(II) gives a mixed oxide MoO_2Ni_4:

$$MoO_4^{2-} + H_2O + 2e^- \rightarrow MoO_2 + 4OH^- \qquad (9.15)$$

and

$$MoO_2 + 4Ni^{2+} + 8e^- \rightarrow MoO_2Ni_4 \qquad (9.16)$$

(ii) Reduction of MoO_2Ni_4 by hydrogen to $MoNi_3$ with included hydrogen, thus

$$MoO_4^{2-} + 3Ni^{2+} + (4+n)H \rightarrow MoNi_3 + 4OH^- + nH_{incl.} \qquad (9.17)$$

The simultaneous hydrogen evolution during electrodeposition of Ni-Mo alloys strongly influences the properties of these deposits. The electrodeposited Ni-Mo alloys show formation of a "cauliflower" surface morphology, high surface area, formation of voids, pits and cracks (Figure 9.3) and a gradient in composition. However, under certain conditions electrodeposition of crack-free, uniform deposits is also possible.

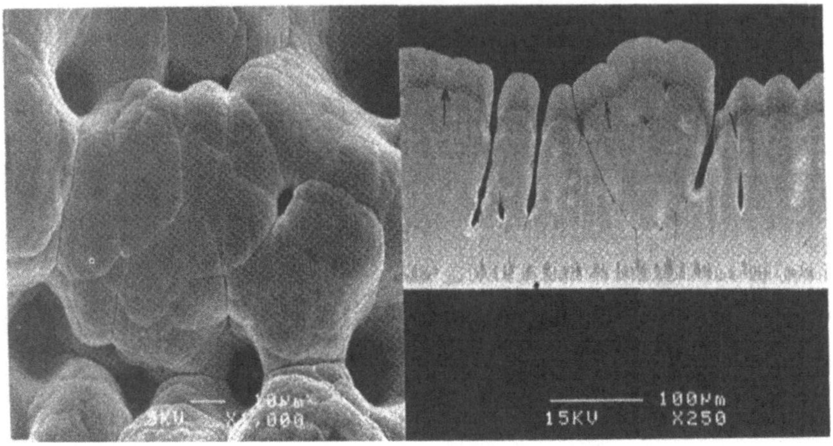

Figure 9.3. SEM micrographs of surface (left) and cross-section (right) of electrodeposited Ni-Mo alloys

Electrodeposited alloys based on the iron group of metals are usually crystalline. Codeposition of P and B with the iron group of metals is especially important because the incorporation of these elements into a deposit influences the structure of the electrodeposited alloys. The incorporation of phosphorus in the alloy deposit is essential factor leading to the production of amorphous or nanocrystalline alloys. [10] Comparison of XRD patterns of "as electrodeposited" Ni-P, Co-P and Ni-Co-P alloys is presented in Figure 9.4. The broad peaks associated with these deposits correspond to an amorphous structure. This result confirms that the alloys based on iron group of metals containing more than 8 % of phosphorus, particularly Ni-P alloys are amorphous. When heat treated at temperatures above 350 °C, these alloys completely devitrify, forming mixtures such as Ni, Ni_3P, Co, Co_2P etc., depending on the overall alloy composition (see Figure 9.5).

Among other alloy plating systems that have attracted the attention of researchers, the multilayer films or compositionally modulated multilayers should be mentioned. [11-13] Compared with pure metals, the compositionally modulated multilayers with distinct sublayers have unusual and enhanced mechanical, electrical, optical and magnetic properties. Examples of these systems include Cu-Ni, Cr-Ni, Cu-Co, Ag-Pd etc.

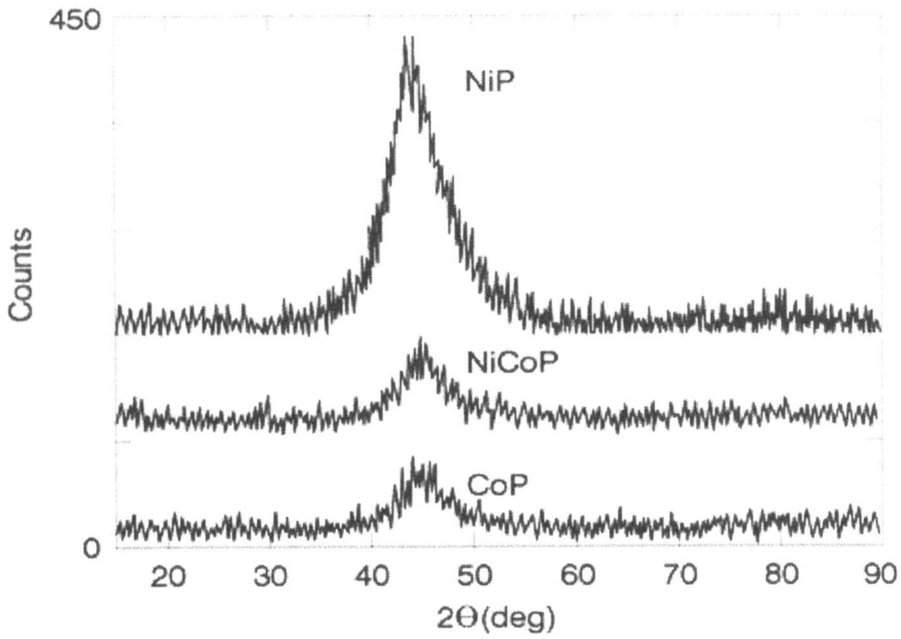

Figure 9.4. A comparison of XRD patterns of electrodeposited Ni-P (92 wt. % Ni, 8 wt. %P, thickness 80 μm), Co-P (92 wt. % Co, 8 wt. % P, thickness 80 μm) and Ni-Co-P (13 wt. % Ni, 77 wt. % Co, 10 wt. % P, thickness 100 μm) alloys[10] (Reprinted with the permission from The Electrochemical Society Inc.)

Two different ways, known as single bath technique (SBT) and double bath technique (DBT) have been proposed for deposition of compositionally modulated multilayers. In the single bath technique, deposition is carried out from one bath (plating solution) which contains ions of both constituents of the multilayer film. This is achieved by periodical variation of the applied potential or current between a value corresponding to deposition of one metal to a value corresponding to predominant deposition of the other component. The thickness of the layer can be modified in a wide range, depending on the time frame and potential or current variation. For the production of multilayeral coatings deposition under periodically changing current conditions is well suited. [13]

The double bath technique involves the use of two different plating solutions containing ions of individual constituents of the multilayer film. In this type of deposition, the substrate is successively moved from one to the other bath and layers of pure metal are deposited in each of the individual baths. The double bath technique is often accompanied by dissolution of plated metal, displacement reaction and passivation of the surface, due to removal from one to another bath. Between the two plating steps, substrates should be rinsed to avoid contamination of plating solutions, which increases the wastewater generation.

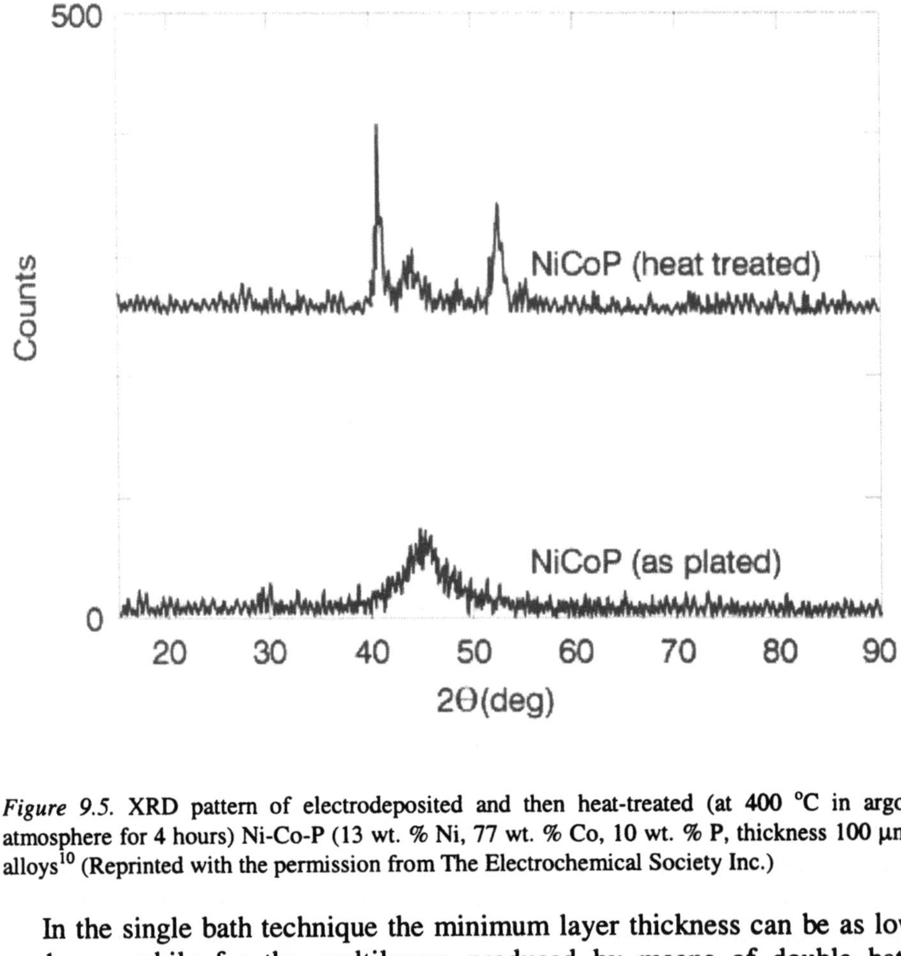

Figure 9.5. XRD pattern of electrodeposited and then heat-treated (at 400 °C in argon atmosphere for 4 hours) Ni-Co-P (13 wt. % Ni, 77 wt. % Co, 10 wt. % P, thickness 100 μm) alloys[10] (Reprinted with the permission from The Electrochemical Society Inc.)

In the single bath technique the minimum layer thickness can be as low as 1 nm, while for the multilayers produced by means of double bath technique the minimum film thickness is estimated at about 25 nm. The individual metal sublayers usually grow epitaxially on top of one another.

Based on the mixed potential theory, in which the measured current in an electrochemical system is equal to the sum of anodic and cathodic partial current, i.e.,

$$I = \sum I_{a,p} + \sum I_{c,p} \tag{9.18}$$

where I is the measured current, $I_{a,p}$ is the partial anodic current and $I_{c,p}$ is the partial cathodic current, Landolt[14] proposed that the composition of electrodeposited alloy differs from that of the electrolyte and depends on both kinetic and thermodynamic quantities. According to the Tafel equation, for the cathodic partial current of species *i*,

$$j_i = i_{o,i} exp(-s_{c,i} \eta_i) \tag{9.19}$$

where $j_{o,i}$ is the exchange current density, $s_{c,j}$ is the inverse of the cathodic Tafel coefficient (V^{-1}), η_i is the overvoltage given by $\eta_i = E - E_{rev,i}$ (E is the applied potential and $E_{rev,i}$ is the equilibrium potential of species *i*), the ratio of partial current densities for two species A and B is given by the equation:

$$ln \frac{j_A}{j_B} = R_o - (s_{c,A} - s_{c,B})E \tag{9.20}$$

where R_o is defined by:

$$R_o = ln \frac{j_{o,A}}{j_{o,B}} + (s_{c,A} E_{rev,A} - s_{c,B} E_{rev,B}) \tag{9.21}$$

The equation (9.20) shows that the composition of an alloy does not depend on potential only if

$$s_{c,A} = s_{c,B} \tag{9.22}$$

On the other hand this equation demonstrates that the composition depends on the exchange current density through the ratio $j_{o,A}/j_{o,B}$. Based on the mixed potential theory and the above equations, Landolt classified electrodeposition of alloys into the following groups:

(i) non-interactive codeposition
(ii) charge-transfer coupled deposition
(iii) transport coupled deposition

In the non-interactive codeposition, the partial currents are independent of each other. A typical example of this type of codeposition is the electrodeposition of nickel-copper alloys. [15]

The charge-transfer coupled is a type in which the partial currents depend of each other. This type of codeposition is divided further into two subgroups designated as inhibited codeposition and catalyzed codeposition.

The examples of the inhibited codeposition include electrodeposition of zinc-nickel or iron-nickel alloys. This type is quite similar to the anomalous deposition as described by Brenner. The partial current density for deposition of more noble metal, e.g., nickel, during the electrodeposition of alloy is much lower than the current density, when this metal is plated alone. On the other hand, the partial current for deposition of less noble metal, e.g., zinc, is not affected by the presence of nickel.

The catalyzed deposition (Landolt's classification) is exactly the same as the induced deposition (Brenner's classification). Examples of this codeposition are electrodeposition of Ni-Mo and Ni-W or Ni-P alloys.

Finally, in the mass transport coupled codeposition the partial current of the component A depends on the transport of component B. This type of codeposition includes systems in which the simultaneous hydrogen evolution occurs during electrodeposition of alloys (e.g. electrodeposition of Fe-Ni or Zn-Ni alloys). Under conditions of simultaneous hydrogen evolution due to consumption of protons, a local increase in pH depends on mass transport conditions and the buffering capacity of the electrolyte, and may effect the mechanism and kinetics of electrodeposition of alloys.

The advantage of Landolt's classification over Brenner's is in the fact that, the former takes into consideration not only thermodynamics, but also charge transfer kinetics and mass transport. However, some systems such as for example electrodeposition of Fe-Ni alloys or Zn-Ni alloys as per Landolt's classification can belong to either charge-transfer coupled or transport coupled codeposition.

Although many systems of electrodeposition of alloys have similarities there are significant differences. These differences arise as a consequence of different conditions of electrodeposition, which includes solution composition and operating conditions. It is obvious that more research is required to evaluate the significance of different parameters for a full explanation of features of alloy electrodeposition.

9.2. ELECTRODEPOSITION OF COMPOSITE MATERIALS

Composites produced by electrodeposition include materials with metallic, oxide or polymer matrices, in which solid particles or fibres are codeposited and uniformly distributed in the deposit.[16] The inert particles used in the electrodeposition of composite materials are usually 0.01 to 1 μm in diameter and are selected from alumina, boron, carbon, silicon carbide, titanium dioxide, tungsten etc., depending on applications. The particles are uniformly dispersed in the plating solution by a mechanical or ultrasonic agitation. During electrodeposition, the inert particles become positively charged and as such, attracted by the cathode and incorporated into the electrodeposited metal, alloy, oxide or polymer. Electrodeposition of composite materials with metallic matrices is usually carried out in order to improve their mechanical and tribological properties, although other characteristics such as corrosion and thermal resistance can also be significantly advanced.

The metal matrices may include nickel, cobalt, copper, zinc, precious metals and related alloys. Solid (inert) particles, which are incorporated into metal deposit, include oxides (Al_2O_3, SiO_2, Cr_2O_3, V_2O_5 etc.), carbides (SiC, Cr_2C_3), graphite, diamond or boron nitride particles, polymer powders (polytetrafluorethylene, polyvinyl chloride) and other components such as salts (MoS_2, $BaSO_4$) or some pigments. The incorporation of submicron powders such as Al_2O_3, $BaSO_4$, Si_3N_4, V_2O_5 and Cr_2O_3 in the nickel-based metallic matrices significantly increases the corrosion resistance. Particulates like WC, diamond and SiC protect metal from abrasion, while MoS_2, and polytetrafluorethylene (PTFE) and graphite reduce the friction coefficient of the composite materials.

For deposition of composite materials with metallic matrices, solutions similar to those in the electrodeposition of metals and alloys are used. The main difference is that solutions used in the electrodeposition of composite materials contain dispersed fine particles or fibres. Codeposition of inert particles into metal matrix is influenced by the adsorption of particles on the cathode. The content of particles in the deposit is influenced by their concentration in solution, additives, pH and current density. Most of the studies show that the content of deposited particles in the metal matrix increases with increasing particle concentration in the solution. In terms of particle size, different results have been reported for the same systems. Additives, such as brightners or wetting agents, influence codeposition of inert particles. Quite opposite observations on the effect of brightners or wetting agents on the amount of solid particles occluded into deposit were reported. In some cases, in the presence of these substances, an increase in the amount of particles is observed. Other researchers, in contrast reported a decrease in the particle content with an addition of wetting agents.

Current density significantly influences the codeposition of particles. Although, some researchers reported no influence of current density on the amount of particles deposited, most commonly, the observed dependence of current density on the particle concentration passes through one or several maximums. Typical examples are presented in Figures 9.6 and 9.7. As shown in Figure 9.6, the dependence of the amount of γ-Al_2O_3 in Au-Al_2O_3 deposit on current density, passes through two maximums for different rotation speeds. [17]

Similar dependencies of the amount of SiC in the Co-SiC deposit on current density, for different concentrations of SiC in the solution are presented in Figure 9.7. [18] A presence of particles in the plating solution increases the current density for the same cathodic potential. This indicates that the presence of particles in the solution causes a depolarization of the cathode.

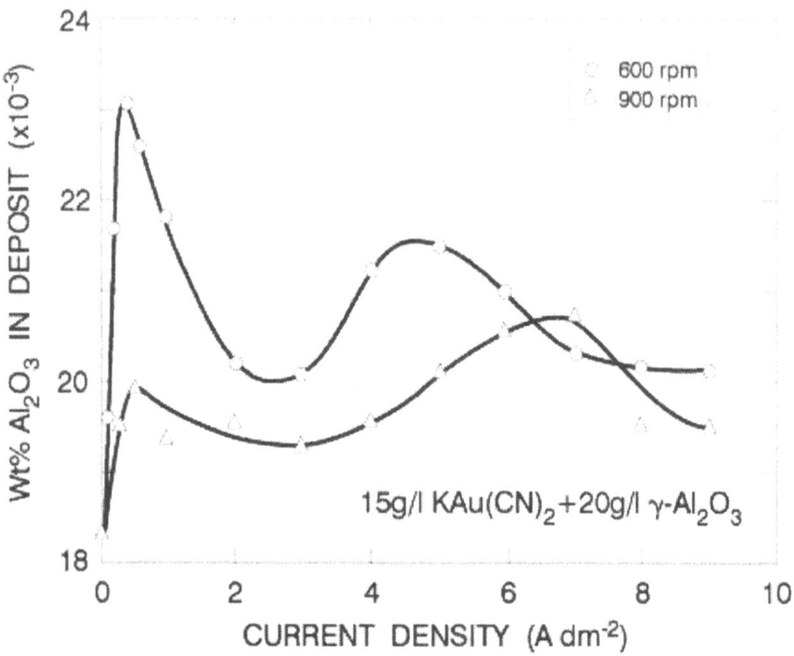

Figure9.6. Dependence of the amount of γ-Al_2O_3 codeposited with gold on current density[17]

The effect of pH on codeposition of solid particles such as Al_2O_3 or SiC into nickel matrix was investigated. The results have shown that an increase in the amount of particles in the deposit is observed with an increase in pH up to 2. With a further in crease in pH, no change in the amount of codeposited particles is observed. [16] Similar results have been reported for other composite materials systems, where electrodeposition is carried out from acidic solutions.

In order to produce coatings with a homogenous composition the solid particles should easily be transported through the solutions and their precipitation on the bottom of the plating cell should be avoided. This is usually achieved with a good agitation and with the addition of corresponding surfactants.

Generally agitation of the solution enhances the particle transport and increases their amount in the deposit. However, a too high agitation causes a decrease in the amount of codeposited particles. Consequently, the dependence of the rate of the solution agitation on the concentration of the particles shows a maximum. The stability of the suspension determines the quality of deposited composite material, and depends on the rate of sedimentation of solid particles, The rate of sedimentation, for an ideal case of spherical particles, is described by the Stokes law with the following formula:

$$v = \frac{2gd_p^2(\rho_p - \rho_e)}{9\upsilon} \tag{9.23}$$

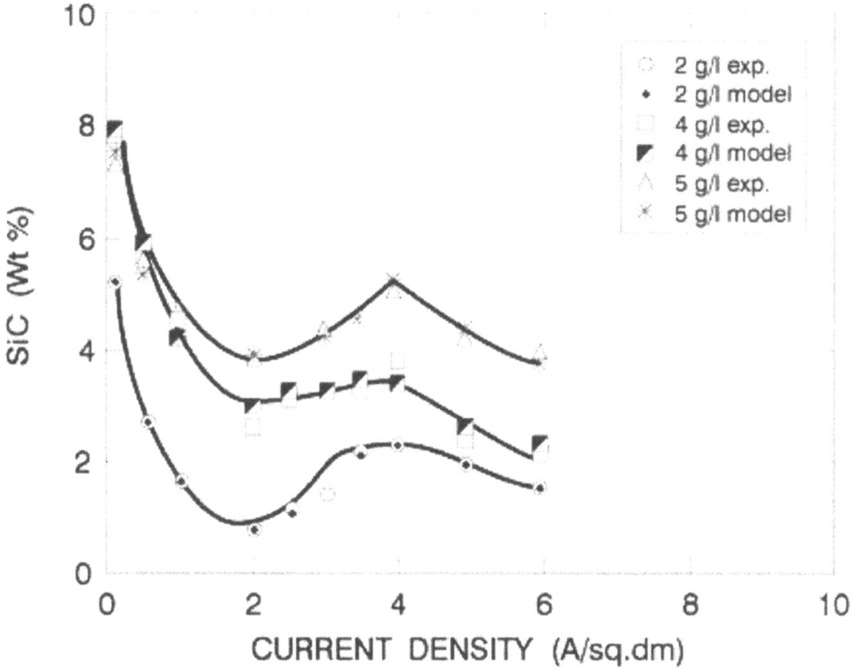

Figure 9.7. Effects of current density on the amount of SiC particles codeposited with cobalt for different particle contents in electrolyte (pH 4.0, t=30 °C, stirring rate 400 rpm) [18] (Reprinted with the permission of The Electrochemical Society Inc.)

The equation (9.23) shows that the rate of sedimentation, v, is directly proportional to the particle size, i.e. their diameter, d_p, particles density ρ_p, density of the electrolyte ρ_e, and indirectly proportional to the viscosity of the solution, υ.

In order to avoid the agglomeration and precipitation of particles in the electrolyte, addition of surfactants such as tannin, gelatine etc. is recommended. The addition of surfactants increases the wettability of particles, and stability of the suspension. Surfactants used in the electrodeposition of composite materials are classified as:

(i) cationic,
(ii) anionic and
(iii) nonionic.

The cationic surfactants confer a net positive charge of the particles, which attracts them electrostatically to the cathode. Contrarily, the anionic surfactants (such as some brightners or wetting agents) confer a negative on the particles, which leads to a decrease in the amount of codeposited-particles. The use of non-ionic surfactants usually promotes codeposition of solid particles. The surfactants, which are used in the electrodeposition of composite materials, can to a certain extent, deteriorate the quality of the coating (high internal stress and brittleness). This occurs due to adsorption of these substances on the electrode surface.

Many authors investigated the mechanistic aspects of electrodeposition of composite materials. The early models suggested that the particles with a positive surface charge are drawn by electrophoresis, or due to agitation, transported to the cathode and mechanically entrapped by the growing metal layer. The idea of mechanical entrapment was rejected, and the attraction of solid particles by the cathode is attributed to the electrostatic force.

One of the most cited models for electrodeposition of composite materials was developed in 1972 by Guglielmi. [19] The model is based on two successive steps, and considers both electrophoresis and adsorption phenomena. According to this model, solid particles are surrounded with a cloud of adsorbed ions. In the first step, when particles approach the cathode they become loosely adsorbed at the surface. The second step involves a strong, irreversible adsorption of these particles at the cathode and their incorporation into the growing metal layer. The strong adsorption of solid particles is preceded by a loosing of their ionic cloud. This model takes into consideration most experimental parameters, however it does not explain the appearance of the maximum in the particle content versus current density curves. Most importantly, this model neglects the mass transfer.

The model proposed by Guglielmi was extensively used by other researchers as a basis for development of other mechanisms, for the electrodeposition of composite materials. As a consequence, several mechanistic models appeared in the literature. [17,18,20] These mechanisms were proposed in an attempt to explain the characteristics of electrodeposition of composite materials, however, further studies are required for a more general understanding of this process.

The generally accepted mechanism for electrodeposition of composite materials involves the transport of particles from a solution to the electrode surface by agitation and their incorporation in the metal matrix by reduction of adsorbed ions. The literature survey shows that the particle concentration in the electrolyte, agitation and metal growth mechanism play important roles in the electrodeposition of composite materials.

The electrochemical impedance spectroscopy study of codeposition of SiC and SiO_2 particles with nickel suggests that particles suspended in a plating bath increase the roughness of the electrode surface. [21] Particles adsorbed, but not embedded in the electrode, remain separated from the electrode surface by a liquid film, which is thicker than the width of the electrical double layer. This liquid film, according to the authors is the key factor controlling the electrolytic deposition of particles.

While the early work was restricted to electrodeposition of composite materials with metallic matrices, the field has recently been extended to the development of other matrix materials (i.e., polymers or ceramics). [22] In order to be used as a matrix, the material must fulfil the following requirements:

(i) must be depositable either on the cathode or on the anode, and
(ii) must be electronically conductive.

Polymeric matrix materials, used in the electrodeposition of composite materials include polypyrrole, polyaniline and polythiophene. These materials were deposited from aqueous or aprotic solutions. The dispersed particles are usually Pt, Pd, PbO_2, RuO_2, TiO_2, etc.. Composite materials with polymeric matrices were mainly investigated as electrode materials in the fuel cell technology for either oxygen reduction or hydrogen oxidation. [23,24] Metal particles, incorporated in polymer films, act as catalytic sites for the electron-transfer processes. Typical examples include platinum nanoparticles incorporated into polypyrrole, or, palladium aggregated into the polyaniline. In the electrodeposition of composites with polymer matrices two main routes are followed. The first, similarly to electrodeposition of metal- or oxide- matrix composite, is based on the electrolysis of suspensions of the dispersed phase in solutions of monomer, which is converted to a solid phase (polymer) by electropolymerization. [25,26] In the second route the electropolymerization is first occurred and then formation of metal clusters within the polymer. [27,28] An example of this type of composite material is polypyrrole film containing highly dispersed platinum particles. $PtCl_4^{2-}$ ions are reduced to Pt^o particles with an average size of about 10 nm, according to the reaction:

$$PPy(PtCl_4^{2-}) + 2e^- \rightarrow PPy/Pt^o + 4Cl^- \qquad (9.24)$$

where PPy denotes polypyrrole.

Oxide matrices in the electrodeposited composite materials include, but are not restricted on PbO_2, Tl_2O_3 and non-stoichiometric W(VI, V) oxides. [29-33]

Composite materials with oxide matrices are usually investigated for the electrocatalysis purposes. PbO_2 and Tl_2O_3 matrices are selected, since they have a high electronic conductivity and they are anodicaly depositable from $Pb(NO_3)_2$ or $TlNO_3$ solutions. Electrodeposition of composites such as $PbO_2 - Co_3O_4$ and $Tl_2O_3 - Co_3O_4$ is investigated for the applications as anodic materials for the oxygen evolution reaction. Codeposition is carried out from $Pb(NO_3)_2$ or $TlNO_3$ electrolytes containing Co_3O_4 suspended particles (less than 1 μm). The incorporation of Co_3O_4 in PbO_2 or Tl_2O_3 matrices leads to an increase in the surface roughness and effective electrode area, which is favorable for the oxygen evolution reaction.

A composite material containing non-stoichiometric W(VI,V) oxides as a matrix and Pt microparticles by the cyclic voltammetry, during the reduction cycle is recommended for the reduction of molecular oxygen in the fuel cells applications. [33]

The features of electrodeposition of composite materials with oxide matrices are similar to electrodeposition of composites with metallic matrices.

9.3 ELECTROFORMING

Traditional electroforming is a method of producing metallic components by electrodeposition over a mandrel or mold, which can subsequently be separated from the deposit. The separated metallic component produced by the electrodeposition, represents itself a finished part. Consequently, in the traditional electroforming process, the surface of a mandrel is prepared in such a way that plated metal does not strongly adhere to the substrate. [34,35] Thus, the metal is readily separated from the substrate after plating. However, the adhesion has to be sufficient in order to avoid the separation of deposited metal before the electroforming process is completed.

For larger articles, electroforming is used in automotive, aerospace, biomedical, jewellery and musical industry applications. The applications of electroformed parts are found in the continuous copper foil used in the printed circuit industry, nickel screen or mesh patterns for printing in the textile industry, mold stampers for the compact audio and video disks, seamless cylinders used in copying machines, components in rocket motors, etc..

Depending on the design of the electroform, and/or, the quantity of parts required, mandrels may be either permanent or disposable. The permanent mandrels are used repeatedly, while disposable mandrels are destroyed after removal from electroform. Permanent mandrels are preferred when the electroform has no undercut surfaces and can easily be separated without damage. In these cases, the mandrel is dissolved or melted to free the electroform.

The material used for mandrel must be dimensionally stable, since its surface morphology is reproduced exactly down to submicroscopic level, with nearly atomic resolution on the electroform. This duplication of the surface details accounts for many applications of electroforming process.

Typical materials used as permanent or disposable mandrels are listed in Table 9.1. It should be noted that every single material used as a mandrel has certain advantages or disadvantages.[34] Consequently, certain precautions such as machinability, scratching, corrosion resistance etc. should be taken into consideration in order to achieve the successful operations of the electroforming process.

In the early approaches, for some applications where formulations based on waxes were used as mandrels, in order to make the surface of these materials conductive, the graphitization was frequently applied.

Table 9.1. Materials used as mandrels in the electroforming process

Permanent Mandrels	Aluminum, Bronze, Brass, Epoxy Resins, Glass, Invar, Kovar, Nickel, Stainless Steel
Disposable Mandrels	Waxes, Low Melting Point Alloys (Pb-Sn-Bi), Plastics (Polystyrene), Zinc and Zinc Alloys

When electroforming is performed on dielectric surfaces, they are usually coated with thin Ag film that has a limited adhesion. For this purpose , either vacuum or electroless deposition techniques can be used. Other metals such as Ni, Cu, Cr, etc., can be applied using available techniques (i.e. vacuum deposition or electroless plating) as long as they provide a limited adhesion and permit a relatively easy separation of the electroformed part from the mandrel. After a proper degreasing and cleaning, the stainless steel or copper mandrels are usually overcoated with a thin flash of chromium, so that the electroformed part is easily removed from them. Copper, nickel and iron are generally used for electroforming purposes. Hard chromium plating is applied on electroforms when wear resistant surfaces are required. In order to obtain a desirable quality of electroforms, electrolytes used in the process should be free of contaminants. Removal of contaminants is realized by continuous filtration through activated carbon.

The application of electroforming in the electronics industries is a very important approach leading in the production of various parts, such as thin film heads, thin film chip carriers, integrated magnetic minimotor, etc.. These parts would be difficult or impossible to make by other methods. In the electronic applications, the electroforming is known as plating through lithographic masks, pattern plating, additive plating etc. In contrast with the traditional electroforming, the metal plated through a resist mask does not represent a finished product by itself. This plated metal is rarely removed from the substrate. It remains on the substrate and is an integral part of the electronic or magnetic device. Electronic or magnetic devices often contain several electroformed layers produced via plating through a mask. These layers may have different patterns and they are usually separated by dielectric substrates. In this way, in contrast with the traditional electroforming metals plated through a mask must have a very good adhesion with the substrate.

Excellent adhesion is achieved by evaporation or sputtering of refractory metals such as Cr, Ti, Ta etc., directly onto dielectric substrate. This layer of refractory metal provides a bridge between the dielectric and layer then overcoated with Cu, Ag, Au or Ni in order to form conductive layer (cathode) for electroplating.

Plating through mask technology is schematically presented in Figure 9.8. The dielectric substrate is metallized with the refractory metal (Cr, Ti, W etc.) to provide an adhesion layer with a thickness of 5 to 50 nm.

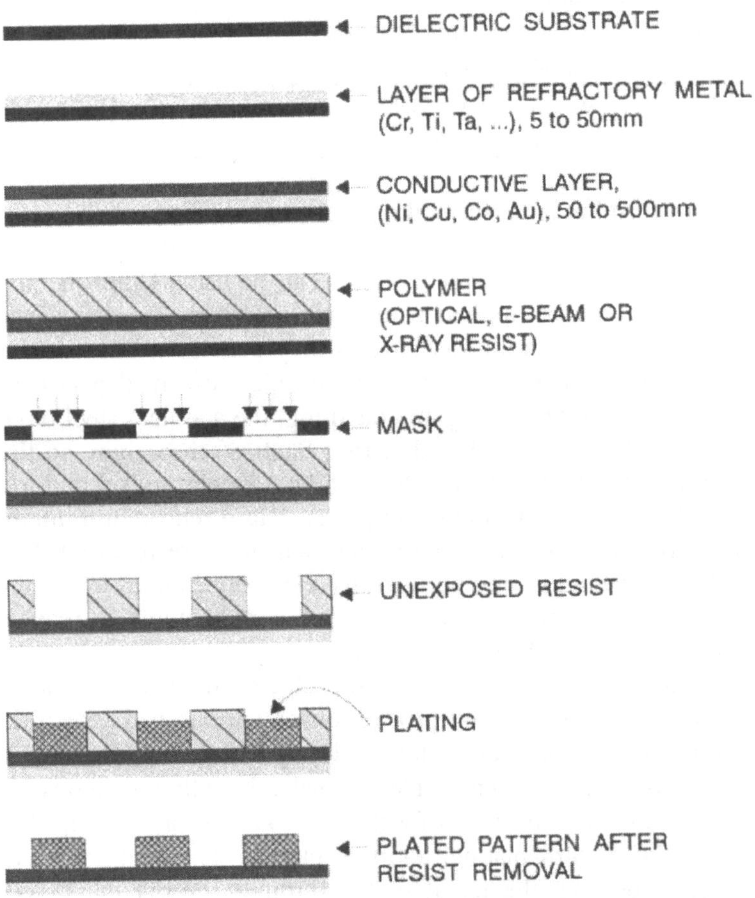

Figure 9.8. Schematic presentation of plating through mask technology

The adhesion layer is overcoated with a conductive metal such as Cu, Ni, Au etc., which usually depends on the applications. In the next step, the surface is coated with an organic polymer, and then through a mask exposed to light, x-ray or e-beam in order to form a pattern. Areas exposed to the light will be depolymerized. Consequently, these areas are dissolved in the developing solution, and after rinsing, drying and removal of residual organic impurities, parts are plated through holes to achieve the desired pattern. After the plating process is completed, the residual resist is removed by a second exposure to the light, x-ray or e-beam and dissolution in the developing solution. The adhesive and conductive layers are removed by chemical etching. By using the e-beam and x-ray lithography instead of optical lithography, pattern dimensions as small as 0.1 μm, have been produced.[36,37]

Although electroplating through mask technology has applications in the production of many devices for computers and other products, it still needs future development.

9.4 ELECTROPLATING FROM NON-AQUEOUS ELECTROLYTES

Electroplating of metals from non-aqueous systems is often referred to as plating from water-free inorganic and organic solutions and does not include deposition from molten salts. It is usually carried out at or below 100 °C. Electroplating from non-aqueous electrolytes is particularly important for metals, which cannot be deposited from aqueous solutions.[38]

If the reduction of metal ions takes place at potentials more negative than potential of discharge of water, the main cathodic process is the hydrogen evolution reaction. In this case, metals with sufficiently negative standard potentials cannot be deposited from aqueous solutions. Due to hydrogen evolution, the alkalinity near the cathode increases, leading in this way to the precipitation of metal hydroxides or the deposition of oxides at the electrode surface.

The analogous processes may occur in organic protic solvents, as a consequence of dissociation and formation of H^+ ions. Protic solvents contain hydrogen that is attached to oxygen or nitrogen and hence is appreciably acidic. In order to avoid hydrogen evolution reaction, aprotic solvents are recommended. These are polar solvents of moderately high dielectric constants, which do not contain acidic hydrogen. They dissolve both organic and inorganic reagents, but in dissolving ionic compounds solvate cations most strongly, and leave the anions relatively encumbered and highly reactive. Aprotic organic solvents have a relatively high electrochemical stability, since their reduction takes place above –3.0 V, and they can be anodically oxidized at 1.0 V to 1.5 V. The electrode material determines the region of the electrochemical stability of these solvents.

Metal ion sources for the electroplating from non-aqueous solutions are selected from suitable inorganic or organic compounds with a good solubility and a high conductivity. The nature of the dissolved salt and structure of cations and anions of non-aqueous electrolytes influences more significantly the electrocrystallization, than is the case for aqueous solutions. This effect is attributed to an increased complexation and specific action between dissolved compound and solvents. In this way the nature of the organic solvent and electrolyte determine the possibility of metal deposition.

Advantages of non-aqueous electrolytes for plating purposes include a larger voltage window of solvent stability, very low or no reactivity with substrates, formation of variety of complex ions in solutions and dissolved salts do not hydrolyze. [39] A larger window allows a greater flexibility in selecting cell-operating voltages. No reactivity of non-aqueous electrolytes with substrate makes possible to plate metals such as for example uranium with nickel or zinc[38], which react with aqueous types of electrolytes.

Disadvantages of non-aqueous electrolytes are associated with toxicity, flammability, explosion, low electrical conductivity, sensitivity to water and a relatively high cost. Electrodeposition of metals from organic solutions requires specially designed systems, which must be protected from oxygen, carbon dioxide and moisture.

In terms of solvents, non-aqueous electrolytes, used in electrodeposition of metals and alloys may be divided into two large groups: organic-solvent based and inorganic-solvent based. Examples of organic solvents are benzene, toluene, ethyl pyridinium bromide, diethyl ether, ethyl benzene, tetrahydrofuran, etc.. The number of inorganic solvents used for plating purposes is significantly smaller. The inorganic solvents include liquid ammonia, thyonil chloride and sulfuryl chloride.

Metals depositable from non-aqueous systems can be divided into two large groups. [38] In the first group are listed metals, which cannot be deposited from aqueous solutions, i.e. metals of the first group of the periodic table (Li, Na, K), metals of the second group (Be, Mg, Ca), metals of the third group (Al) and metals of the fourth group (Ge, Ti, Zr). To this group are also added metals of the fifth and the sixth groups of the periodic table (i.e. V, Nb, Mo and W). Note that metals such as Mo and W are listed into the first group, although they can be deposited from the aqueous solutions, but not in the pure state. Metals such as Mo and W are readily deposited from aqueous solutions, however only in the presence of iron group of metals (i.e. Ni, Co, etc., see the section related to the electrodeposition of alloys).

In the second group of metals, which can be plated from non-aqueous solutions are listed metals usually deposited from aqueous systems (i.e., Cu, Zn, Co, Sn, Ni etc.). Although these metals are commonly deposited from aqueous solutions, for some specific requirements they can also be deposited from non-aqueous solutions.

Very little or no success is achieved in deposition of metals of the first group in their pure state from non-aqueous solutions. This may be due to their limited industrial applications. Metals that are not deposited so far from non-aqueous electrolytes in their pure state include calcium, strontium, barium, the lanthanides and actinides, titanium, zirconium, hafnium, vanadium, niobium, molybdenum, tungsten and tantalum. However, literature shows that these metals were deposited from non-aqueous solutions as alloys, although at low current efficiency. [40]

The most important metal from the first group, in terms of platability from non-aqueous solutions, is aluminum. [41-43]. Deposition of aluminum from non-aqueous solutions has attracted researchers and industry for the two simple reasons. First, it cannot be deposited from aqueous solutions and second it has immense applications in various technical fields. Electroplating of aluminum is carried out industrially, although to a limited extent due to technical difficulties, and a relatively high cost of operation. [44] This bath consists of aluminum alkyl and sodium or potassium fluoride dissolved in toluene and a high purity aluminum used as the anode. The cell operates at 100 °C with 100 % of anodic and cathodic current efficiencies. The bath has an excellent throwing power.

Aluminum is electroplated in an enclosed plating cell to prevent reactions with oxygen, carbon dioxide or water from air, which would degrade the electrolyte and shorten its useful life. However, with the special process control and plating equipment, the electrolyte is stable, and is not consumed during the plating for one year.

Different types of electrolytes that may be used in the electroplating of aluminum from non-aqueous solutions are listed in Table 9.4.1. Practically, all these electrolytes with exception of aromatic solvents work under extremely dry conditions (no presence of water). All these solutions are unstable up to a certain degree, which is disadvantageous in their industrial applications. On the other hand specific precautions should be taken since some of these solutions are very flammable. While the aluminum is plated on the cathode, the main anodic process is dissolution. In this process aluminum anodes are used.

In the alkyl benzene electrolytes the cathodic current efficiency is estimated at about 50 to 80 %, while the anodic efficiency is close to a 100 %. Due to anodic dissolution of aluminum, an increase in the aluminum-ion concentration is observed. The excess of aluminum ions reacts with bromide ions, which are introduced in the solution with an addition of HBr.

Electrodeposition of aluminum in alkyl benzene electrolytes is described by the reaction:

$$AlBr_2 \cdot nC_6H_4(CH_3)_2^+ + 3e^- \rightarrow Al + nC_6H_4(CH_3)_2 + 2Br \quad (9.25)$$

A positive influence of water is attributed to a formation of dischargable hydroxo-complexes, according to the reaction:

$$2(Al_2Br_6 \cdot C_6H_5C_2H_5) + H_2O \rightarrow (Al_2Br_4OH \cdot C_6H_5C_2H_5)^+ + Br^- + Al_2Br_7^- + C_6H_5C_2H_5H^+ \quad (9.26)$$

Discharge of these hydroxo-complexes is then given by the equation:

$$C_6H_5C_2H_5H^+ + (Al_2Br_4OH \cdot C_6H_5C_2H_5)^+ + 6e^- \rightarrow 2Al + 4Br^- + H_2O + 2C_6H_5C_2H_5 \quad (9.27)$$

In the real systems, however, the discharge mechanism is more complicated and involves side reactions such as evolution of H_2, HBr formation and a possible polymeraization of the solvent and impurities. Electrodeposited aluminum is of a higher purity (99.5 to 99.999 %) than the anode, since the impurities and alloying elements are insoluble in the electrolyte. Impurities are continuously eliminated by filtration of the electrolyte. The pure aluminum layer becomes pore-free at a thickness greater than 8 μm. The thickness of electrodeposited aluminum can reach 500 μm, if some special applications are required. The aluminum coating has the very good ductility, corrosion resistance, and when electropolished it is convenient for the production of mirrors of high optical quality. [45]

As mentioned above, the second group includes metals, which can be electrodeposited from aqueous solutions (i.e., Cu, Zn, Ni, Co, Ag, Au etc..). The electroplating of these metals from non-aqueous solutions at the present does not have significant industrial applications, and therefore is mostly of academic value. The research shows that electrodeposition of this group of metals takes place in polar types of solvents (solvents that have active functional groups such as OH^-, CO, NH_2^- and $CONH_2$) in which, the same type of salts used in the aqueous solutions are dissolved (providing they have sufficient solubility). As examples of studied systems, electrodeposition of various metals, such as Zn, Pb, In and Sn, from formamide, acetamide, glycerol, ethanol, acetone etc. should be mentioned. [38] The results do not show any advantage over electroplating of these metals from aqueous solutions. Usually, deposition of these metals from non aqueous solutions produces deposits of poor quality and maintaining of solution chemistry encounters experimental difficulties.

Electrodeposition of metals from liquid ammonia has also been studied (Ag, Cu, Pb, Hg etc.). However, no results of a practical value were obtained. Any metals from this group can readily be deposited from aqueous solutions.

Table 9.2. Electroplating of Aluminum from Organic Electrolytes

Electrolyte	J [A/m^2]	CE [%]	Remark
$AlBr_3$, Ethyl Bromide, NaBr, KBr	200	60 – 70	Thin Al films with a good adhesion are deposited
$AlBr_3$, Dimethyl Aniline		~95	Smooth, bright and ductile coatings
$AlCl_3$, Ethyl-Pyridine-Bromide, Toluene	120 – 220		Stirring of the electrolyte is carried out in the nitrogen atmosphere. The thickness of plated aluminum can approach up to 1 mm
NaF, $Al(C_2H_5)_3$, Toluene	60 – 80		Aluminum coatings with a high corrosion resistance
LiH or $LiAlH_4$, $AlCl_3$ in Diethyl Ether	500	~100	The thickness of aluminum can approach 2 mm with an addition of methyl borate. The coatings can be used for the corrosion protection of uranium in atomic reactors
$AlBr_3$, Xylene or Other Aromatic Solvents (Alkyl Benzene, Toluene)		50 – 80	The aluminum can be plated in the presence of small amounts of water. Electroplating is successful even when the electrolyte is in equilibrium with atmospheric humidity

Electrodeposition from non-aqueous electrolytes is attractive for several metals and alloys that cannot be deposited from aqueous solutions. Significant results have been obtained with aluminum and its alloys. Improvements in electrodeposition from non-aqueous electrolytes will continue to grow due to desirable physico-chemical characteristics of coatings (e.g., aluminum or its alloys) or reactivity of substrate with aqueous solutions (e.g., uranium). Developments in other areas of technology (energy conversion, advanced batteries, electronics etc..) may lead to requirements for materials with specific properties and to advancement in the field of electrodeposition of metals and alloys from non-aqueous electrolytes.

9.5 ELECTROPLATING FROM ROOM TEMPERATURE MOLTEN SALTS

As with plating from non-aqueous electrolytes, electrodeposition of metals from molten salts has attracted the attention of researchers from two different reasons. The first is related to the search of platability of metals from molten salts, especially those that cannot be plated from aqueous solutions or non-aqueous types of solutions. For those metals which are platable from aqueous solutions *e.g.* zinc, the search for a convenient molten salts electroplating is often related to a target to avoid the simultaneous hydrogen evolution reaction and to obtain coatings of more desirable properties. In the electrodeposition of zinc from aqueous solutions, the simultaneous hydrogen evolution reaction significantly affects the embrittlement and sometimes reduces current efficiency of the process. As a result an aprotic-plating bath is required.

Although there are well-established molten salts electrolyses, particularly those related to the electrowinning of metals such as aluminum, magnesium, sodium etc., attention in this section is directed towards the room temperature molten salts electrodeposition. In this case, the hydrogen reaction can be avoided, or deposition of metals that are not platable from aqueous solutions may occur. The room temperature molten salts are based on anhydrous $AlCl_3$. They are analogous to the high temperature melts ($AlCl_3$ – NaCl), with a difference that the NaCl is replaced with an aromatic organic chloride. This replacement of NaCl results in lowering the melting point well below room temperature, sometimes as low as – 50 °C. Several types of organic aromatic chlorides have been investigated. This includes 1-methyl-3-ethyl-imidazolium chloride (MEIC), 1-(1-butyl) pyridinium ammonium chloride (BPAC), 1, 2-dimethyl-3-propyl-imidazolium chloride (DMPIC) etc. The most studied aromatic organic chloride for the room temperature molten salts has been the MEIC. Room temperature molten salts can be obtained from the combination of anhydrous $AlCl_3$ and MEIC. [46] These chloroaluminate salts have a high ionic conductivity, good thermal stability, a wide electrochemical window and adjustable Lewis acidity. The MEIC/$AlCl_3$ mixtures are liquid at room temperature (about 25 °C) over the range of composition from 40 to 67 mol. % $AlCl_3$.

In the acidic melts, metal ions are believed to be only weekly complexed or solvated by $AlCl_4^-$ ions and thus can be reduced at more positive potentials. These melts contain a molar excess of $AlCl_3$. In the alkaline melts, the metal ions are strongly complexed by chloride ions and exist as discrete anionic chloride complexes, e.g. $[MCl_p]^{(p-z)-}$, where z is the valence of the metal ion. The alkaline melts contain a molar excess of MEIC. The formation of strong anionic chloride complexes in alkaline region makes the metal ions more difficult, or in some cases impossible to reduce within the electrochemical window of the melt. Consequently, most of the studies on the electrodeposition of metals from room temperature chloroaluminate melts are carried out in the Lewis acidic composition region of the melt, which contains a molar excess of $AlCl_3$.

The Lewis acidity of $AlCl_3$ and organic chloride donor (RCl) mixtures is a function of the $AlCl_3$/RCl molar ratio, N, according to the equation:

$$2AlCl_4^- \rightarrow Al_2Cl_7^- + Cl^- \quad (9.28)$$

Melts with the ratio N>1 (molar excess of $AlCl_4^-$) are Lewis acidic, while those with N<1 (molar excess of RCl) are Lewis basic. The equilibrium constant for the above equation is estimated from the potentiometric titration[46] at 2.0 x 10^{-19}. In acidic melts, the dominant anionic species are $AlCl_4^-$ and $Al_2Cl_7^-$. The electrochemical potentials are determined by $AlCl_4^-$ oxidation and $Al_2Cl_7^-$ reduction. Deposition of aluminum proceeds according to the reaction:

$$4Al_2Cl_7^- + 3e^- \rightarrow Al + 7AlCl_4^- \quad (9.29)$$

Pure metals such as palladium, gold, tin, mercury, lead and zinc, have been electrodeposited from the acidic chloroaluminate melts.[47] The electrodeposition of some transition metals is complicated by the codeposition of aluminum. The formation of alloys such as Co-Al, Cu-Al, Ni-Al, Cr-Al, Fe-Al is observed at potentials several hundreds milivolts more positive than the potential at which the bulk deposition of aluminum occurs.[48-52]

Experiments are performed in a glove box with an inert atmosphere (usually nitrogen). To remove the protonic or other impurities, the $AlCl_3$-MEIC melt is purified by pre-electrolysis of the melt at constant current density for several days, while the electrolyte is stirred. The electrolyte is then filtered to remove aluminum particles, which are produced during the pre-electrolysis step.

Metal-ions (i.e. Zn, Co, Cu etc.) are introduced in the melt by an anodic dissolution of corresponding metals. For the electrodeposition process materials such as glassy carbon, platinum, tungsten etc. are used as working electrodes (cathodes). To maintain the concentration of the metal ion in the melt constant, the counter electrodes (anodes) are usually prepared from the same metal. For the electrochemical measurements, the aluminum reference electrodes are commonly used.

The current efficiency of the plating process frequently achieves 100 %. The surface morphology of deposited films can vary from smooth and dense to nodular, porous and dendritic deposits, depending on the experimental conditions.

Due to experimental difficulties there are not, at the present, commercial applications of the room temperature molten salts for the electroplating purposes. Although very promising results are obtained by far, significant amount of research and engineering should be carried out in the future, in order to apply these processes on industrial scale.

9.6 ELECTROPOLISHING

Electropolishing is defined as a process of anodic dissolution of metals or alloys in an appropriate solution resulting in production of improved morphology and geometry of the surface and a shiny, bright and smooth appearance. Technical advantages of the electropolishing include a reduction in coefficient of friction, an increase in the magnetic susceptibility of some magnetic materials, an increase in corrosion resistance and excellent reflectance. In addition, electropolishing is widely used in the metallography for the microscopic investigation of crystallographic structure of metals and alloys. Some theoretical aspects of electropolishing are discussed in the section 3.2.3.

Electropolishing as an anode process, is similar to electromachining, however there are significant differences between them. For instance, electropolishing is usually carried out from unstirred, concentrated acidic solutions as electrolytes, at lower current densities, and with the electrode separations of at least 1 cm. The quality of electropolished surfaces depends on electrochemical conditions including anodic polarization, electrolyte composition and microgeometry. It is determined by the appearance, measurements of profiles with optical profilometers, and also using microscopic techniques. In terms of the surface roughness, the two types of electropolishing are distinguished. The first, commonly called anodic levelling or smoothing, refers to the elimination of the surface roughness with a height of more than 1 μm. The second type is called anodic brightening and is referred to the elimination of surface roughness less than 1 μm. However, this distinguishing between the smoothing and brightening is a very approximate simplification, since there is no simple relationship between measurements of profile and brightness.

A schematic presentation of the anodic current density – potential relationship, during the electropolishing process is given in Figure 9.9. Four distinguishable regions on this curve can be seen: AB, BC, CD and DE. In the region AB, the anode dissolves. Under these conditions the surface microroughness does not disappear. The electropolishing takes place under mass transfer conditions, at limiting current density and in the potential range between E_2 and E_3. The surface of the electrode becomes smooth and microroughness decreases. When the potential approaches the value of E_3 the surface of the metal becomes bright. An increase in the potential above E_3 leads to a rapid increase in the current density, causing in this way an increase in the roughness of the electrode surface, due to metal dissolution and simultaneous oxygen evolution reaction.

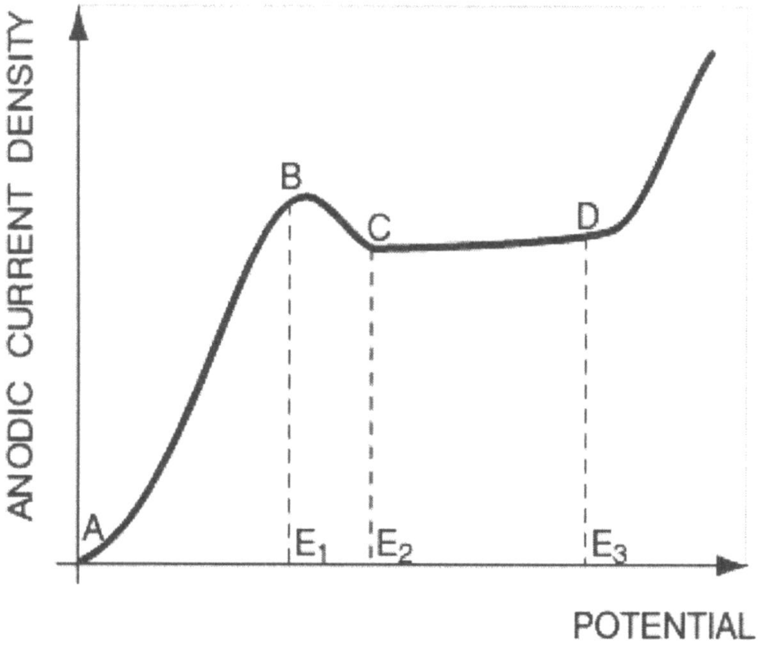

Figure 9.9. Schematic presentation of anodic current density – potential relationship during electropolishing

A decrease in the microroughness during electropolishing is a consequence of current distribution at the electrode surface. The surface of an electrode profile, with distinguished protrusions and depressions is schematically presented in Figure 9.10. It seems that the protrusions are more accessible to the current flow than depressions. Therefore, under the conditions of electropolishing, protrusions will dissolve faster, thus leading in this way to a decrease in the microroughness. Thermodynamically, it is more probable that the preferential transformation of solid phase (i.e., metal crystals) into the ionic (solvated) phase would take place at protrusions than in depressions. This is due to smaller energy of transformation of ions from solid to solvated phase at protrusions than in the depressions. In practice, the anodic dissolution frequently deteriorates microtopography due to unequal localized etching.

The formation of a passive oxide layer at the anode surface plays crucial role during the electropolishing. [53-55] Electrolytes used for electropolishing contain substances (for example CrO_3), which form an oxide passive film, and acids, i.e. H_3PO_4 or H_2SO_4, which dissolve this passive film. Typical formulations used in the electropolishing of aluminum, copper and stainless steel are given in table 9.3.

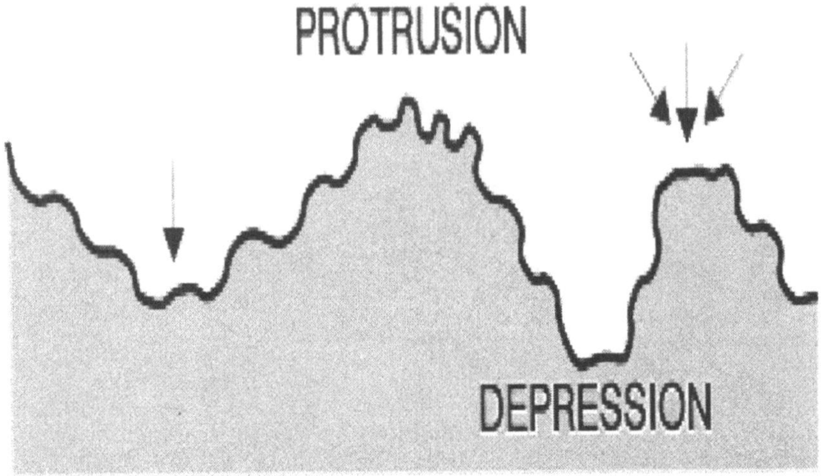

Figure 9.10. Electrode profile with distinguished protrusions and depressions

Table 9.3. Composition of electrolytes and conditions in electropolishing of aluminum, copper and stainless steal

Metal	Electrolyte	T [°C]	j[A/cm²]	Time [min]
Al and its alloys	H_2SO_4 60 mL HF 10 mL H_2O 690 mL	60	0.1	5 – 10
	H_2SO_4 45 mL H_3PO_4 750 mL CrO_3 65 g H_2O 200 mL	80	0.15	5 – 10
	H_3PO_4 200 mL H_2SO_4 200 mL HNO_3 200 mL	Room	0.15	5 – 10
Cu and its alloys	H_3PO_4 200 mL n-Butyl alcohol 90 mL	Room	0.015 – 0.03	5 – 10
Stainless Steal	H_3PO_4 520 mL H_2SO_4 420 mL H_2O 300 mL	70	0.5 – 0.6	10 – 15

Electropolishing is carried out in concentrated aqueous solutions of phosphoric acid, sulfuric acid or their mixtures, and sometimes in combinations of perchloric and acetic acids. There are also formulations in which, methanol is used instead of water. [56]

If the rate of passive film formation is less then the rate of dissolution, the anode surface is rather etched, leading to an increase in the microroughness, due to non-uniform coverage with oxide passive film. On the other hand, when the rate of passive oxide film formation is more than the rate of dissolution, the film thickness increases covering the anode surface. In this way the electropolishing is not achieved. It is obvious that the electropolishing occurs when the rates of the passive film formation and dissolution are comparable.

The levelling of the electrode surface during electropolishing is a consequence of a non-uniform passivation of protrusions and depressions. It seems that the depressions are better covered with passive films than the protrusions, leading in this way to faster dissolution of protrusions than depressions. Passivation of protrusions to a lesser degree is explained by an increased chemical activity, due to a faster rate of diffusion of metal ions on protrusions than on depressions.

The rate of anodic levelling is equal to the difference in dissolution rate between protrusions and depressions on a rough surface. It is determined by the predominant current distribution on the surface profile. Consequently, the rate of anodic levelling is dependent on geometrical, electrochemical and hydrodynamic parameters.[53,57]

While the influence of the geometry of the surface remains qualitatively the same, the charge transfer overvoltage (secondary current distribution) tends to reduce the rate of anodic levelling. [57] When the concentration overvolatge is present (tertiary current distribution), two cases are distinguished. Below the limiting current, potential and mass transport influence the current distribution, and, therefore, the rate of levelling. This situation is of a little interest for the practical applications. At the limiting current density, the current distribution depends only on mass transport.

For the anodic levelling and anodic brightening terms macrosmoothing and microsmoothing were introduced by Edwards. [53] The macrosmoothing results from local differences in the current distribution on the surface profile or the concentration of the transport limiting species, and is preceded by microsmoothing. Macrosmoothing is a consequence of a higher current distribution on protrusions, which causes higher distribution rates.

Microsmoothing results from the surface kinetics and passivation behavior, due to suppression of the influence of surface defects and crystallographic orientations on the dissolution process. The microsmoothing occurs when dissolution of metal is mass transport controlled, which corresponds to the limiting current plateau. In most electropolishing systems, the rate of transport of dissolution products from the anode into the bulk solution determines the limiting current.

The rate of levelling at the limiting current density may theoretically be predicted on the basis of Nernst diffusion layer model. The Nernst diffusion layer, δ, for a triangular profile with the height, ε_o, is presented in Figure

9.11. The broken line represents the outer limit of the Nernst diffusion layer for an ideal microprofile. In an ideal case, when $\delta \ll \varepsilon_o$, the diffusion layer should follow the surface profile as indicated by the broken line. Under these conditions, the current density is uniform and only the geometrical levelling should occur. However, due to local hydrodynamics, the situation is more complicated, since the ratio between δ and ε_o can change significantly. For this case a detailed modelling of the hydrodynamic perturbation, caused by the surface microprofile is required for a quantitative prediction of the rate of levelling.

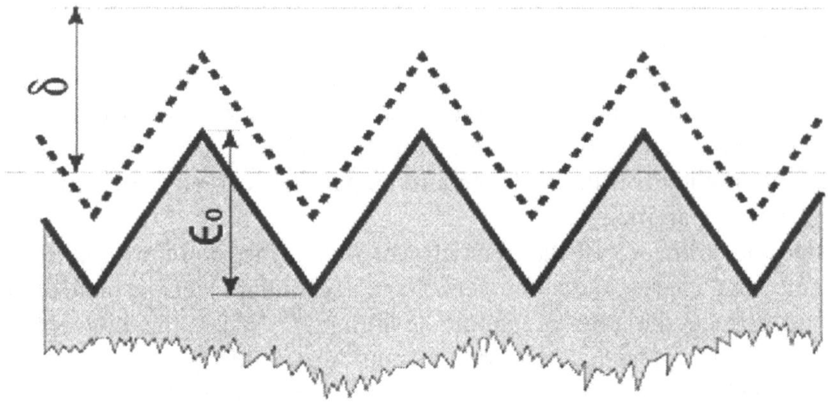

Figure 9.11. Triangular surface profile with a height ε_o and the Nernst diffusion layer δ. The broken layer represents the outer limit of Nernst diffusion layer for an ideal microprofile

In the aqueous solutions, during the high dissolution rate, the surface concentration of dissolution products is in a reasonable agreement with the saturation concentration of the corresponding metal salt. In the concentrated acid type solutions (H_2SO_4, H_3PO_4) the surface concentration of products of dissolution significantly exceeds the saturation concentration, which probably causes formation of metastable species at the anode surface. During the electropolishing, a low water concentration may reduce the limiting current by decreasing the saturation of metal ions produced due to dissolution.

There is general agreement among researchers that electropolishing occurs when the reaction rate is controlled by mass transfer. In an attempt to explain the electropolishing process the following mechanisms appeared in the literature:

(i) salt – film mechanism, and
(ii) acceptor mechanism.

The salt – film mechanism is based on an assumption that the surface concentration of the metal ions due to dissolution is very high that exceeds solubility and causes the precipitation of a salt film. The rate of reaction is then determined by the rate of diffusion of metal ions away from the electrode surface.

In the acceptor mechanism, the metal ions produced from the dissolution remain on the electrode surface until they are complexed by an "acceptor" species, which include either an anion or water. The rate of reaction, for this case, is determined by mass-transfer of the acceptor to the electrode surface. Consistent with the acceptor mechanism, the reactions describing dissolution of copper may be presented as follows:

$$\mathrm{Cu} \rightarrow \mathrm{Cu}^{2+}_{(\mathrm{ads.})} + 2\mathrm{e}^- \tag{9.30}$$

and

$$\mathrm{Cu}^{2+}_{(\mathrm{ads.})} + \mathrm{nA}^{\mathrm{m}-}_{(\mathrm{ads.})} \rightarrow (\mathrm{CuA_n})^{2-\mathrm{nm}}_{(\mathrm{aq.})} \tag{9.31}$$

The experimentally obtained limiting current plateaus for electropolishing of copper in concentrated phosphoric acid (85 % solution) at different rotation speed are presented in Figure 9.12. [58] As this figure shows, the limiting current plateaus extend over a potential range of 1 V. Above the current plateau, the oxygen evolution reaction most probably takes place. According to the results presented in Figure 9.12, the value of the limiting current density depends on the rotation speed. An increase in the rotation speed leads to an increase in the limiting current density.

In Figure 9.13 are presented dependencies of limiting current density on the square root of rotation speed for temperatures 30 °C, 40 °C and 90 °C.

Based on their experimental results , Vidal and West supported the acceptor mechanism theory, in which the limiting current density is given by the Levich equation:

$$j_L = \frac{2}{n} F D_A^{2/3} \nu^{-1/6} \omega^{1/2} C_A \tag{9.32}$$

where j_L is the limiting current density, F is the Faraday constant, D_A is the diffusion coefficient of the acceptor species (probably water), ν is the kinematic viscosity, ω is the rotation speed and C_A is the concentration of the acceptor species. They assumed that the concentration of the acceptor is constant over a wide range of temperatures. In this way, the validity of above equation is experimentally confirmed by the results presented in Figure 9.12. These authors rejected the salt-film mechanism, since it is unlikely that the saturation concentration of salt is independent on temperature.

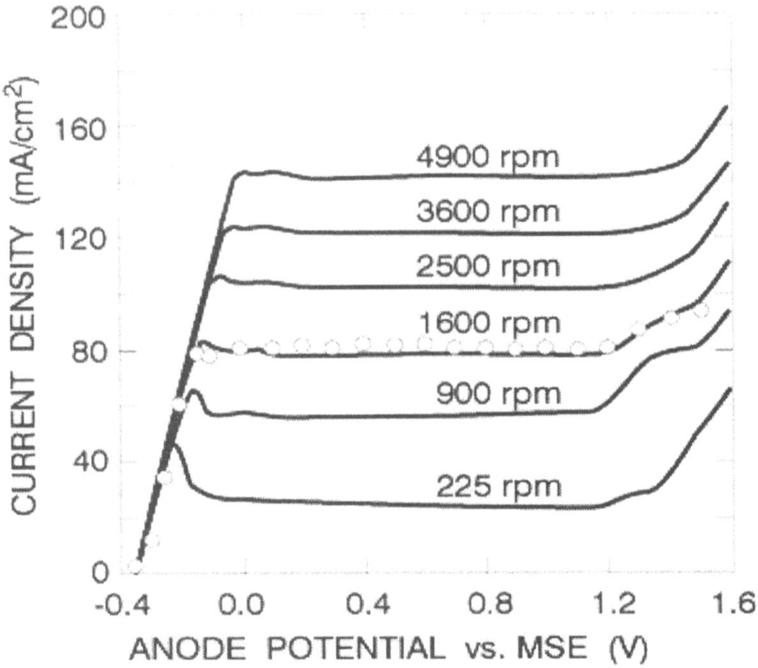

Figure 9.12. Anodic polarization curves for six rotation speeds. The hollow circles correspond to potentiostatic measurements. All sweeps were done with a scan rate 5 mV/s at 30 °C [58] (Reproduced with the permission from The Electrochemical Society)

If the salt film– mechanism were true the physical properties of this film and its thickness are not well known. It is not clear if this film is a solid oxide type film[59] or an anhydrous film[60]. Difficulties in determining the nature of this film arise due to its disappearance after the current is switched off. The role of this salt film in microsmoothing is not clear yet.

The presence of a current plateau, associated with an anodic film on the surface, is a very important condition for the microsmoothing. However, the necessary condition for the microsmoothing is that the reaction rate is mass transport controlled, which is achieved only in a specific concentration and temperature range, as shown with well-defined current plateaus for transpassive dissolution of nickel in sulfuric acid. [53]

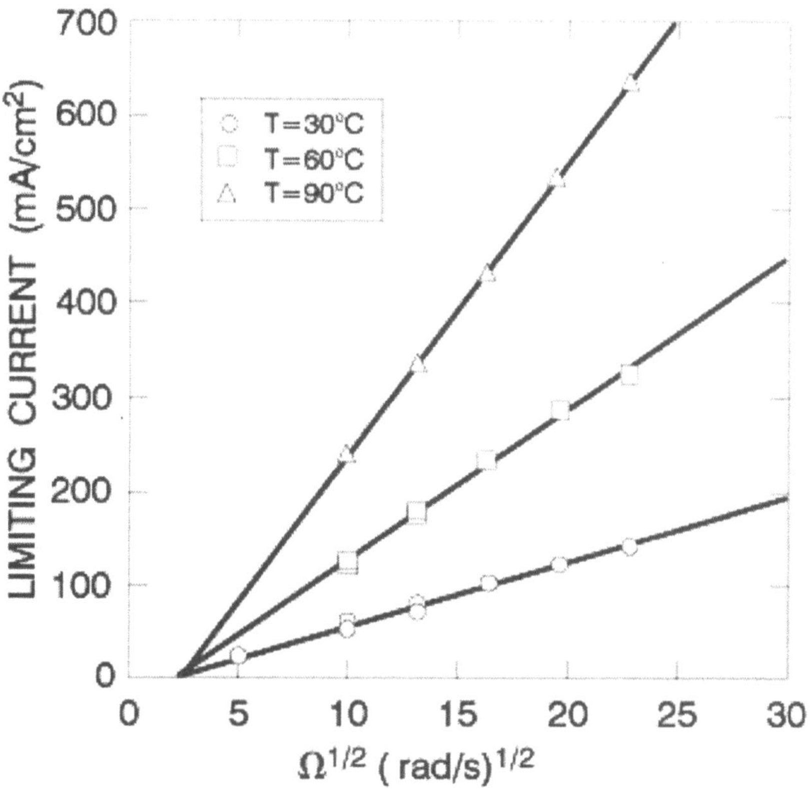

Figure 9.13. Levich plot (limiting current *vs.* square root of rotation speed)at three temperatures. For all cases, the limiting current densities were measured at an applied potential of 0.6 V vs. reference (Hg/mercurous sulfate electrode) [58] (Reproduced with the permission from The Electrochemical Society)

Anodic dissolution in the transpassive potential region below the limiting current leads to crystallographic etching or pitting. The pitting is a local attack of a passive metal, which is induced by certain ions under the effect of high anodic potential. In order to achieve the uniform electropolishing the pitting must be avoided. In systems where pitting may occur electropolishing can be established with a sufficient increase in the anodic potential in order to break down the passive film. Many electropolishing electrolytes are based on a limited amount of water, which renders the formation of passive oxide films more difficult.

9.7 ELECTROMACHINING

Electromachining is an electrochemical process in which metal removal is achieved by the anodic dissolution. This process, frequently called electrochemical machining (ECM) is investigated as a method for shaping high strength, heat-resistant metals and alloys, which are difficult to cut by other established techniques. At the end of the last century, the electromachining became a method employed in different industries including automotive, offshore petroleum and medical engineering. Electrochemically speaking, electromachining is a process very similar to the electropolishing, since both processes are based on the anodic dissolution reactions. However, the rate of metal removal for an electromachining process should be considerably higher than that in the electropolishing. Therefore in the electromachining, higher current densities are required, the electrode separation is less than 1 cm and the process is carried out in diluted electrolytes with stirring.

An anodic dissolution reaction is usually represented by:

$$M \rightarrow M^{z+}_{(aq)} + ze^- \tag{9.33}$$

The electrolyte and the material of the cathode are chosen in a way that the cathodic process is usually hydrogen evolution reaction. Consequently, the shape of this electrode does not change during the electromachining process. Depending on the metal or alloy being electromachined, hydrolysis reaction may occur due to dissolution:

$$M^{z+}_{(aq)} + H_2O \rightarrow Me(OH)^{+}_{z,(aq)} + H_2 \tag{9.34}$$

To achieve reasonable results the precipitated hydroxides must be removed from the electrolyte, and this is usually carried out by filtration.

A schematic presentation of electrochemical machining is given in Figure 9.14. During the electrolysis the cathode is moved simultaneously towards the anode and a shape complementary to that of the cathode is produced on the anode. The rate of the anode dissolution (metal removal) is in inverse proportion to the distance between the cathode and the anode. Typical rates of movement of the cathode towards the anode are about 0.02 mm/s.

A choice of the electrolyte in the electromachining process is crucial in order to keep the shape of the cathode unchanged and to achieve desirable current efficiencies of the process. This depends on the metal or alloy used in the processing, and also on the nature of the cathode. The electrolyte is usually pumped through the gap between the electrodes in order to remove the products of machining (i.e., hydroxides and other solid particles, hydrogen-gas bubbles accumulated at the cathode etc..), and also to reduce the heating due to current flow.

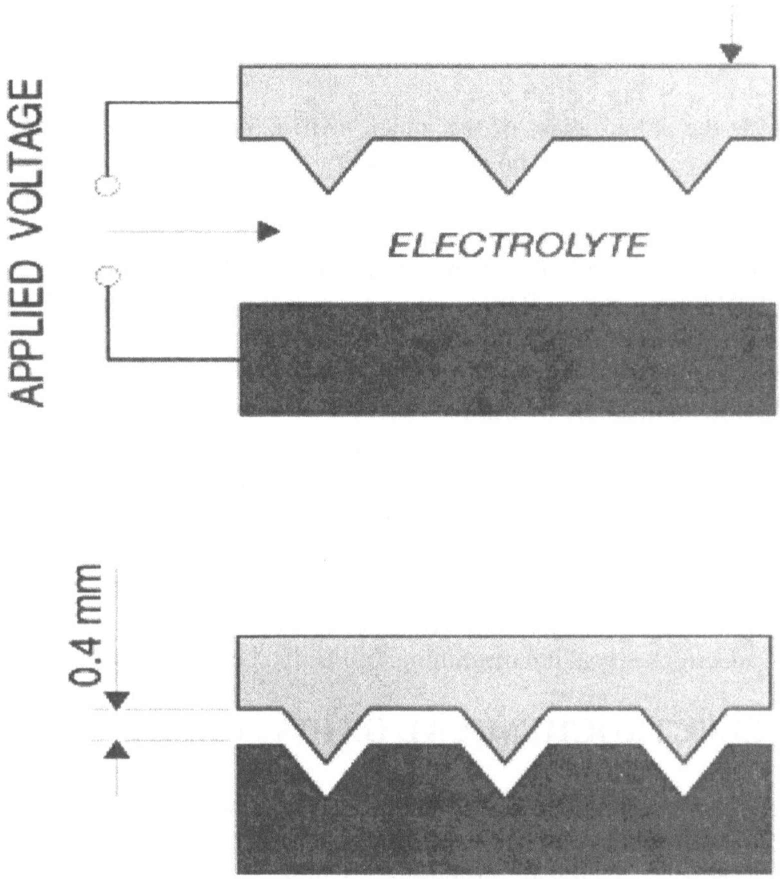

Figure 9.14. Schematic presentation of electrochemical machining

The cathode, usually produced from a metal softer than the anode, with a designed complementary shape is used as a tool. A workpiece is the anode. Electrolytes based on NaCl, $NaNO_3$, $NaClO_3$ etc. are used for the electromachining process. [61,62] The results show that the surface brightening depends on the concentration and the temperature of the electrolyte. After passing through the gap between the electrodes the electrolyte is carefully filtered to remove the products of electrolysis, and then heated in a reservoir to the working (electromachining) temperature. The gap between the electrodes is estimated at about 0.8 to 0.4 mm. The rate of metal machining does not depend on the hardness of the material (i.e. anode), and any types of profiles can be reproduced on hard metals, without the wearing the tool (cathode).

The rate of electrochemical machining can be determined on the basis of Faraday's laws. The mass of metal dissolved (removed), m, is determined according to the equation:

$$m = \frac{M}{nF} It \tag{2.16}$$

where M is the atomic mass of the metal, n is the number of electrons, F is the Faraday's constant, I is the current and t is the time. The average current densities depend on metal, and their values are usually between 50 and 150 A/cm^2, while the voltage is about 10 to 20 V. Typical tolerances of about 0.127 mm are reported, however, under special circumstances they can achieve 0.013 mm, or even 0.002 mm under pulsating current conditions. [63]

During the electromachining the formation of surface oxide films frequently occurs. To break the oxide film, higher voltages should be applied. Due to oxygen evolution reaction at the anode the gas bubbles rupture the oxide film causing localized pitting. The process variables can significantly influence the surface finish. The smoother surface finish is generally observed with higher current densities or with higher velocities of the electrolyte.

Electrochemical machining processes have various applications such as smoothing of rough surfaces, hole drilling, full form shaping, electrochemical grinding, electrochemical arc machining, biomedical engineering etc. [64-67]

9.8 ELECTROCHEMICAL OXIDATION OF METALS

Electrochemical oxidation of metals is an anodic process in which thin oxide films are produced. These oxide films may have different color and attractive physico-chemical properties. Color and other properties of anodically produced oxide films are determined by the conditions of electrochemical oxidation, which include composition of the electrolyte, temperature, current density, voltage and duration of the process.

Thin oxide films can be produced anodically on many metals. Metals of interest include so-called "valve" metals (i.e., metals such as aluminum, tantalum, niobium, titanium, zirconium etc., which form adherent electrically insulating anodic films,). At the present, this process is commercially applied only to aluminum. Anodic oxidation of aluminum is often called "anodizing". Oxide films produced on aluminum surfaces as a consequence of anodizing, have a very good hardness, abrasion- and corrosion-resistances and unique columnar and porous structure.

Applications of anodized aluminum include protection against corrosion and abrasion, decorative surfaces which provide color and base for paints etc. These anodized surfaces are used in aggressive environments, permanent external and architectural constructions, automotive, aircraft and electronics industries.

The anodic oxidation of aluminum is carried out under both periodically changing and direct current conditions. Typical electrolytes used for the electrochemical oxidation of aluminum are given in Table 9.4.

The most widely used electrolytes for anodizing of aluminum are based on sulfuric acid. In this case tanks are lead-lined with lead acting as the cathode. So called "hard anodizing" (where extremely hard and abrasion-resistant coatings are required) is carried out using sulfuric acid solutions at higher voltages (above 25 V) and lower temperatures (about – 5 to + 5 °C)

Oxalic acid anodizing is mainly used for the wear resistance applications, while anodic films produced by the anodic oxidation in the chromic acid solutions have excellent corrosion properties.

General reaction for the anodic oxidation of aluminum is presented as:

$$2Al + 3H_2O \rightarrow Al_2O_3 + 6H^+ + 6e^- \qquad (9.35)$$

Table 9.4. Typical Electrolytes and Conditions Used in the Anodic Oxidation of Aluminum

Electrolyte	T [°C]	j [A/dm²]	U [V]	Time [min]	Thick. [μm]
H_2SO_4 10 – 20 %	20 – 30	1 – 2	10 – 20	10 – 20	3 – 35
$(COOH)_2$ 3 – 5 %	30 – 60	2 – 3	20 – 120	10 – 240	10 – 20
H_3PO_4 35 – 65 %	20 – 50	1 – 3	10 – 15	3 – 5	3 – 5
CrO_3 2 – 10 %	35 – 40	0.1 – 0.5	40 – 50	50 – 60	2 – 15

Depending on the electrolyte and conditions of electrochemical oxidation, the reaction products may be:

i) soluble in the electrolyte,
ii) almost insoluble in electrolyte,
iii) sparingly soluble in the electrolyte, and
iv) moderately soluble in the electrolyte.

When the reaction products are soluble in the electrolyte, the metal is dissolved until the solution is saturated with its ions. This type of reaction occurs in strong inorganic acids and bases. When the reaction products are almost insoluble (electrolytes based on borates or tartrates) strongly adherent and practically non-conductive very thin oxide films are formed. Sparingly soluble oxide films are usually produced in the electrolytes based on sulfuric, chromic or oxalic acid. In this case, the film growth is accompanied by its dissolution at the surface. The rate of film growth is obviously higher than the rate of dissolution. Pores formed in the film are wide enough to permit continuous access of the current to the metal, which leads to its further oxidation. Finally, when the reaction products are moderately soluble, the electropolishing is also possible if a proper electrolyte is used (e.g. addition of sodium hydroxide to a sodium citrate bath.[68]

The mechanism of anodic oxidation of aluminum is very complex by its nature and still not well understood. Formation of thin oxide films and their composition depend on the electrolytes and conditions of electrochemical oxidation.

There is a general agreement among researchers that the anodic oxide film mainly consists of anhydrous aluminum oxide, which is either amorphous or in the γ - state. [45] In the amorphous anodic alumina material, Al^{3+} cations are both octahedrally and tetrahedrally coordinated to O^{2-} ions.

Fresh films formed at in H_2SO_4 solutions are composed of amorphous Al_2O_3 with a small amount of water (about 1 %) and 2 – 16 % sulfate. [69] Coatings produced in oxalic or chromic acid solutions contain about 3 % oxalate or up to 0.7 % chromate. When freshly formed porous type anodic coatings are boiled (a process known as "sealing"), the alumina is converted to a crystalline monohydrate, by take-up of about 5 to 6 % water.

Formation of thin oxide film during anodization of aluminum is schematically presented in Figure 9.15.

As shown in Figure 9.15., the anodic films formed on aluminum during the anodic oxidation consist of two layers, an inner, thin, dense, dielectrically compact and the outer thick, porous layer. The inner layer is called the active, barrier or dielectric layer. The thickness of this film may vary from approximately 0.01 μm to 0.1 μm and represents only 0.1 to 2 % of the total film. The barrier layer formed in the anodic oxidation of aluminum is similar to the natural oxide layer formed at the surface in the air atmosphere. It is non-porous, and conducts current only because of faults in the skeleton and the fact that is very thin. The thickness of the barrier coating is proportional to the voltage of the cell and is given with the following empirical formula:

$$d = k \times U \tag{9.36}$$

where d is the thickness, U is the voltage and k is a constant, with an approximate value of 1.4×10^{-7} cm/V.

It is assumed that the barrier type of coating is produced as a result of migration of mobile species across the pre-existing air-formed film. The precise nature of mobile species and their mechanism of transport is not clear yet, however the results indicate that Al^{3+} ion engress and O^{2-}/OH^{-} ingress proceed through the air-formed film. [70]

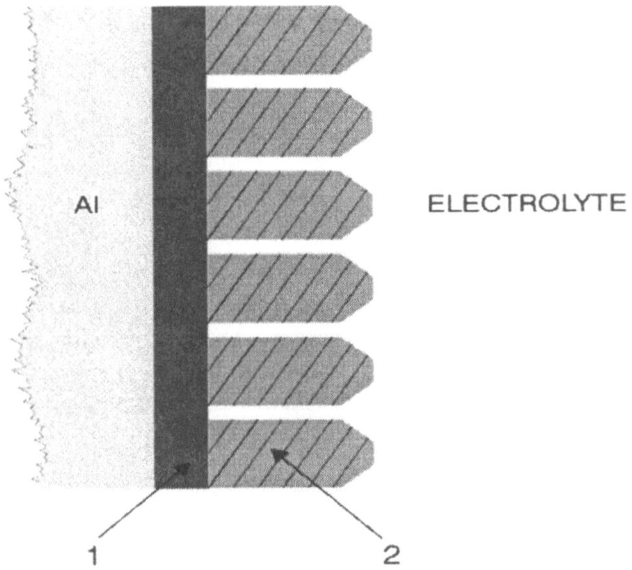

Figure 9.15. Formation of oxide film on aluminum during anodic oxidation: 1- barrier coating; 2 – porous film

The thickness of the barrier layer is mostly influenced by the type of electrolyte (H_2SO_4, $(COOH)_2$, CrO_3, H_3PO_4 etc.) and its concentration. A decrease in the barrier film thickness is observed with an increase in the concentration of the electrolyte, however the reasons for this are not clear yet.

During electrolysis, the outer film dissolves according to the reaction:

$$Al_2O_3 + 6H^+ \rightarrow 2Al^{3+} + 3H_2O \quad (9.37)$$

As a consequence a porous film is formed (Figure 9.15). In this way, the two parallel reactions at the anode during electrochemical oxidation of aluminum are described by the equations (9.35) and (9.37). The main cathodic reaction during anodic oxidation of aluminum is attributed to the hydrogen evolution.

Porous anodic films, with their relatively regular morphology characteristics have received significant attention from researchers. A change in the thickness of the porous layer with time is presented schematically in Figure 9.16. This, typically obtained thickness vs. time curve shows that after the initial increase, the film thickness rapidly decreases. This decrease in change of thickness is a consequence of an increase in film dissolution and oxygen evolution reaction.

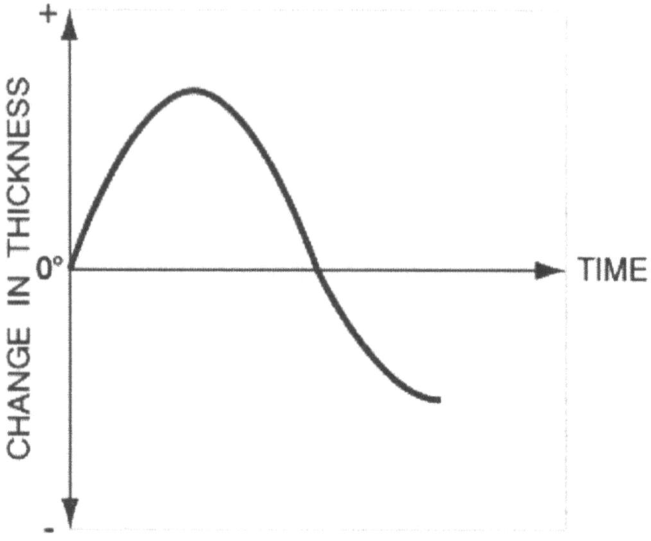

Figure 9.16. A change in film thickness during the anodic oxidation of aluminum

With an increase in H_2SO_4 concentration, when other parameters (i.e. temperature, current density etc.) are constant, the rate of dissolution of oxide increases leading to a decrease in film thickness and an increase in the porosity. Temperature shows a similar effect on the growth of oxide film and an increase in porosity. In order to produce thicker films, electrolysis in H_2SO_4 solutions should be carried out at lower temperatures. Another approach is to use less aggressive electrolytes.

Theoretically, if the rate of electrochemical film formation is proportional to the current density, the rate of chemical dissolution should be constant. In this way, an increase in the current density leads to an increase in the film growth. However, in practice an increase in the current density leads to an increase in temperature and, consequently, to an increase in the rate of dissolution.

During the anodic dissolution, the total charge is consumed by the following processes:

(i) oxygen evolution reaction,
(ii) oxide formation, and
(iii) transfer of aluminum ions into electrolyte, due to film dissolution.

Depending on the nature of the electrolyte involved in the anodic oxidation, various secondary reactions may take place at the anode, which affects the properties of the oxide film. As results show, sulfate, chromate or oxalate are incorporated in the barrier layer. Reaction mechanism for incorporation of these anions is not yet clarified.

The structure of porous Al_2O_3 formed during anodic oxidation of aluminum is described as a close-packed array of approximately hexagonal columnar cells, which contain elongated pores normal to the Al substrate surface. The schematic illustration of the porous anodic film is presented in Figure 9.17. The processes inside the barrier layer are of a electrochemical nature, while the processes inside the porous layer are of a chemical and physical nature. [71-72]

Anodic films produced in sulfuric acid solutions are semi-transparent and colorless. In this way, they provide very good substrates for coloring. Coloring of anodized aluminum is usually carried out with inorganic and organic compounds and also electrolytically. [73,74]

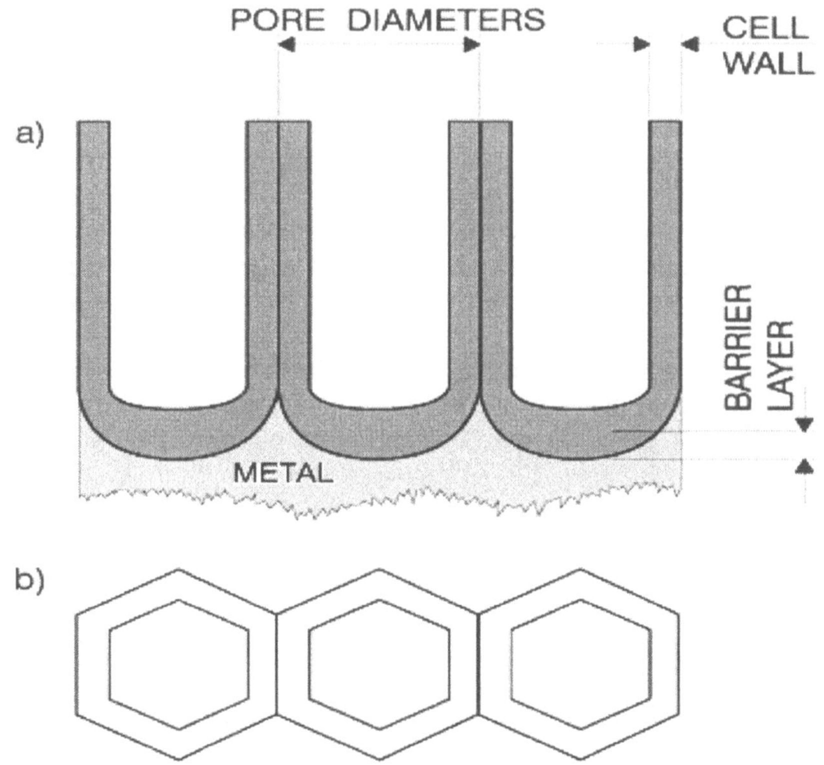

Figure 9.17.. Schematic illustration of the porous anodic film: (a) cross section, (b) top view

Inorganic coloring techniques are based on immersion of anodized aluminum into solutions containing ions, which will produce a color. For example, anodized aluminum is successively immersed into solution containing cobalt acetate and then a solution of potassium carbonate, at 30 to 50 °C, for blue color. A successive immersion of anodized aluminum, first in a solution of lead nitrate and then in a solution of potassium chromate, at 40 to 50 °C, is carried out to produce yellow color. A green color is produced by an immersion of anodized aluminum in a solution containing copper sulfate and then a solution of ammonium sulfide. Mechanistic aspects of coloring are not well understood, but results suggest this process is based on absorption and diffusion of anions and cations into pores.

Among organic compounds dyestuff such as Alizarin Red S, Chrome Fast Orange R, Aluminum Black MLW, Aluminum Turqoise PLWS etc., are used for coloring.

Electrolytic coloring is carried out from Ni(II), Cu(II) or Sn(II) electrolytes. Under the conditions of direct current (d.c.), anodized aluminum acts as the cathode. However, since d.c. processes are sensitive to contaminants present in the electrolyte, electrolytic coloring is frequently carried out under alternating current conditions. The counter electrode materials include graphite, nickel, stainless steal and lead.

As mentioned earlier, the oxide films on anodized aluminum are porous and they have very good adsorption properties. In order to improve the protection properties films are treated with boiling water, chromate, Ni(II) solutions etc., which causes a further hydration of oxide and sealing of pores.

Sealing is a process in which major pore blockage and an increase in water content in the anodic films occurs. This preventive method is practically very important for the improvement of properties of anodized films.

In the sealing with non-aqueous sealants, organic substances provide a physical blockage of the coating pores. For this process lubricating oil, silicone polymers, electrophoretic coatings with pigmented paint etc. are used.[75]

Sealing based on aqueous solutions is more widely used in commercial applications. In this process, after a careful water-rinse, oxidized parts are treated in saturated steam at atmospheric pressure or in hot water, nickel acetate, chromate and various cold sealing solutions.

While "as produced" oxidized aluminum contains about 0.5 % H_2O, during sealing the amount of water increases rapidly in the initial stages of the process, and later much more slowly. The amount of water in the sealed coating is estimated at approximately 8 to 13 %.

In the sealing process, the porous structure of the anodic films becomes less distinct and pores are blocked starting at outer surface. The aluminum oxide is transformed into various forms of hydrated oxides. The most important form of these hydrated oxides is böhmite.

Sealing methods used in practice are divided into following groups:

i) water sealing,
ii) steam sealing and
iii) sealing in metal-salt solution.

Sealing in water depends on temperature (usually the working temperature is 96 to 100 °C), presence of various ions (Cu^{2+}, Fe^{2+}, F^-, Cl^- etc.) and pH (between 5.5 and 6.5). [76] The time of sealing is about 2 to 3 minutes per 1 μm of coating thickness.

The steam sealing presents considerable engineering difficulties, and there is no evidence for superiority of this process over hot water sealing. As mentioned above, the process is carried out with saturated steam and at atmospheric pressure, although some plants use very large chambers, which are hermetically sealed and the work is steamed under pressure. [77]

Sealing in metal-salt solutions is based on nickel acetate or potassium dichromate formulations. These processes are carried out at elevated temperatures (e.g. sealing in the nickel acetate solution at 75 - 80 °C, and sealing in the chromate solutions between 94 and 98 °C).

There are other sealing systems that were developed in order to reduce the higher costs associated with hot water, or environmental concerns related to chromate solutions. Among other systems cold sealing in nickel fluoride solutions, sealing in sodium silicate solutions, or use of ammonia vapor under pressure, should be mentioned.[78]

9.9 FURTHER READINGS

1. Brenner A., *Electrodeposition of Alloys, Principles and Practice*, Vol.1, New York: Academic Press, 1963.
2. Ilyushenko L.F. *Elektroliticheski Osazhdenie Magnitnie Plenki*, Minsk : Nauka i Tekhnika, 1979.
3. Dahms H., Croll I.M. The Anomalous Codeposition of Iron-Nickel Alloys. J. Electrochem. Soc. 1965; 112:771-775.
4. Horkans J.. Effect of Plating Parameters on Electrodeposited NiFe. J. Electrochem. Soc. 1981; 128:45-49.
5. Matlosz M. Competitive Adsorption Effects in the Electrodeposition of Iron-Nickel Alloys. J. Electrochem. Soc. 1993; 140:2272-2279.
6. Vaes J., Franser J., Celis J.P. The Role of Metal Hydroxides in NiFe Deposition. J. Electrochem. Soc. 2000; 147:3718-3724.
7. Djokić S.S., Maksimović M.D.. "Electrodeposition of Nickel-Iron Alloys" In *Modern Aspects of Electrochemistry*, Vol. 22, pp. 417-466, J.O'M. Bockris, B.E. Conway and R.E. White, eds., New York: Plenum Press 1992.
8. Podlaha E.J., Landolt D. Induced Codeposition. J. Electrochem. Soc. 1996; 143:885-892.

9. Chassaing E., Quang K.Vu., Wiart R.. Mechanism of Nickel-Molybdenum Alloy Electrodeposition in Citrate Electrolytes. J. Appl. Electrochem. 1989; 19:839-843.
10. Djokić S.S. Electrodeposition of Amorphous Alloys Based on the Iron Group of Metals. J. Electrochem. Soc. 1999; 146:1824-1828.
11. Haseeb A.S.M.A., Celis J.P., Ross J.R. Dual-bath Electrodeposition of Cu/Ni Compositionally Modulated Multilayers. J. Electrochem. Soc. 1994; 141:230-237.
12. Rousseau A., Benabeu P. Single-bath Electrodeposition of Chromium-Nickel Compositionally Modulated Multilayers (CMM) from a Trivalent Chromium Bath. Plat. And Surf. Finish. 1999; 86 (9):106-110.
13. Kelly J.J., Bradley P.E., Landolt D. Additive Effects During Pulsed Deposition of Cu-Co Nanostructures, J. Electrochem. Soc. 2000; 147:2975-2980.
14. Landolt D. Electrochemical and Materials Science Aspects of Alloy Deposition. Electrochim. Acta. 1994; 39:1075-1090.
15. Ying R. Electrodeposition of Copper-Nickel Alloys from Citrate Solutions on Rotating Disk Electrode. I Experimental Results. J. Electrochem. Soc. 1988; 135:2957-2964.
16. Hovestad A., Janssen L.J.J., Electrochemical Codeposition of Inert Particles in a Metallic Matrix. J. Appl. Electrochemistry. 1995; 25:579-527.
17. Buelens C., Celis J.P., Roos J.R. Electrochemical Aspects of the Codeposition of Gold and Copper with Inert Particles. J. Appl. Electrochem. 1983; 13:541-548.
18. Hwang B. J. Hwang C. S. Mechanism of Codeposition of Silicon Carbide with Electrolytic Cobalt. J. Electrochem. Soc. 1993; 140:979-984.
19. Guglielmi N. Kinetics of the Deposition of Inert Particles from Electrolytic Baths. J. Electrochem. Soc. 2000; 119:1009-1012.
20. Celis J., Roos J. R.. Kinetics of the Deposition of Alumina Particles from Copper Sulfate Plating Baths. J. Electrochem. Soc. 1977; 124:1508-1511.
21. Nowak P., Socha R.P., Kaisheva M., Fransaer J., Celis J. P., Stoinov Z., Electrochemical Investigation of the Codeposition of SiC and SiO_2 Particles with Nickel. J. Appl. Electrochem. 2000; 30:429-437.
22. Musiani M., Electrodeposition of Composites: An Expanding Subject in Electrochemical Materials Science. Electrocim. Acta 2000; 45:3397-3402.
23. Chen C.C., Bose C.S.C., Rajeshwar K.. The Reduction of Oxygen and Oxidation of Hydrogen at Polypyrrole Film Electrodes Containing Nanodispersed Platinum Particles. J. Electroanal. Chem. 1993; 350:161.
24. Alonso-Vante N., Cattarin C., Musiani M. Electrocatalysis of O_2 Reduction at Polyaniline + Molybdenum-Doped Ruthenium Selenide Composite Electrodes. J. Electroanal. Chem. 2000; 481:200-207.
25. Beck F., Dahlhaus F., Zahedi N. Anodic Codeposition of Polypyrrole and Dispersed TiO_2. Electrochim. Acta.1992; 37:1265.
26. Beck F., Dahlhaus M. Anodic Formation of Polypyrrole/Tungsten Trioxide Composites. J. Appl. Electrochem. 1993; 23:781.
27. Li H.S., Josowicz M., Baer D.R., Engelhard M.H., Janata J. Preparation and Characterization of Polyaniline – Palladium Composite Films, J. Electrochem. Soc. 1995; 142:798-805.
28. Hapel M. The Electrolytic Oxidation of Methanol at Finely Dispersed Platinum Nanoparticles in Polypyrrole Films. J. Electrochem. Soc, 1998; 145:124-134.
29. Musiani M. Anodic Deposition of PbO_2/Co_3O_4 Composites and Their Use as Electrodes for Oxygen Evolution reaction. J. Chem. Soc. Commun. 1996; 21:2403-2404.
30. Cattarin S., Frateur I., Guerriero P., Musiani M. Electrodeposition of PbO_2 + CoO_x Composites by Simultaneous Oxidation of Pb^{2+} and Co^{2+} and Their Use as Anodes for Oxygen Evolution. Electrochim. Acta. 2000; 45:2279-2288.

31. Musiani M., Guerriero P. Electrodeposited Tl_2O_3 – Matrix Composites I: Effect of the Dispersed Phase on Nucleation and Growth of the Matrix. J. Electrochem. Soc. 1998; 145:549-554.
32. Musiani M.M., Furlanetto F., Guerriero P. Electrodeposited Tl_2O_3 – Matrix Composites II Electrocatalysis of Oxygen Evolution reaction on Tl_2O_3/Co_3O_4 Composites. J. Electrochem. Soc. 1998; 145:555-560.
33. Kulesza P.S., Grzybowska B., Malik M.A., Galkowski M.T. Tungsten Oxides as Active Supports for Highly Dispersed Platinum Microcenters: Electrocatalytic Reactivity Toward Reduction of Hydrogen Peroxide and Oxygen. J. Electrochem. Soc. 1997; 144:1911-1917.
34. Spencer L.F.. Modern Electroforming. I Requirements and Mandrels. Metal Finishing. 1973; February:64-72.
35. Wearmouth W.R. Applications and Developments in Nickel Electroforming and Toolmaking. Metal Finishing. 1980; November:35 – 39.
36. Mehdizadeh S., Dukovic J., Andricacos P.C., Romankiw L.T., Cheh H. Y. The Influence of Lithographic Patterning on Current Distribution in Electrodeposition: Experimental Study and Mass-Transfer Effects. J. Electrochem. Soc. 1993; 140:3497-3505.
37. Romankiw L.T. Electroforming of Electronic Devices. Plat. And Surf. Finish. 1997; 84 (1):10-15.
38. Brenner A. "Electrolysis of Nonaqueous Systems." In *Advances in Electrochemistry and Electrochemical Engineering*. C.W. Tobias, ed. Vol. 5: pp.205-248. New York: Interscience Publishers, 1967
39. Popovych O., Tomkins R.P.T., *Nonaqueous Solution Chemistry*, New York: John Wiley & Sons, 1981.
40. Lowenheim F. A., *Modern Electroplating*, Third Edition, New York: John Wiley & Sons, 1974.
41. Capuano G.A., Davenport W.G. Electrodeposition of Aluminum from Alkylbenzebe Electrolytes. J. Electrochem. Soc. 1971; 118:1688 – 1695.
42. Capuano G.A., Davenport W.G. Cathodic Polarization of Aluminum in Alkylbenzene Electrolytes. J. Electrochem. Soc. 1984; 131:2595 – 2600.
43. Biallazor S. Lisowska-Oleksiak A. The Modification of Aromatic Electrolytes for Electrodeposition of Aluminum. J. Appl. Electrochem. 1990; 20:590-595.
44. Mc Chesney M. Electrodeposited Aluminum. Plating and Surface Finishing. 1995; 82 (10):42.
45. Safranik W.H., *The Properties of Elctrodeposited Metals and Alloys*, Second Edition, Orlando, FL: American Electroplaters and Surface Finishers Society, 1986.
46. Wilkes J.S., Levisky J.A., Wilson R.A., Hussey C.L. Dialkylimidazolium Chloroaluminate Melts: A New Class of Room temperature Ionic Liquids for Electrochemistry, Spectroscopy and Synthesis. Inorg. Chem. 1982; 21:1263-1265.
47. Lin Y.F., Sun I.W. Electrodeposition of Zinc from a Mixture of ZincChhloride and Neutral Aluminum-Chloride-1-Methyl-3-Ethylimidazolium Chloride Molten Salt. J. Electrochem. Soc. 1999; 146:1054-1059.
48. Moffat T.P. Electrodeposition of Al-Cr Metallic Glasses. J. Electrochem. Soc. 1994; 141:L115-L117.
49. Mitchell J.A., Pitner W.R., Hussey C.L.. Stafford G.R. Electrodeposition of Cobalt and Cobalt-Aluminum Alloys from a Room temperature Chloroaluminate Molten Salt. J. Electrochem. Soc. 1996; 143:3448-3455.
50. Pitner W.R., Hussey C.L. Electrodeposition of Zinc from the Lewis Acid Aluminum Chloride-1-Methyl-3Ethylimidazolium Chloride Room Temperature Molten Salt. J. Electrochem. Soc. 1997; 144:3095-3103.

51. Lee J.J., Bae I.T., Sherson D.A., Miller B., Wheeler K.A. Underpotential Deposition of Aluminum and Alloy Formation on Polycrystalline Gold Electrode from $AlCl_3$/EMIC Room Temperature Molten Salts. J. Electrochem. Soc. 2000; 147:562-566.
52. Chen P.Y., Lin M.C., Sun I.W. "Electrodeposition of Cu-Zn Alloy from Acidic $ZnCl_2$-EMIC Molten Salt", J. Electrochem. Soc. 2000; 147:3350 – 3355.
53. Landolt D. Fundamental Aspects of Electropolishing. Electrochim. Acta. 1987; 32:1-11.
54. Mathieu J.B., Mathieu H. J., Landolt D. Electropolishing of Titanium in Perchloric Acid – Acetic Acid Solution. I. Auger Spectroscopy Study of Anodic Film, J. Electrochem. Soc. 1978; 125:1039-1043.
55. Mathieu J. B., Landolt D. Electropolishing of Titanium in Perchloric Acid – Acetic Acid Solution. II. Polarization and Stoichiometry. J. Electrochem. Soc. 1978; 125:1044-1049.
56. Piotrowski O., Mandore C., Landolt D., Electropolishing of Tantalum in Sulfuric Acid – Methanol Electrolyte. Electrochim. Acta. 1999; 44:3389-3399.
57. Sautebin R., Landolt D. Anodic Leveling Under Secondary and Tertiary Current Distribution Conditions, J. Electrochem. Soc., 1982; 129:946-953.
58. Vidal R., West A.C., Copper Electropolishing in Concentrated Phosphoric Acid, J. Electrochem. Soc. 1995; 142:2682-2694.
59. Novak M., Reddy A.K. N., Wroblowa H. An Ellipsometric Study of Surface Films on Copper Electrodes Undergoing Electropolishing. J. Electrochem. Soc. 1970; 117:733-737.
60. Glarum S.H., Marshall J.H. The Anodic Dissolution of Copper into Phosphoric Acid. I. Voltammetric and Oscillatory Behavior. J. Electrochem. Soc. 1985; 132:2872-2878.
61. Datta M., Landolt D. On the Influence of Electrolyte Concentration, pH and Temperature on Surface Brightening of Nickel Under ECM Conditions. J. Appl. Electrochem. 1997; 7:247-252.
62. Datta M, Landolt D., On the Role of Mass Transport in High Rate Dissolution of Iron and Nickel in ECM Electrolytes – II. Chlorate and Nitrate Solutions. Electrochim. Acta. 1980; 25:1263-1271.
63. Datta M., Shenoy R.V., Romankiw L.T. Recent Advances in the Study of Electrochemical Manufacturing. Journal of Engineering for Industry, 1996; 118 (2):29.
64. Jain V.K., Dixit P.M., Pandey P.M. On the Analysis of the Electrochemical Spark Machining Process. International Journal of Machine Tools and Manufacturing, 1999; 39(1):165.
65. Ni X, Mc Geough J.A., Greated C.A. A Study of Electrical Discharges in Electrolyte by High Speed Photography. J. Electrochem. Soc. 1993; 140 (12):3505-3512.
66. Butterfield D, Sgarzi A. Method and Apparatus for Electrochemical Machining of Spray Holes in Fuel Injection. US Patent, 5,026,462 (1991).
67. Finneran M.T., Finneran R.J. Electromyographic Electrode. U.S. Patent, 6,047,202 (2000).
68. Neufeld P., Ali H.O. The Influence of Anions on the Structure of Porous Anodic Al_2O_3 Films Grown in Alkaline Electrolytes. J. Electrochem. Soc. 1973; 120:479-484.
69. Farnann I., Durpee R., Jeong Y., Thompson G.E., Wood G.C., Forty A. Structural Chemistry of Anodic Alumina. Thin Solid Films. 1989; 173:209-215.
70. Thompson G.E., Furneaux R.C., Goode J.S., Wood G. C. Porous Anodic Films Formation on Aluminum Substrates in Phosphoric Acid. Trans. Inst. Met. Finish. 1978; 56:159-167.
71. Patermarakis G., Moussoutzanis K. Mathematical Models for the Anodization Conditions and Structural Features of Porous Anodic Al_2O_3 Films on Aluminum. J. Electrochem. Soc. 1995; 142:737-743.

72. Patermarakis G. Development of a Theory for the Determination of the Composition of the Anodizing Solution Inside the Pores During the Growth of Porous Anodic Al_2O_3 Films on Aluminum by a Transport Phenomenon Analysis. J. Electroanal. Chem., 1998; 447:25-41.
73. Patermarakis G., Papandreadis N., Effect of the Structure of Porous Anodic Al_2O_3 Films on the Mechanism of Their Hydration and Pore Closure During Hydrothermal Treatment. Electrochim. Acta. 1993; 38:1413-1418.
74. Patermarakis G., Karayannis H.S. The Mechanism of Growth of Porous Anodic Al_2O_3 Films on Aluminum at High Film Thickness. Electrochim. Acta. 1995; 40:2647.
75. Young L., *Anodic Oxide Films*, London: Academic Press, 1961.
76. Brace A.W., Sheasby P.G., *The Technology of Anodizing Aluminum.*, Stonehouse, Glos. Technology Limited, , 1979.
77. Henley V.F., *Anodic Oxidation of Aluminum and its Alloys,* Oxford: Pergamon Press, 1982.
78. Mansfeld F., Zhang G., Chen C., Evaluation of Sealing Methods for Anodized Aluminum Alloys with Electrochemical Impedance Spectroscopy (EIS), Plat. and Surf. Finish., 1997; 84 (12):72-81.

Chapter 10

METAL DEPOSITION WITHOUT AN EXTERNAL CURRENT

Deposition of metals and alloys without an external current source is very important in modern technology, especially in the production of new materials for applications in electronics, wear and corrosion resistant materials, medical devices, battery technologies, etc.

These processes supplement and in some cases replace electrodeposition for several practical reasons. The solutions for deposition of metals without an external current source have excellent throwing power and allow plating on articles of very complex shapes and plating through holes. Deposits are denser and exhibit better properties for corrosion and electronics applications. Other important advantages of this type of deposition over electrodeposition include applicability for metallization of non-conductive surfaces (glass, ceramics, polymers, etc.) and the ability to selectively deposit thin metal films only on catalyzed areas of the substrate. Finally, for this type of deposition, an external current source is not needed.

It seems that all metals electrochemically depositable from aqueous solutions can also be deposited chemically under proper conditions (bath composition, pH, temperature, and corresponding catalytic surface), using suitable reducing agents.

Table 10.1 presents a survey of metals and alloys that have been deposited without an external current source hitherto. In the first group are listed commonly deposited single metals such as Ni, Co, Cu, Ag, Au and Pd. Other metals from this group do not have significant applications at the present, but it should be noted that there are reports on their deposition in the published literature.

The second group lists elements, which cannot be deposited alone. However, they can easily be codeposited with nickel or cobalt. Typical examples are Mo and W. The phenomenon is somehow analogous to induced electrodeposition as Brenner defined it.

Table 10.1. Metals and alloys deposited without an external current source

Single metals	Elements codeposited with Ni or Co	Other alloys
Ni, Co, Cu, Cd, Pb, Sb, Bi, Ag, Au, Pt, Pd, Rh, Ru, Sn, In	P, B, V, Mo, W, Mn, Re, Fe, Zn, Tl	CuNi, CuCo, CuCd, CuAu, PdNiP, PdCoP, PdZnP, AuAg, AuSn, PbSn, AuIn

The third group represents alloys based on the first and/or second group of elements. These alloys have been deposited for various applications, mainly in the electronics industry. There is a high probability that other alloys, which are commonly electrodeposited, can also be electrolessly deposited, but there is no published data so far.

10.1 BASIC DEFINITIONS

Brenner and Riddell were the first authors to introduce the term *electroless* metal deposition when describing an autocatalytic process of depositing a metal in the absence of an external source of electrical current. Since there are other metal depositions from aqueous solutions that are carried out without an external current, this process can be divided into three main groups:

1. displacement deposition
2. contact deposition
3. autocatalytic deposition

10.1.1 Displacement Deposition

Displacement deposition is a heterogenous galvanic process in which the noble metal ions are reduced and deposited at the surface of an active metal, as a consequence of dissolution of that metal. The process is sometimes called immersion plating, although this term is not a specific description, and therefore should be avoided, or cementation. The overall displacement reaction is quite simple[1]:

$$M_1^{n_1^+} + \frac{n_1}{n_2} M_2^o = \frac{n_1}{b_2} M_2^{n_2^+} + M_1^o \qquad (10.1)$$

and involves the displacement half-reaction of a more active metal M_2:

$$M_2^o = M_2^{n_2^+} + n_2 e^- \qquad (10.2)$$

by a more noble metal, M_I:

$$M_I^{n_I^+} + n_I e^- = M_I^o \quad (10.3)$$

Typical cementation systems in practice are Ag/Zn, Ag/Cu, Cu/Zn, Cu/Fe, Cu/Al, Sn/Cu etc. The displacement reaction stops immediately after the reduced metal (more positive metal) covers the surface of the immersed metal (more negative metal). Accordingly, the thickness of the deposited metal is always limited. The time of immersion is particularly critical for achieving a uniform coating layer. Very often, the adhesion of the deposited films is not as good as that of films prepared by electrodeposition or by autocatalytic deposition. The displacement deposition differs from all other plating processes from aqueous solutions without an external source of electrical current, because it does not require a reducing agent. Because of lower quality and thinner coatings, displacement deposition has found applications mainly in the refining metals. To a certain extent, however, there are other applications such as coatings for porcelain enamelling, zincate coatings, decorative finishing, soldering, purification of electrolytes before electrowinning, environmental purposes, etc.. [2]

10.1.2 Contact Deposition

Contact deposition is equivalent to electrochemical deposition with the exception that the current is derived from the chemical reaction and not from an outside source. The metal on which deposition takes place, and the auxiliary metal with which it is in contact, form a galvanic element. In this galvanic element, the auxiliary metal acts as an anode and dissolves; the other metal is a cathode. Consequently, the dissolved metal is deposited on the cathode (metal on which deposition takes place) at a mixed potential.

The importance of contact deposition for industrial applications is relatively small. Sometimes the process is used with autocatalytic Ni to initiate Ni deposition on copper and its alloys.[2] This is achieved by coupling the Cu or Cu-alloys with Al , Fe or Ni. The contact deposition is applicable only to a limited extent, and uniform thicker deposits cannot be obtained. On the other hand, the constant increase of dissolved metal concentration in the solution may cause instability of the solution.

10.1.2 Autocatalytic Deposition

Autocatalytic deposition is the most commonly used chemical method for the deposition of metallic films from aqueous solution without an external source of electrical current. The metal films are formed only on catalytically active surfaces without an external source of electrical current and by the chemical reduction of metallic ions in an aqueous solution containing a reducing agent. Autocatalytic deposition is defined as a process for deposition of metallic films by a controlled chemical reaction that is catalyzed by the metal or alloy being deposited.[3] If the metal ion, M^{z+}, is reduced by the reducing agent ion, R^{n-}, the process can be simply described by the following reaction:

$$M^{z+} + R^{n-}(catalyst) = M + R^{z-n} \quad (10.4)$$

Although the term *electroless deposition* broadly describes all processes of metal and alloy deposition without an external source of electrical current, it should be noted that this term is commonly used the for autocatalytic deposition process. Consequently, in this chapter, the term *electroless deposition* is used only for the autocatalytic deposition processes.

The development of electroless deposition is mainly connected with Ni or Cu deposition. However, other electrolessly depositable metals and/or alloys such as Ag, Au, Co, Sn, AuSn, NiWP, etc. have also been studied because of their important applications.

10.2 SOLUTIONS FOR ELECTROLESS DEPOSITION

All solutions for electroless metal deposition have many similarities, but depending on the metal or alloy to be deposited, there are also some differences. Typically, a solution for electroless metal deposition is consisted of the following components:

(i) Source(s) of metal ions
(ii) Complexing agent(s)
(iii) Reducing agent(s)
(iv) Stabilizer(s) and inhibitor(s)

Table 10.2. presents sources of metal ions in electroless deposition of common metals. Generally speaking, the metal ion sources can be any water-soluble salts such as sulfates, chlorides, acetates, cyanides, etc. The nature of the metal ion source is usually determined by the stability of the solution, properties of the deposited films, and also by environmental issues.

The majority of complexing agents used in electroless metal deposition are organic acids or their salts, with a few exceptions of inorganic ions such as CN^-, NH_4^+ or $P_2O_7^{4-}$. Ammonia and NH_4^+ ions, in the case of nickel solutions, are mainly used for pH control. The choice of the complexing agents is dependent, first of all, on the nature of the metal ion used for deposition. The principal functions of complexing agents are: buffering action, prevention of precipitation of hydroxides and salts, and reduction of the concentration of free (aquo) metal ions. In addition, complexing agents affect the rate of reduction and the properties of metal deposits. In some cases, complexing agents apparently form strong complexes with metallic contaminants, thereby making them less susceptible to react with reducing agents.

Table 10.2. Sources of Metal Ions in Electroless Deposition of Common Metals

Metal	Salt
Ni	$NiSO_4$, $NiCl_2$, $Ni(H_2PO_2)_2$,$Ni(CH_3COO)_2$
Cu	$CuSO_4$
Co	$CoSO_4$, $CoCl_2$
Au	$KAu(CN)_2$, $KAuCl_4$, $Na_3Au(SO_3)_2$
Ag	$AgNO_3$, $NaAg(CN)_2$
Pd	$PdCl_2$, $Pd(NH_3)_4Cl_2$
Pt	$Na_2Pt(OH)_6$, $(NH_3)_2Pt(NO_2)_2$,Na_2PtCl_6

Complexing agents used for the electroless deposition of common metals are listed in Table 10.3. Commercial solutions for nickel electroless deposition operate in the pH range 4.5 to 6. The complexing agents are most effective in this pH range. However, in the electroless deposition of Cu, Au, Ag, Pd and in some cases Ni, solutions with pH > 8 are used.

Table 10.3, Complexing agents used in the electroless deposition of common metals

Metal	Complexing agent
Ni, Co	Acetate, Propionate, Succinate, Hydroxyacetate, Aminoacetate, Ethylenediamine, Malonate, Pyrophosphate, Malate, Citrate, Tartrate,
Cu	Tartrate, Glycolic acid, Triethyl amine, EDTA, Cyanide
Au	Cyanide, Sulfite, Ethanolamine, Citrate, Chloride
Ag	Cyanide, Ammonium
Pd, Pt, Ru	Citrate, Succinate, Acetate, Ammonium, Ethylenediamine

The choice of reducing agent depends on conditions of electroless deposition and, of course, on the metal or alloy being deposited, including their physico-chemical properties. Use of reducing agents containing phosphorus or boron leads unavoidably to the incorporation of these elements, which can dramatically affect the properties of the metal deposit. On the other hand, electroless deposition of pure metals is also possible using reducing agents such as hydrazine or formaldehyde.

Hypophosphite is mainly used for the electroless deposition of Ni, Co, Pd and their alloys. The deposits are not purely metallic as they usually contain phosphorus. Utilization of hypophosphite in electroless metal deposition is considerably less than 100 %. The reduction reaction takes place only at certain surfaces such as metals of the group VIII (Fe, Co, Ni, Rh, Pd, and Pt). It also takes place on Au.

The most studied reaction among electroless processes is definitely deposition of Ni with hypophosphite. The overall reaction for nickel deposition with hypophosphite can be represented as:

$$Ni^{2+} + 2H_2PO_2^- + 2H_2O = Ni + 2H_2PO_3^- + H_2 + 2H^+ \quad (10.5)$$

Generally, if the concentration of hypophosphite is increased, the phosphorous content in NiP alloy is increased.

Electroless deposition of Cu with hypophosphite is still doubtful. In the presence of nickel, however, NiCuP alloy films have been deposited successfully. [4]

Boron-containing reducing agents used in electroless Ni deposition are mainly borohydrides and amine boranes. Deposits usually contain 90 to 99 % metallic phase, depending on the composition of the solution and operating conditions. The rest is usually boron and other occluded reacting agents. The-boron containing reducing agents are used for electroless deposition of common metals, such as Ni, Co, Pd, Pt, Au, Ag and their alloys.

The electroless deposition of Ni with borohydride takes place in alkaline solutions. Theoretically, each borohydride ion can reduce four nickel ions:

$$4Ni^{2+} + BH_4^- + 8OH^- = B(OH)_4^- + 4Ni + 4H_2O \quad (10.6)$$

However, experimental results show that one mole of borohydride reduces approximately one mole of nickel ion.

Gorbunova *et al.* investigated the conditions for electroless nickel-boron deposition using sodium borohydride as a reducing agent.[5] They found that an increase in the $NaBH_4$ concentration in solution without a stabilizer or with stabilizers such as lead chloride, 2-mercaptobenzothiazole or thallium nitrate, leads to an increase in the rate of Ni-B deposition (Figure 10.1). Using a solution containing $TlNO_3$ as a stabilizer gives a faster rate of Ni-B deposition.

Using borohydride as the reducing agent, gold-based alloys (Au-Ag, Au-In) and metals such as Pt, In and Co were deposited. [5]

Whereas borohydrides such as $NaBH_4$ are completely ionic, the amine boranes are covalent compounds. The electrons in the amine boranes are displaced toward the boron atom, while the nitrogen atom displaces positive charge as is illustrated by the following formula:

$$(CH_3)_2HN^{\delta+} : {}^{\delta-}BH_3$$

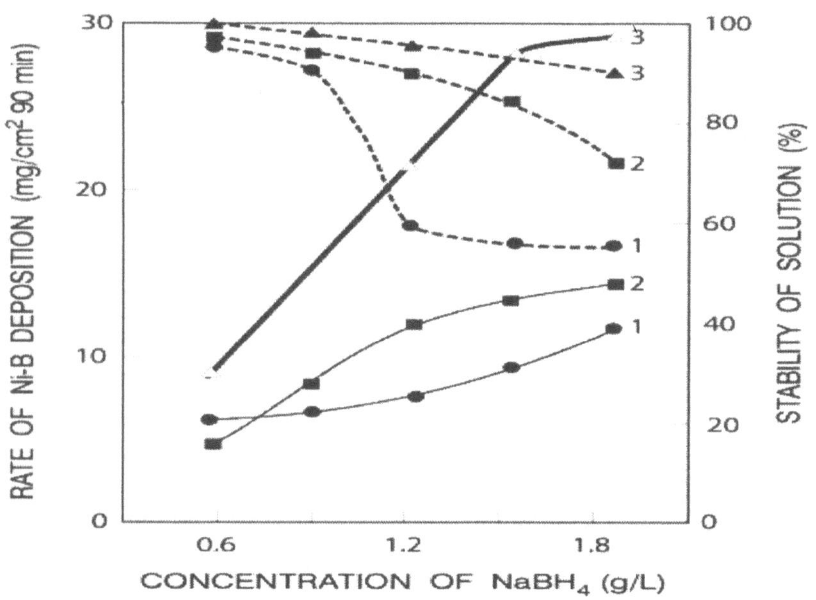

Figure 10.1. Influence of $NaBH_4$ concentration on the rate of Ni-B deposition (solid lines) and stability of solution (dashed lines): 1- solution without stabilizer; 2-solution with $PbCl_2$ and 2-mercaptobenzothiazole; 3 - solution with $TlNO_3$.[5] (Reproduced by the permission of The Electrochemical Society, Inc.)

In practice, the application of aminoboranes is limited to dimethylamine borane, $(CH_3)_2NHBH_3$. Dimethylamine borane (DMAB) is used for the electroless deposition of Ni, Cu, Co and Ag. In alkaline and neutral solutions, the preceding chemical reaction of dimethylamine borane with OH^- ions can be represented as:

$$(CH_3)_2NHBH_3 + OH^- = (CH_3)_2NH + BH_3OH^- \qquad (10.7)$$

The acid-catalyzed hydrolysis of dimethylamine borane occurs according to the following equation:

$$(CH_3)_2NHBH_3 + 3H_2O + H^+ = (CH_3)_2NH_2^+ + H_3BO_3 + 3H_2 \qquad (10.8)$$

Based on experimental results, in the electroless nickel deposition the molar ratio of nickel ions reduced to DMAB molecules consumed during the process is approximately 1:1.

Parlstein and Weightman investigated electroless deposition of Co with DMAB from acid solutions. [6] Dependence of deposition rate on DMAB concentration is presented in Figure 10.2. As illustrated in this figure, the rate of Co deposition increases almost linearly up to DMAB concentration of about 4 g/ dm^3. A further increase in DMAB concentration results in a rapid decomposition of the solution.

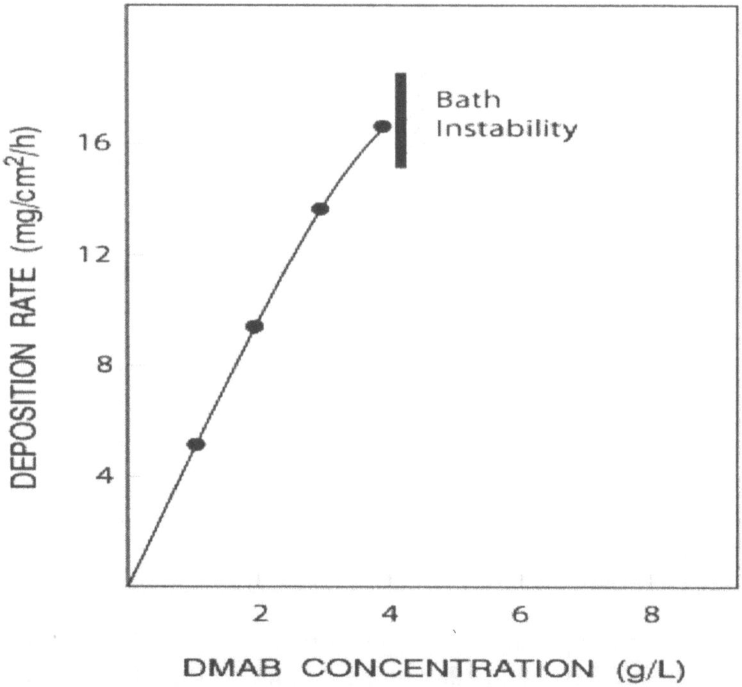

Figure 10.2. The effect of DMAB on electroless deposition of cobalt from a solution containing 25 g/ dm^3 $CoSO_4 \cdot 7H_2O$, 15 g/ dm^3 $Na_2C_4H_4O_6$, pH 5.0, temperature 80 °C. [6] (Reproduced by the permission of The Electrochemical Society, Inc.)

Formaldehyde is mainly used for electroless copper and silver deposition; however, there are reports that this reducing agent can also be used for electroless deposition of AuCu alloy or Co.

An overall reaction for electroless copper deposition with formaldehyde is described as follows:

$$Cu^{2+} + 2HCHO + 4OH^- = Cu + 2HCOO^- + 2H_2O + H_2 \qquad (10.9)$$

Dumesic *et al.* studied electroless copper deposition from an EDTA alkaline solution using formaldehyde as a reducing agent.[7] They reported that an increase in the formaldehyde concentration from 0.03 to 0.4 mol/dm^3 leads to a linear increase in the initial deposition rate (Figure 10.3.).

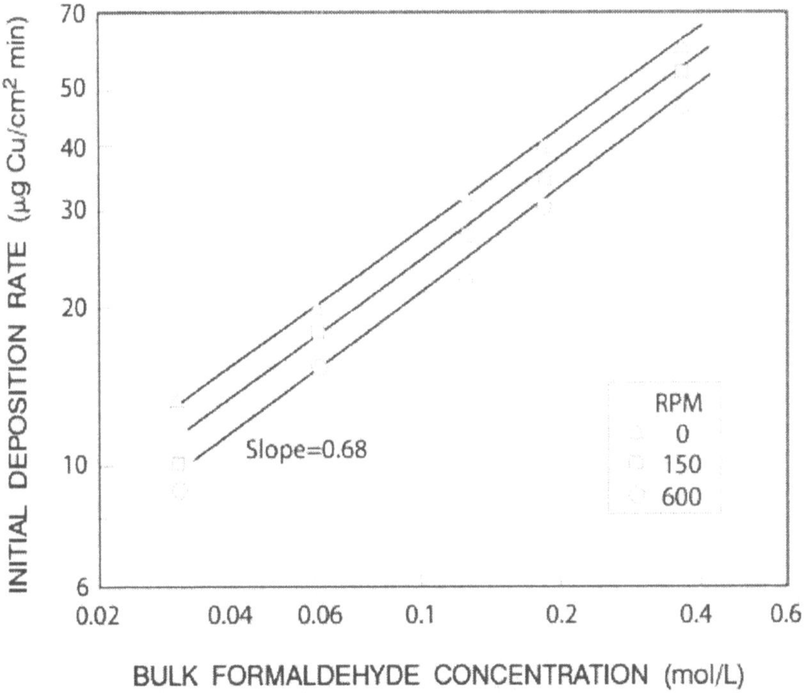

Figure 10.3. Dependence of the initial rate of Cu deposition on HCOH concentration for various stirring speeds.[7] (Reproduced by the permission of The Electrochemical Society, Inc.)

Electroless Ag deposition with formaldehyde is fast, but either a cloudy film of silver metal is obtained or peeling occurs. From other metals, as mentioned earlier, the electroless Co deposition is carried out using formaldehyde as a reducing agent.[8]

Hydrazine has long been recognized as a very powerful reductant of metallic ions, and has been used for electroless deposition of metals and alloys. Examples include electroless Cu, Ni, Co, Au, Ag, Pt-group of metals and their alloys, NiSnW, NiFe and alloys resembling stainless steel.

The rate of electroless deposition of Co with hydrazine, increases with an increase in N_2H_4 concentration, which is presented in Figure 10.4.[9] The net reaction for electroless deposition of Co with hydrazine is described as:

$$Co^{2+} + N_2H_4 + 2OH^- = Co + H_2 + N_2 + 2H_2O \quad (10.10)$$

Similarly to other reducing agents, increasing hydrazine concentration leads to an increase in the rate of electroless deposition.

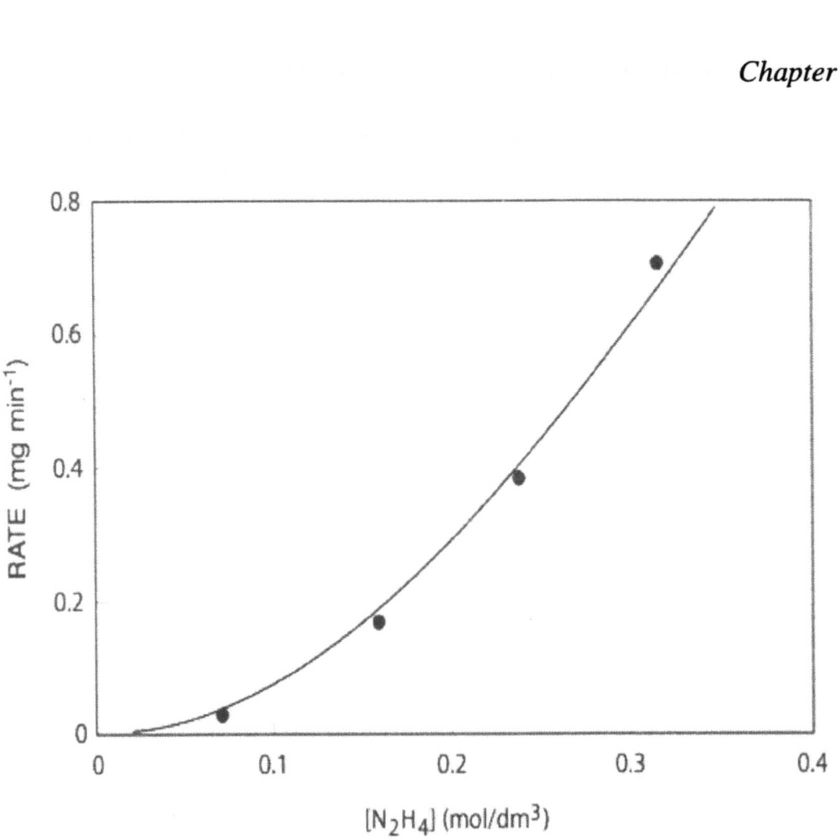

Figure 10.4. The effect of N_2H_4 concentration on the rate of Co deposition (72 °C, pH 12.2, [Co(II)] = 0.08mol/dm^3).[9] (Reproduced by the permission of The Electrochemical Society, Inc.)

Hydrazine is often used in the spray method for mirror production as the deposition rate is fast. The deposition of Ag from an $Ag(NH_3)_2^+$ complex solution can be described with the following reaction:

$$4Ag(NH_3)_2^+ + N_2H_4 = 4NH_4^+ + 4NH_3 + N_2 + 4Ag \tag{10.11}$$

Stabilizers are chemical compounds used in electroless deposition of metals in order to avoid the decomposition of the solution. Addition of these compounds to the plating solution assures, under proper conditions, operations over an extended period of time. Bath decomposition occurs as a precipitation of metallic particles in the bulk solution. These particles act as a highly efficient catalyst for further metal reduction because of their large surface area. The choice of a stabilizer depends on the metal being deposited and its compatibility with the process.

Stabilizers, used in the electroless deposition of Ni, have been divided into the following classes:[2]

I) compounds containing elements such as S, Se, Te;
II) compounds containing oxygen (AsO_2^-, IO_3^-, MoO_4^{2-});
III) heavy metals cations (Sn^{2+}, Pb^{2+}, Hg_2^{2+}, and Sb^{3+}),
IV) unsaturated organic acids (maleic, itaconic acid, etc.)

The concentration of stabilizers is very important since it determines the rate of deposition. An increase in the concentration of stabilizers of classes I or II above 2 ppm may completely inhibit the deposition reaction. The concentration of class III stabilizers is in the range 10^{-5} mol/ dm^3 to 10^{-3} mol/ dm^3, and the concentration of class IV stabilizers is in the range 10^{-3} mol/dm^3 to 10^{-1} mol/ dm^3.

10.3 MECHANISTIC ASPECTS OF ELECTROLESS DEPOSITION

In spite of relatively intensive study of the electroless deposition of metals and alloys, there is still some disagreement in the treatment of mechanistic aspects of these processes. In order to explain electroless deposition of metals and alloys, five different mechanisms have been proposed, as follows:

1. "atomic hydrogen" mechanism
2. "hydride ion" mechanism
3. "pure electrochemical" mechanism
4. "metal hydroxide" mechanism
5. "uniform" mechanism

These mechanisms involve various attempts to explain electroless deposition. However, according to some experimentally observed characteristics, it is difficult to use any one of these mechanisms for a general explanation of an electroless deposition process.

10.3.1 The Atomic Hydrogen Mechanism

The atomic hydrogen mechanism was developed for electroless Ni deposition with hypophosphite. Brenner and Riddell[2] postulated that the atomic hydrogen reduces Ni^{2+} ions and acts by heterogenous catalysis at the catalytic Ni surface. Atomic hydrogen is generated by the reaction of hypophosphite with water, and is then desorbed at the catalytic surface according to the equation below:

$$H_2PO_2^- + H_2O = H_2PO_3^- + 2H_{ad} \tag{10.12}$$

At the catalytic surface, the adsorbed hydrogen reduces Ni^{2+} ions:

$$2H_{ad} + Ni^{2+} = Ni + 2H^{+} \qquad (10.13)$$

The atomic hydrogen mechanism fails to explain many of the features of electroless deposition such as the simultaneous hydrogen evolution reaction. In this mechanism, deposition of phosphorus and involvement of hydrogen evolution reactions are explained as side reactions. Furthermore, this scheme does not explain why the stoichiometric utilization of hypophosphite is always less than 50 %.

10.3.2 The Hydride Ion Mechanism

In the hydride ion mechanism, the hypophosphite acts as the donor of hydride (H^-) ions. The hydride ion is the reducing agent of both Ni^{2+} and H^+ ions. This mechanism, was modified by Lukes [10] who applied it to both acidic and alkaline solutions. In acidic solutions, formation of the hydride ion was described by the reaction:

$$H_2PO_2^- + H_2O = HPO_3^{2-} + 2H^+ + H^- \qquad (10.14)$$

In alkaline solutions, the formation of H^-, Lukes described by the following reaction:

$$H_2PO_2^- + 2OH^- = HPO_3^{2-} + H_2O + H^- \qquad (10.15)$$

Two hydride ions from the above reactions can then react with Ni^{2+}, or one ion with either a hydrogen ion or water, to form Ni metal and H_2:

$$Ni^{2+} + 2H^- = [Ni + 2H] = Ni + H_2 \qquad (10.16)$$

In broad terms, the hydride ion mechanism can then be described by the following general equations:

$$2RH + 2OH^- = 2ROH + 2H^- \qquad (10.17)$$

$$M^{2+} + 2H^- = [M + 2H] = M + H_2 \qquad (10.18)$$

$$H_2O + H^- = H_2 + OH^- \quad (10.19)$$

where RH is formaldehyde, hydrazine or hypophosphite as previously.

From Lukes' theory arises a question of the reality of a hydride ion formation having a standard reduction potential of - 2.08 V in a hypophosphite solution with standard potential of -1.57 V. [11] Both potentials are reported for pH=0. The change-over from standard conditions to those in which metals are reduced by hypophosphite does not alter the difference between these potentials. On the other hand, the existence of hydride ions in an alkaline medium, even in an intermediate state, appears very unlikely.

10.3.3 The Electrochemical Mechanism

The so-called electrochemical mechanism was first proposed by Brenner and Riddell[2] and later modified by other researchers. In this mechanism, electroless deposition is considered to result from mixed anodic and cathodic reactions. In the case of electroless Ni deposition, the oxidation of hypophosphite with water generates electrons, and is considered as the partial anodic process:

$$H_2PO_2^- + H_2O = H_2PO_3^- + 2H^+ + 2e^- \quad (10.20)$$

with E^o =0.50 V.

The electrons formed in the above reaction are utilized in the coupled cathodic processes for deposition of Ni and P:

$$Ni^{2+} + 2e^- = Ni \quad (10.21)$$

with E^o = - 0.25 V, and

$$H_2PO_2^- + 2H^+ + e^- = P + H_2O \quad (10.22)$$

with E^o= - 0.50 V.

According to the electrochemical mechanism, the evolution of hydrogen gas is a result of the secondary reaction, which follows:

$$2H^+ + 2e^- = H_2 \quad (10.23)$$

with E^o=0.00 V.

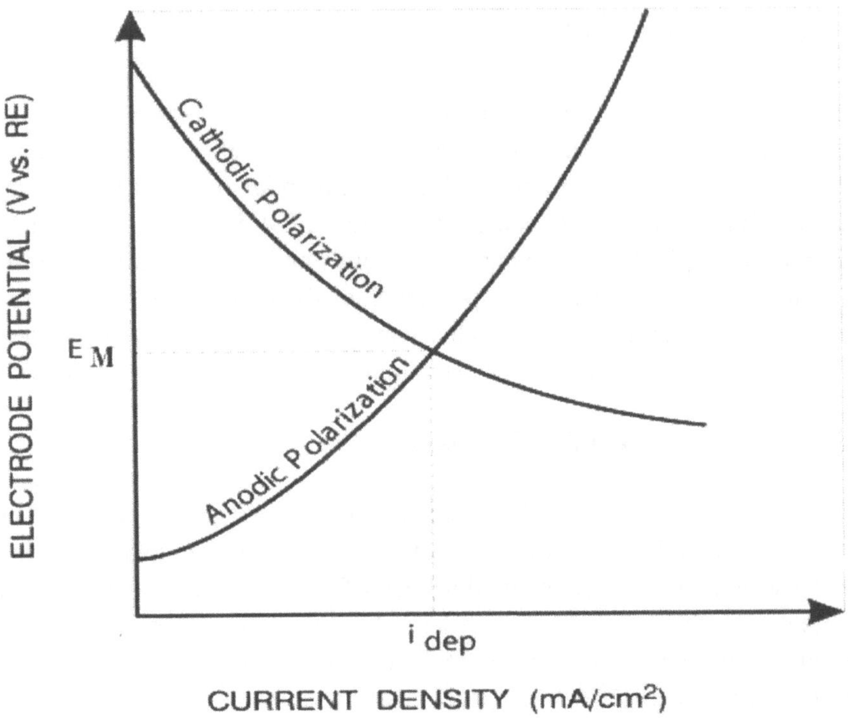

Figure 10.5. Schematic diagram of partial cathodic and partial anodic polarization curves in terms of mixed potential theory

The electrochemical mechanism does not explain reduction of metal ions in the bulk solution (*i.e.* without the presence of a metallic substrate). It also does not explain the reduction of metal hydroxides (formed as precipitates) to a metallic state. As experimental results show, the presence of any metallic surface is not a sufficient condition to start electroless deposition.

In terms of mixed potential theory, electroless deposition was first described by Paunovic. [12] According to this theory, electroless metal deposition can be considered as the superposition of anodic and cathodic curves crossing at the mixed potential, E_M. Electroless deposition of metals takes place at the mixed potential. The mixed potential, E_M, and the deposition current, i_{dep}, are obtained by the intersection of the partial anodic and cathodic polarization curves, as it is schematically shown in Figure 10.5.

This theory predicts that the rates of anodic reactions do not depend on the cathodic processes occurring simultaneously at the cathodic surface. The rates of separate reactions (anodic and cathodic) depend only on the mixed potential at which they have the same values.

By the applying the mixed potential theory it was suggested that the mechanism can be predicted from the polarization curves for the partial processes. However, the extrapolation of partial polarization curves and application of the mixed potential theory is not often realized, since the two partial processes are independent of each other. [13]

10.3.4 Metal Hydroxide Mechanism

The metal hydroxide mechanism was originally proposed by Salvago and Cavallotti. [14] This mechanism can be described briefly by the following scheme.

At the catalytic Ni surface, the ionization of water takes place according to the reaction:

$$H_2O = H^+ + OH^- \tag{10.24}$$

Hydrolysis of Ni^{2+} and formation of hydroxo-complexes takes place as follows:

$$Ni^{2+} + OH^- = NiOH^+_{ads} \tag{10.25}$$

$$Ni^{2+} + 2OH^- = Ni(OH)_{2,(ads.)} \tag{10.26}$$

and

$$Ni(OH)^+_{ads.} + H_2O = Ni(OH)_{2,aq.} + H^+ \tag{10.27}$$

The hypophosphite ions interact directly with hydrolyzed species, as is indicated below:

$$Ni(OH)_2 + H_2PO_2^- = NiOH + H_2PO_3^- + H \tag{10.28}$$

$$NiOH + H_2PO_2^- = Ni + H_2PO_3^- + H \tag{10.29}$$

Deposition of phosphorus is explained in terms of the reaction:

$$Ni_{cat.} + H_2PO_2^- = P + NiOH_{ads.} + OH^- \tag{10.30}$$

The hydrolyzed Ni(I) species interact directly with water:

$$NiOH + H_2O = Ni(OH)_2 + H \tag{10.31}$$

Evolution of hydrogen can is explained as:

$$H + H = H_2 \tag{10.32}$$

or by the reaction of hypophosphite ions with water:

$$H_2PO_2^- + H_2O = H_2PO_3^- + H_2 \tag{10.33}$$

According to the reactions (10.29) and (10.30) Savago and Cavallotti[14] explained lamellar morphology of electroless NiP deposits. It is obvious that any periodicity between the reactions (10.29) and (10.30) will produce deposits having layers richer with P, and then layers richer with Ni (lamellar morphology). Cavallotti and Savago reported that when nickel hydroxide is precipitated, inhibition phenomena are evident.

Experimental results, obtained for electroless deposition of Co, using hydrazine as the reducing agent, support the metal hydroxide mechanism.[9] The observations in this work[9] support the metal hydroxide mechanism as a means of explaining the electroless deposition of Co by hydrazine. Hydrolyzed species of Co^{2+} can react directly with hydrazine producing metal powder. The reaction occurs in bulk electrolyte, when a precipitate of cobalt hydroxide is formed and production of Co powder takes place. The SEM micrographs show that the Co powder produced under these conditions was dendritic in terms of surface morphology (Fig. 10.6).

Figure 10.6. SEM micrograph of Co powder obtained during electroless deposition of Co with hydrazine.[9] (Reproduced by the permission of The Electrochemical Society, Inc.)

XRD patterns in Figure 10.7. show that the Co powder contained the 70 % Co having the hcp structure and 30 % Co the fcc. Although the metal hydroxide mechanism explains most of the characteristics of electroless Co deposition by hydrazine, particularly the reduction of precipitated cobalt hydroxides and deposition of dendritic Co powder, there are still some doubts about this mechanism. For example, it does not explain the oxidation of N_2H_4 at a Pd-activated surface in Co(II)-free solutions. In order to explain deposition of shiny and smooth coatings at flat surfaces, Djokić [9] suggested that contributions from both the electrochemical and the metal hydroxide mechanisms should be considered.

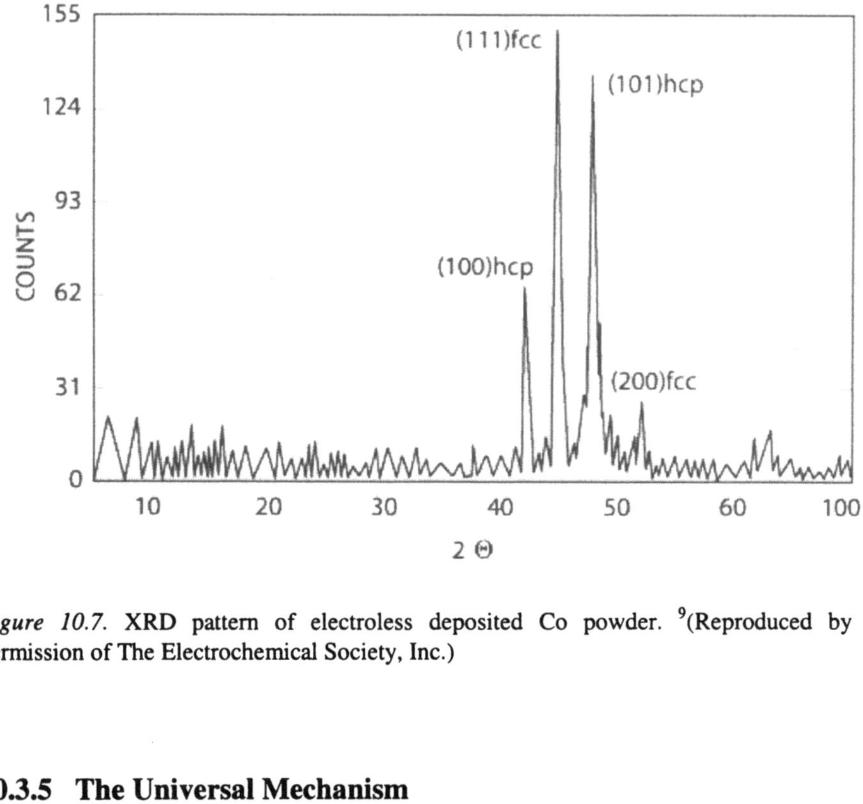

Figure 10.7. XRD pattern of electroless deposited Co powder. [9](Reproduced by the permission of The Electrochemical Society, Inc.)

10.3.5 The Universal Mechanism

Based on similarities among electroless processes, van den Meeraker proposed a mechanism that accounts for both the electrochemical and the catalytic nature of the process.[15] This mechanism was developed according to the following features, which are common to different electroless systems:

(a) The electroless deposition process proceeds only on certain catalytic metals that are known as effective hydrogenation-dehydrogenation catalysts;

(b) Electroless deposition is always accompanied by evolution of hydrogen gas;

(c) Poisons for hydrogenation-dehydrogenation reactions, such as thiourea and mercaptobenzothiazole, act as stabilizers in practically all electroless processes; and

(d) The deposition rate increases with an increase in pH.

The reactions taking place during electroless deposition were described as follows:

Anodic:

Dehydrogenation: $RH = R + H$

Oxidation: $R + OH^- = ROH + e^-$

Recombination: $H + H = H_2$

Oxidation: $H + OH^- = H_2O + e^-$

Cathodic:

Metal Deposition: $M^{n+} + n\,e^- = M$

Hydrogen Evolution: $2H_2O + 2e^- = H_2 + 2OH^-$

In this scheme, RH represents the reducing agent. It dissociates to a radical R and atomic hydrogen. The electrons for reduction of metal ions are supplied by the oxidation of R and/or reaction of H with OH^-.

The universal mechanism is not adequate for the explanation of all electroless processes. It fails to explain electroless deposition of metals on non-conductive surfaces, and also deposition of metal particles in solution.

The proposed mechanisms explain most of the characteristics of electroless deposition. However, as discussed above, there are some characteristics, which cannot be explained using these mechanisms. It seems that major problems arise when attempting to generalize the proposed models for electroless deposition. A more realistic approach would be to look for specific reactions, for particular conditions and substrates. It is very unlikely, in spite of the similarities of electroless processes, that a general mechanism will be developed explaining features for all electroless deposition of metals.

10.4 APPLICATIONS AND PROPERTIES OF ELECTROLESS DEPOSITED FILMS

Development of electroless deposition of metals and alloys in the past years has been remarkable and still continues. This process was investigated for various applications such as magnetic disks, printed circuits, selective plating on semiconductors, batteries, medical devices, etc. Most of these applications are related to electroless deposition of copper or nickel. Considering the fact that many of these applications use similar approaches and in a way overlap, the further discussion in this section is presented as follows:

1. metallization of non-conductive surfaces
2. electroless deposition of composite coatings
3. electroless deposition of gold and other metals

10.4.1 Metallization of Non-Conductive Surfaces

Metallization of non-conductive surfaces (polymers, ceramics and glass) requires specific treatments prior to electroless plating. Usually, these surfaces are first etched, then sensitized by a simple immersion in a $SnCl_2$/HCl solution. During the sensitization process, the adsorption of Sn^{2+} ions takes place. Senzitized surfaces are then exposed to a solution containing $PdCl_2$ and HCl. This process is called activation. The activation process can be described by the following equation:

$$Sn^{2+} + Pd^{2+} = Sn^{4+} + Pd^{o} \tag{10.34}$$

The Pd sites formed during the activation step allow chemical deposition of Ni or Cu. In some cases, sensitization and activation steps are combined in one step. In other words, solutions for the sensitization and activation are combined and represent mixtures of $SnCl_2$, $PdCl_2$ and HCl.

Other processes recommended for the activation of non-conductive surfaces for metal deposition from aqueous solutions (electroless deposition or electrodeposition) are carbon/graphite systems, conductive polymers, and non-formaldehyde based electroless processes.

For applications where metallization of advanced devices on non-conductive substrates takes place, Cu, Ni, Ag etc. are deposited by chemical or physical vapor deposition. These substrates are then ready for further metallization by the electroless deposition.

Metallization of polymers and plastics has also attracted significant attention from researchers because of its various industrial applications. Polymers of interest included polyimides, polystyrene, polycarbonates, polyetherimides, ABS plastics, aramid fibres, fluoropolymers etc.

10.4.2 Electroless Deposition of Composite Coatings

Codeposition of solid, inert particulates within a metal matrix during electroless deposition of that metal matrix (single metal or alloy) leads to production of composite coatings. In typical composite coatings, the fine particulates' diameter size ranges from 0.1 to 10 μm. [2] Their content in the coating can exceed up to 40 vol.%. The metal matrix in this class of composite coatings is usually electroless NiP or NiB. These materials are used for the improvement of wear and corrosion resistances, friction coefficient and hardness. Inert particulates, depending on applications, may include chromium carbide, alumina, titanium carbide, silicon carbide, boron carbide, diamonds, PTFE etc.

There is renewed interest in coatings with exceptional hardness, wear and friction properties for automotive and other mechanical applications. High performance coatings include alloys of cobalt and tungsten, composites with fluoropolymers etc. These coatings have potential for replacement of hard chromium.

Commercial acceptance of composite coatings has increased in a number of applications, especially since productivity, quality and environmental concerns continue to expand at increasing rates.

10.4.3 Electroless Deposition of Gold

Electroless deposition of Au is used for applications in the electronics industries (deposition on semiconductors and circuit patterns), as well as for decorative purposes. Solutions for electroless deposition and properties of deposited Au films were reviewed recently. [16]

The classical electroless Au solutions utilize $KAu(CN)_2$ as a source of Au and KBH_4 or DMAB as reducing agents. These solutions are autocatalytic, and it is possible to deposit sufficiently thick layers of gold.

In order to replace cyanide-based solutions, two different systems for electroless deposition of Au were developed in recent years. Both use gold thiosulfate as their source of gold. [17] The main difference is the reducing agent. The first system uses thiourea, while the second uses ascorbic acid.

Deposition of Au using ascorbate as the reducing agent, is explained as the combination of an anodic reaction (oxidation of ascorbic acid):

$$C_6H_8O_8 = C_6H_6O_6 + 2H^+ + 2e^- \qquad (10.35)$$

coupled with the cathodic reaction (reduction of Au(I) to Au):

$$Au(S_2O_3)_2^{3-} + e = Au + 2SO_3^{2-} \qquad (10.36)$$

In deposition of Au with thiourea as the reducing agent, the anodic reaction is:

$$CS(NH_2)_2 + 5H_2O = CO(NH_2)_2 + H_2SO_4 + 8H^+ + 8e^- \qquad (10.37)$$

while the cathodic reaction is described by equation (10.36).

Other developments include improvements of non-cyanide solutions (gold(I) thiosulfate with ascorbic acid as a reducing agent) for the electroless deposition of gold, which prevents the formation of any precipitates during the storage of the bath[18], and solutions with chelating agents, such as diethylenetetraaminepentaacetic acid[19], dimethylamine[20], etc.

10.4.4 Electroless Deposition of Other Metals

Other metals, studied from the aspect of electroless deposition, include Ag, Sn, Sb-Pb, Bi, Sn-Bi solder, Pd, Ni-Sn-P alloys, etc.

It is first reported by Rutkevich *et al.*, that Bi can be reduced by Ti(III) complexes in an autocatalytic mode. [21] The main characteristics of the process are explained in terms of pure electrochemical mechanism. The reduction of Bi(III) to Bi is represented as:

$$Bi^{3+} + 3Ti^{3+} = Bi + 3Ti^{4+} \tag{10.38}$$

The authors also claimed a possibility of using Ti(III) complexes to reduce Ni(II) and Co(II), as well as application of V(III) for autocatalytic Cu(II) reduction.

NiSnP and NiSnB alloys are deposited from alkaline solutions containing $Na_3C_6H_5O_7$ as a complexing agent, using sodium hypophosphite and DMAB as reducing agents, respectively. For electroless deposition of NiSnP a source of Sn is Na_2SnO_2; and for NiSnB, a source of Sn is Sn(IV) gluconate.

The maximum contents of Sn in the deposit are estimated at 30 at.% for NiSnP and 42 at.% for NiSnB alloys. The crystallinity of alloys increases as the Sn content increases.

In spite of development of other competitive technologies, it is obvious that applications of electroless deposition of metals and alloys will continue to grow in the future. More work is required to understand fundamental issues related to the reaction mechanisms of electroless deposition, the influence of processing parameters on properties of deposited coatings, etc. This knowledge is needed to ensure the successful operation of the process.

The new electroless deposition-based technologies should improve selectivity, satisfy quality requirements of deposited coatings, and assure the consistency of the process. The environmental concerns related to the solutions used for electroless deposition must also be investigated, in order to develop environmentally-friendly technologies, and to allow successful competition with other available processes.

10.5 FURTHER READINGS

1. Djokić S.S. Cementation of Copper onto Aluminum in Alkaline Solutions. J. Electrochem. Soc. 1996; 143:1300-1305
2. Mallory G.O., Hajdu J.B., eds. *Electroless Plating: Fundamentals and Applications* Orlando, Florida: AESFS, 1990.
3. Gawrilov G.G., *Chemical (Electroless) Nickel Plating*, Surrey, U.K.: Portcullis Press Limited, 1979.
4. Chassaing E., Cherakaoni M., Srhiri A. Electrochemical Investigation of the Autocatalytic Deposition of Ni-Cu-P Alloys. J. Appl. Electrochem., 1993; 23:1169-1174.
5. Gorbunova K.M., Ivanov M.V, Moiseev V.P. Electroless Deposition of Nickel Boron Alloys. J. Electrochem. Soc., 1993; 120:613-618.
6. F. Pearlstein F, Weightman R.F. Electroless Cobalt Deposition from Acid Bath. J. Electrochem. Soc., 1974; 121:1023-1028.
7. Dumesic J., Koutsky J.A., Champan T.W. The Rate of Electroless Copper Deposition by Formaldehyde Reduction. J. Electrochem. Soc, 1974; 121:1405-1411.
8. Saranov E.I, Bulatov K.N., Lundin A.B. Deposition of Thin Cobalt Coatings by Autocatalytic Reduction of Its Salts with Formaldehyde, Zashch. Metall., 1970; 6:612-614.
9. Djokić S.S. Electroless Deposition of Cobalt Using Hydrazine as a Reducing Agent. J. Electrochem. Soc., 1997; 144:2358-2363.
10. Lukes M. The mechanism for the Autocatalytic Reduction of Nickel by Hypophosphite Ion. Plating, 1964; 51:969-971.
11. Ivanovskaya, T.V., Gorbunova K.M. On the Mechanism of Catalytic reduction of Metals by Hypophosphite. Zashch. Metall., 1966;2:477-481.
12. Paunovic M, Electrochemical Aspects of Electroless Deposition of Metals. Plating, (1968); 51:1161-1167.
13. Donahue F.M. Interactions in Mixed Potential Systems. J. Electrochem. Soc. 1972; 119:72-74.
14. Salvago G, Cavallotti P. Characteristics of the Reduction of Nickel Alloys with Hypophosphite. Plating. 1972; 59(7):665-671.
15. van den Meerakker J.E.A.M. On the Mechanism of Electroless Plating. II One Mechanism for Different Reactions. J. Appl. Electrochem. 1981; 11:395-400.
16. Okinaka Y, Osaka T. "Electroless Deposition Processes: Fundamentals and Applications" In *Advances in Electrochemical Science and Engineering*, Gerischer H. and Tobias C., eds., p.55-116, New York: VCH Publishers Inc., 1994.
17. Kato M., Yazawa Y., Okinaka Y., Proceedings of AESF SUR/FIN'95, Annual Technical Conference Electroless Gold Plating Bath Using Ascorbic Acid as Reducing Agent – Recent Improvements. pp. 805 – 813, 1995 June 26-29; Baltimore, MD: AESFS Orlando, FL, 1995.
18. Kato M., Yazawa Y., Hoshino S. Electroless Gold Plating Solution. USA Patent (1995); 5,470,381.
19. Wachi H, Otani Y. Electroless Gold Plating Solution. USA Patent (1997); 5,660,619.
20. Wachi H, Otani Y. Electroless Gold Plating Solution. USA Patent (1996); 5,560,764.
21. Rutkevich D.L., Shevchenko G.P., Svirido N.V., Osipovich N.P. Electrochemical Study of Autocatalytic Bismuth (III) Reduction with Ti(III) – Complexes., J. Electrochem. Soc. 1993; 140:3473-3478.

Chapter 11

ELECTRODEPOSITION OF METALS FROM MOLTEN SALTS

Molten salts electrolysis is an attractive method for production of metals that cannot be deposited from aqueous solutions. Although there are many metals which can be deposited from fused-salt systems, on the industrial scale this method is used for the production of only a few metals.

Molten salts electrolysis is used for electrodeposition of alkaline and alkaline earth metals, Al, Ti, Zr, Ta, Mo, W etc. Among these metals, aluminum, magnesium, alkaline and some refractory metals have the most significant industrial importance from an electrometallurgy point of view.

Sodium is won by electrolysis of the fused hydroxide (Castner process), or a fused mixture of NaCl and $CaCl_2$ (Down process). At the present, in western countries the electrolytic production of calcium is completely replaced by the alumothermic processes. However, in China and Russia, it is believed that calcium is produced by the electrolytic method.

Magnesium is usually produced by electrolysis of the molten electrolyte containing alkali metal and magnesium chloride. In this process anhydrous $MgCl_2$ as the feed material is used, which requires a relatively high cost. More recently, electrowinning of magnesium from MgO in a melt containing neodymium chloride is suggested.[1] This process is based on the reaction of MgO with $NdCl_3$. Magnesium is then electrowon from a resulting $MgCl_2/NdCl_3/NdOCl$ melt. Aluminum is produced by the electrolysis of a molten cryolite, in which alumina (as a source of aluminum) is dissolved.

In Table 11.1 are presented some metals, electrowon from their fused salts. This table includes Li, Na, K, Mg, Mn and Al, although manganese is at the present mainly produced by the electrolysis of aqueous-based electrolytes.

Other metals that can be produced by the electrolysis from their fused salts include Be, Ti, Ta, Nb, W etc.

Procedures for deposition of beryllium involve electrolysis of $BeCl_2$ in a variety of fused salts. Tantalum or niobium can be produced by electrolysis of chloride-fluoride melts such as LiF-NaF-KCl-NaCl at 700 °C, in which K_2TaF_7 or K_2NbF_7 are added.[2,3] The refractory metals, such as for example tungsten, can be produced from $ZnBr_2$-NaBr or $ZnCl_2$-NaCl melts containing WBr_5 or WCl_6 at 350 to 450 °C.[4]

Table 11.1. Some metals produced by the electrolysis from their fused salts

Metal	Electrolyte	Cathode	Anode	T [°C]	CE[%]	Remark
Li	55 wt.% LiCl 45 wt.% KCl	Carbon Steel	Graphite	~460	80-85	
Na	Fused NaOH	Copper	Nickel	320	40-50	Castner Process
	42 wt.% NaCl 58 wt.% $CaCl_2$	Steel	Graphite	580-600	80-90	Down Process
Mg	$MgCl_2$, KCl, NaCl	Steel	Graphite	~700	75-90	
Mn	MnO, CaF_2, CaO	Steel	Carbon	1300	-	Discont. in (1985)
Al	Na_3AlF_6, Al_2O_3	Carbon	Carbon	960-1000	80-95	

Principally, the electrolysis of molten salts is very similar to the electrolysis from aqueous solutions. However, there are significant differences.

Decrease in the current efficiency during electrowinning from fused salts is often observed, and attributed to the following factors:

(i) evaporation of the products of electrolysis,
(ii) appearance of secondary reactions,
(iii) dissolution of metals in fused salts, and
(iv) dissolution of anodic products in fused salts.

In order to reduce the evaporation of the products of electrolysis, the process is usually carried out at lowest possible temperatures. The appearance of secondary reactions may be avoided using proper cell design and proper electrode materials. The most significant decrease in the current efficiency is related to the dissolution of the cathodic and anodic products in the melt, their diffusion into the bulk electrolyte, formation of intermediate compounds and oxidation of metal dissolved with the oxygen from air.

Molten salts electrochemical studies include aspects of melt handling, cell design, materials choice, selection of electrodes, properties of melt, its purification etc. Purification of melts is a very important operation and it always depends on the nature of a melt involved in the study. For example, water is a very critical contaminant of lithium and magnesium based melts. It is usually removed by long-term drying over P_4O_{10} followed by pumping, but only in the range of low concentrations. On the other hand P_4O_{10} could be a very dangerous contaminant of alumina-cryolite melts. With higher concentrations of water more sophisticated techniques are required. Another, commonly used pretreatment is pre-electrolysis. This step is usually applied for removal of some heavy metals and is carried out carefully to avoid undesirable side reactions of some anionic impurities.

Materials employed in studies or industrial processes are selected on the basis of the nature of a melt and operating conditions. The use of dried and deoxygenated, inert atmosphere is a general requirement in order to reduce oxidation and corrosion as well as to prevent the generation of electroactive oxygen-containing species in the melts. Crucibles are usually made of ceramic materials, graphite and refractory metals.

Reference electrodes in molten electrolytes have considerable problems related to the very strong ionic interactions at high temperatures. Consequently, redox series can only be defined for single melts.

Since metals are electrodeposited on cathodes, the choice of cathodes is an important factor for normal electrolysis. In the many instances, various carbon materials or stainless steel are used as the cathode. Studies on electrodeposition of metals from fused salts at electrodes such as platinum, gold, tungsten and molybdenum have also been performed in spite of difficulties associated with their use.

In the electrodeposition of metals molten salts, counter electrodes (anodes) are very important theoretically and practically. They will be discussed later in a separate section, but it is to be noted that a choice of an anode material depends on the melt composition and electrolysis conditions. Anode materials used for this purpose include various types of graphite materials, platinum, gold, refractory metals etc.

11.1 IONICALLY CONDUCTING MELTS

Electrolytic conductivity of molten salts is a very important property from both theoretical and practical points of view. This information is useful for a better understanding of the mechanisms of the transport processes and electrodeposition of metals from fused salts, and also for the reduction of ohmic drop through the electrolyte and an improvement of the efficiency of the electrowinning process.

The electrical conductivity, κ, of a molten electrolyte is related to its resistance, R, according to the expression that follows:[5]

$$\kappa = \frac{1}{R} G \tag{11.1}$$

where G is the cell constant defined as the ratio of length to area. In an uniform electric field with a potential gradient of 1 V/cm, the specific conductivity is determined by the concentration of each ionic species (c_i), the charge of each ion (z_i) and the mobility (u_i), i.e.,

$$\kappa = \frac{F}{1000} \sum c_i z_i u_i \tag{11.2}$$

where F is the Faraday's constant. In the case of binary, fully dissociated electrolyte, κ can be written as:

$$\kappa = \frac{F}{1000} cz(u_{cation} + u_{anion}) \tag{11.3}$$

This equation can be rearranged to:

$$\frac{1000}{cz} \kappa = (\lambda_{anion} + \lambda_{cation}) = \Lambda \tag{11.4}$$

where λ_{anion} and λ_{cation} are the ionic conductivities of the anion and cation, and Λ is the equivalent conductivity of the melt. The electrical conductance depends on temperature according to the basic Arrhenius type equation:

$$\kappa = A_\kappa \exp(-E_\kappa/RT) \tag{11.5}$$

$$\Lambda = A_\Lambda \exp(-E_\Lambda/RT) \tag{11.6}$$

The relationship between E_κ and E_Λ is given as:

$$E_\Lambda = E_\kappa + RT^2 \alpha \tag{11.7}$$

where α is the coefficient of expansion. This equation shows that the two forms of the activation energy are equal only for systems with very small temperature coefficients of expansion (for example silicate melts) or at relatively low temperatures (the maximum temperature suggested is about 125 °C.

Measurements of electrical conductivity are similar to those in aqueous solutions, however it should be noted that temperature and corrosion resistant materials must be used. Combined with other transport properties (viscosity, diffusivity, thermal conductivity) and their temperature coefficients, the electrical conductivity is of particular value for postulating mechanisms for the transport processes in terms of lattice geometry, molecular force fields and molecular motion.

11.2 ELECTROCHEMICAL STUDIES IN ELECTRODEPOSITION FROM MOLTEN SALTS

The electrochemical measurements are not basically different from those in the aqueous solutions. However, practical difficulties often arise from the fact that these systems are very corrosive and that measurements should be carried out at relatively high temperatures, above 600 °C for magnesium chloride melts and about 1000 °C for the alumina-cryolite melts.

Electrochemical studies of molten salts systems have led to determination of important parameters such as reversible potentials, diffusion coefficients etc., which advance fundamental understanding of molten salts electrolysis and facilitates process control. However, there are some discrepancies in the published results. These discrepancies arise due to extremely difficult experimental conditions (high temperature and very corrosive environment).

Electrochemical techniques are easily applied for studies of molten salts since they have very low ohmic resistance, the interference from surface-active materials is relatively small and the rates of the charge transfer reactions and secondary chemical processes are high. Most of the reactions in the molten salts systems are under mass-transfer conditions.

Electrochemical studies in electrodeposition from molten salts are very useful for obvious reasons. They allow an in-depth understanding of anodic and cathodic reactions involved in the process. The knowledge gained through these measurements is important for a better maintenance and control of industrial cells, an increase in the process efficiency, etc..

Almost all electrochemical techniques developed in the aqueous solutions, with specific adjustments, have been used in the molten salt electrochemistry for studying electrodeposition of metals. Among these techniques, steady state, open circuit potential, chronoamperometry, chronopotentiometry, cyclic voltammetry, a.c. impedance etc., should be mentioned. The application of these methods requires specific attention on choice of materials, which have to be able to withstand extremely corrosive conditions and elevated temperatures. These electrochemical processes have been used in studying of both, cathodic and anodic processes.

11.2.1 Electrolysis of Alumina-Cryolite Melts

Taking into consideration the fact that the most important metal produced by the electrodeposition from molten salts is aluminum, further discussions in this chapter are mainly restricted to the achievements related to this field.

The principal basis for production of aluminum by the Hall-Héroult process can simply be described in terms of the following reactions, for which the respective thermodynamically calculated E^o values are indicated as shown:

$$Al_2O_3 + 3/2C \rightarrow 2Al + 3/2CO_2 \tag{11.8}$$

$E^o = 1.163$ V at 1010 °C

and

$$Al_2O_3 + 3C \rightarrow 2Al + 3CO \tag{11.9}$$

$E^o = 1.024$ V at 1010 °C

The electrolyte consists of fused cryolite (Na_3AlF_6) in which alumina (Al_2O_3) in the concentration range from 2 to 6 wt.% is dissolved. The melt always contains about 4 to 8 % CaF_2, arising from low levels of calcium oxide impurities in the alumina. The concentration of Al_2O_3 in the melt is adjusted automatically in the smelter cell by means of periodic mechanical-feed system. The relative ratio of molar contents of NaF and AlF_3 in the electrolyte is the so-called cryolite ratio. Industrial cells operate at cryolite ratios between 2 and 3 and temperatures from 940 to 980 °C.

Ionic conductivity of alumina-cryolite melts arises from the migration of Na^+ toward the cathode and complex alumofluoride anions toward the anode.[6] There is no evidence of existence of any uncomplexed aluminum cations.

The primary ionization of cryolite melt is represented by the following equations:

$$Na_3AlF_6 \rightarrow 3Na^+ + AlF_6^{3-} \tag{11.10}$$

$$AlF_6^{3-} \rightarrow AlF_4^- + 2F^- \qquad (11.11)$$

The degree of dissociation of hexafluoraluminate ion at the melting point of cryolite is about 0.3. In this way, cryolite melts contain mainly Na^+, AlF_6^{3-}, F^- and AlF_4^-. The addition of alumina to a cryolite melt and a consequent Al_2O_3 dissolution cause properties of the melt such as surface tension, vapor pressure, electrical conductivity and density to change rapidly, with a tendency to decrease while the viscosity increases with increase of alumina content.

The nature and the number of the ionic species formed in the solution of Al_2O_3 in cryolite are still incompletely understood, although this matter has been studied by means of a variety of modern physico-chemical methods.[7] The complex oxyfluoroaluminates, with a general formula $Al_xO_yF_z^{(3x-2y-z)+}$ are probably formed in the dissolution process of alumina in cryolite, rather than simpler aluminates such as $Al_xO_y^{(3x-2y)+}$.

The effective cathodic reaction in the electrolysis of alumina-cryolite melt is reduction of Al^{3+} ions and production of aluminum metal:

$$Al^{3+} + 3e^- \rightarrow Al \qquad (11.12)$$

with the following side reactions:

$$Al^{3+} + 2e^- \rightarrow Al^+ \qquad (11.13)$$

$$Al^+ + e^- \rightarrow Al \qquad (11.14)$$

The real situation is, however, much more complicated. Taking into account the fact that the free Al^{3+} ions are not present in the melt as well as the fact that Na^+ ions are the principal current carriers it seems that the discharge of Na^+ could be the primary process at the cathode:

$$Na^+ + e^- \rightarrow Na \qquad (11.15)$$

$$2Na^+ + e^- \rightarrow Na_2^+ \qquad (11.16)$$

and discharge of aluminum metal results from secondary reactions. In industrial cells, however, the cathodic product is mainly aluminum, with sodium present at very low activity. It appears that the cathodic process can be considered as a reversible three-electron transfer. One possible mechanism for the cathodic discharge of Al^{3+} ions is based on the hypothesis that this process is proceeded by dissociation:

$$AlF_6^{3-} \rightarrow Al^{3+} + 6F^- \tag{11.17}$$

$$AlF_4^- \rightarrow Al^{3+} + 4F^- \tag{11.18}$$

In the existing literature, however, there is no evidence for a chemical reaction preceding electron transfer. Other three-electron-transfer processes are also possible, such as, for example, a process involving oxyfluoroaluminate ions. In Houpins and Frak's opinion[6], the most probable cathodic reactions are:

$$AlF_4^- + 3e^- \rightarrow Al = 4F^- \tag{11.19}$$

and

$$AlF_6^{3-} + 3e^- \rightarrow Al + 6F^- \tag{11.20}$$

Anodic processes in the electrolysis of alumina-cryolite melts include the discharge of oxygen-containing species and consequent formation of CO and CO_2. The primary gas found at carbon anodes is CO_2, although chemical analysis shows that only about 60 % of the gaseous products is CO_2. This is a result of secondary process that arise on the account of the reaction of aluminum metal with CO_2 forming Al_2O_3 and CO.

The anode processes probably involve electrosorptive formation of oxygen-carbon compounds to CO and CO_2, and their desorption from the electrode surface. In general, in the intermediate compounds C_xO_y, the ratio x/y is a function of time, temperature, nature of the carbon anode material, current density etc. The rate of CO formation is slow so that, at commercial current densities, the composition of anodically produced gases approaches about 100 % CO_2. The following anodic reactions are suggested:

a) at low alumina concentrations and high cryolite ratios:

$$2Al_2OF_6^{2-} + 2AlF_6^{3-} + C \rightarrow 6AlF_4^- + CO_2 + 4e^- \tag{11.21}$$

b) at low alumina concentrations and low cryolite ratios:

$$2Al_2OF_6^{2-} + 4F^- + C \rightarrow 4AlF_4^- + CO_2 + 4e^- \tag{11.22}$$

c) at high alumina concentrations and low cryolite ratios:

$$AlOF_2^- + 4F^- + C \rightarrow 2AlF_4^- + CO_2 + 4e^- \qquad (11.23)$$

d) at high alumina concentrations and high cryolite ratios:

$$2AlOF_2^- + 2AlF_6^{3-} + C \rightarrow 4AlF_4^- + CO_2 + 4e^- \qquad (11.24)$$

With increase in current density and/or decrease in alumina concentration (in other words, oxygen-containing ionic species), the anode becomes passivated, leading to discharge of fluoride anions according to the reaction:

$$4F^- + C \rightarrow CF_4 + 4e^- \qquad (11.25)$$

and/or

$$4AlF_6^{3-} + 3C \rightarrow 4AlF_3 + 3CF_4 + 12e^- \qquad (11.26)$$

Investigation of the cathodic processes in the electrowinning of aluminum from alumina-cryolite melts has received far less attention than the anodic processes. This comes as a consequence of a general opinion that the cathodic reaction was considered to be simple. However, research in this field show that cathodic reactions are very complex and the exact mechanism has not yet been postulated. Cathode reactions in alumina-cryolite melts have been studied on molybdenum, platinum, graphite and molten aluminum electrodes, which has led to significant discrepancies in the published results. On industrial scale, carbon lining is exclusively used as a cathode material. Under these conditions, carbon linings exhibit significant disruption. This disruption is a consequence of the sodium intercalation and crystal growth of the electrolyte in the carbon. In order to avoid these problems, composite materials based on TiB_2, ZrB_2, Al_4C_3, TiC and similar, have been proposed. The requirements from those materials include physical properties such as low porosity, high electrical conductivity, excellent corrosion and chemical resistance under the production conditions and good wettability by the liquid aluminum. However, proper solutions have not been found by far, since carbon cathodes are still in use.

At 1000 °C, aluminum is for about 250 mV more positive than sodium. The sodium ions are principal carriers of the electricity, and as a consequence, the enrichment of NaF is observed in the vicinity of the cathode. This causes the appreciable diffusion overvoltage (from 50 to 400 mV, at current densities ca. 0.5 to 1.0 A/cm^2). The change in the overvoltage is observed with a change in the cryolite ratio. An increase in the cryolite ratio, causes a decrease in the cathodic overvoltage. These findings are related to the laboratory experiments.

In industrial cells, the diffusion overvoltage is significantly lower, due to convection. Thonstad estimated the cathodic overvoltage at about 100 mV. Due to enrichment of the cathodic diffusion layer with sodium fluoride, it is expected that electrowon aluminum metal contain higher concentration of sodium than the amount corresponding to that of the equilibrium data for the electrolyte (bulk melt). Most researchers have estimated that an average of sodium is about 80 ppm, which is below the equilibrium data.

At current densities, which are normally used in the production of aluminum, the primary gas evolved at the anode is CO_2. At lower current densities (below 0.05 A/cm^2) formation of CO in high contents may occur.

Anodic processes in the production of aluminum metal have received far more attention than the cathodic processes. This comes as a consequence of the complexity of anodic processes. However, despite the relatively active research that has gone on the study of these processes, there is no general agreement among published results nor between explanations of the behavior observed, which is often associated with poor reproducibility.

For the purpose of studying the anodic reactions involved in alumina-cryolite melts, the following electroanalytical procedures have been investigated: chronoamperometry, chronopotentiometry, cyclic voltammetry, impedance spectroscopy and related electrochemical methods.[8] Materials used for the study of anodic reactions include various types of carbon, platinum, gold and refractory metals.

The anodic overvoltage on various types of carbon electrodes in cryolite-alumina melts was studied by steady-state measurements. Tafel slopes and exchange current densities evaluated from these experiments depend on the nature of the carbon materials. The reported overvoltages are very high. At 1 A/cm^2 overvoltage values are 1.4 V, 1.0 V and 0.8 V for glassy carbon, graphite and baked carbon, respectively.[7] The overvoltage increases with decreasing porosity, which is attributed to a decrease in the wetted area.

The anodic overvoltage is higher on large anodes due to the shielding effect of gas bubbles. Dewing and van der Kouwe[9], as shown in Figure 11.1 for the graphite ATJ, found Tafel plots with slopes of 0.29.

As this figure shows, the exchange current varies from 1.5 mA/cm^2 for current densities below 0.1 A/cm^2, to 5 mA/cm^2 for current densities above 0.5 A/cm^2. The break between 0.1 A/cm^2 and 0.5 A/cm^2 is attributed to changing CO/CO_2 ratio in the gas generated. At lower current densities, low exchange current is due to adsorption of CO on the electrode surface. The CO, which is produced by reaction of CO_2 with dissolved aluminum, acts as a catalytic poison. The fraction of CO in the gas will depend on how fast CO_2 is being generated electrochemically. At higher current densities, more CO_2 is produced, which leads to a dilution of CO and to a higher exchange current. For baked carbons the same value of the Tafel slope is obtained, although it applies over a restricted range of current density and overvoltages are lower than for graphite anodes.

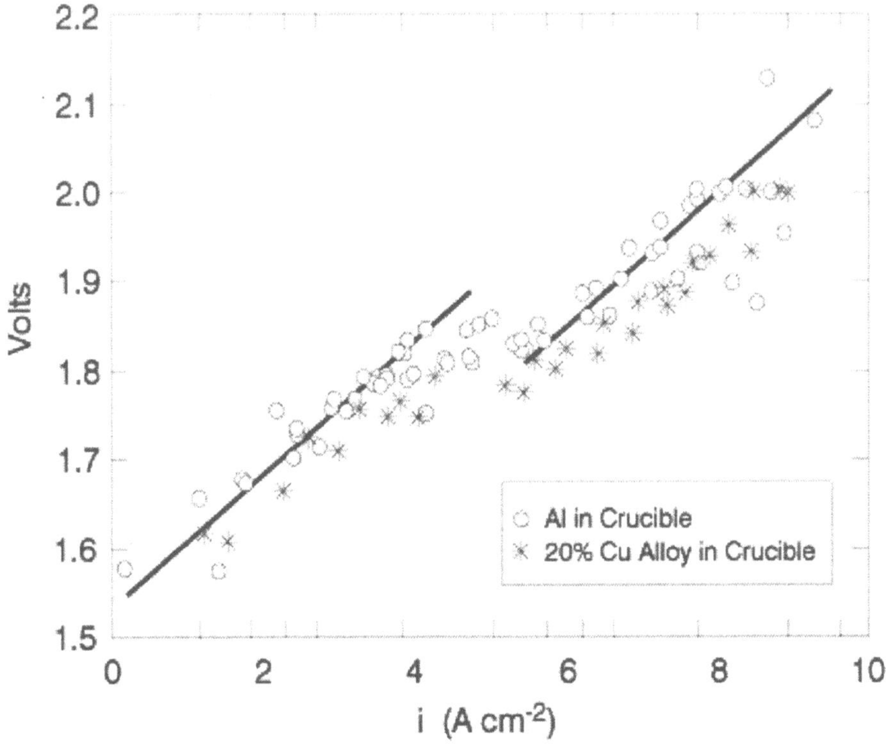

Figure 11.1. Tafel plot for ATJ graphite anode[9] (Reprinted with the permission of The Electrochemical Society Inc.)

The chronoamperometric method provides a known and convenient method for study of electrochemical reactions under diffusion control in aqueous solutions at room temperatures. Determination of the concentration of an electroactive species in chronoamperometric experiment is based on the well-known Cottrell equation:

$$I = \frac{nAFD^{1/2}C_o^{\infty}}{\pi^{1/2}t^{1/2}} \tag{11.27}$$

where I is the time-dependent current, C_o^{∞} is the bulk concentration, A is the surface area, F is the Faraday's constant and D is the diffusion coefficient of electroactive species. While this approach is well established in aqueous solutions for analytical purposes under conditions of semi-infinite linear diffusion to a planar electrode, much less work has been carried out on the corresponding problem in molten salts at higher temperatures, when many new serious experimental problems arise.

Anodic behavior in alumina-cryolite melts was studied on carbon, gold and platinum electrodes by means of chronoamperometry. At the graphite materials, the anode processes are not fully diffusion controlled, nor are the results adequately reproducible. At the glassy carbon electrode, anodic processes are diffusion controlled. This is illustrated in Figure 11.2. As this figure shows with increasing concentration of Al_2O_3 in the melt, the corresponding current function increases in a satisfactory linear way.

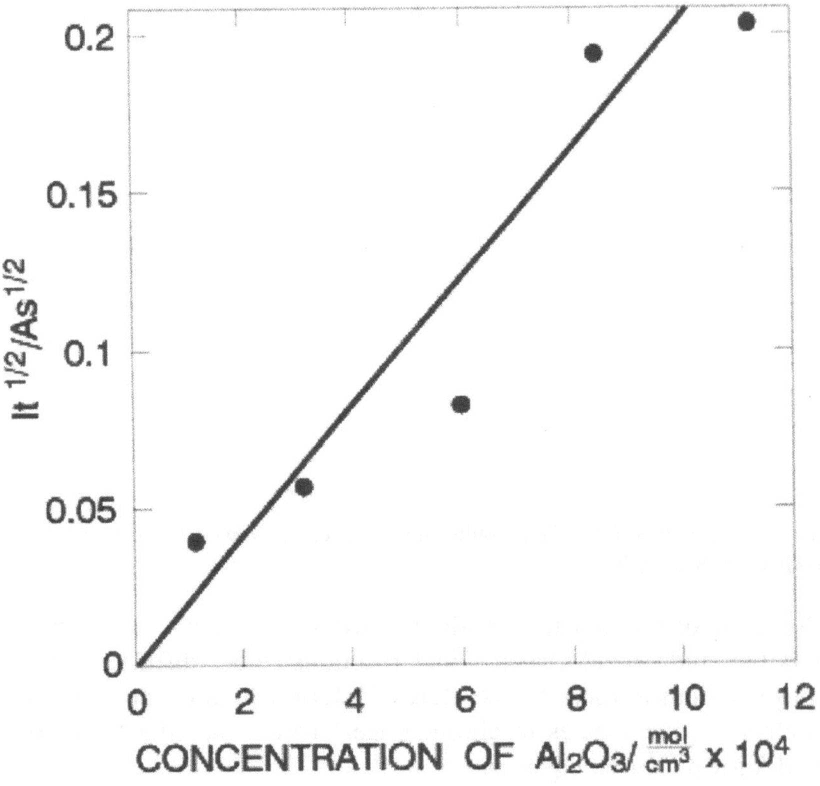

Figure 11.2. Dependence of $It^{1/2}$ on Al_2O_3 concentration (evaluated from chronoamperometry experiments) in alumina – cryolite melts for glassy carbon working electrode (1000 °C)[8] (Reprinted with the permission of The Electrochemical Society Inc.)

The departure from classically expected behavior at graphite materials is due to the reduction of the electrode surface in contact with the melt, which is caused by the build-up of evolved adherent gas bubbles or gas films at the electrode surface.

The results indicate kinetic difference between the anodic reactions of the graphites and glassy carbon. Evidently, the kinetic reactivity of glassy carbon for anodic oxidation in the melt is more facile than that for graphite where their electro-oxidation is not fast enough to become limited by diffusion control, implying a relative slow electrode process. This difference could be attributed to the relative stability of the structure of graphite associated with its multiple conjugated bonding in contrast to that in glassy carbon, especially on the basal-plane exposures of the graphite crystals of the materials.

At platinum electrode anodic reactions are limited by the formation of an oxide film, which obeys, kinetically, Wagner's parabolic growth law:

$$y^2 = Kt \tag{11.28}$$

where y is the film thickness, K is a constant depending on the diffusion coefficient oxygen-containing species. For platinum electrode the current response function does not depend on the Al_2O_3 concentration.

To diminish the effect of convection on the electrode-kinetic behavior of the electrochemical reactions, as well as to minimize disturbances at the interface due to evolution of gas, the method of fast cyclic voltammetry for study of the anodic processes in cryolite-alumina melts can be profitably used. Using very fast cyclic voltammetry researchers found four to five current peaks in the range 1 to 4 V. The shape and the peak current values in cyclic voltammograms depend on the kind of carbon material used as the working electrode in the investigation of anodic reactions in alumina-cryolite melt.

Typical examples of cyclic voltammograms obtained at different carbon materials, are presented in Figures 11.3. and 11.4. For the example of graphite ATJ, four distinguishable anodic current peaks appeared at approximately 1.1, 1.95, 2.45 and 3.3 V.

In the case of glassy carbon, on the negative going sweep curves of the voltammogram, between 2 and 3.2 V relatively high anodic currents were recorded (Figure 11.4.). The sharp decrease in the current at 3.5 V, and a subsequent absence of current flow on the reverse sweep between 4 and 3.2 V clearly indicate the occurrence of an "anode effect" (see later discussions). Around 3.2 V, the anode comes out of the "anode effect" permitting current to flow again. The discussion of the cyclic voltammetry results is restricted on the second peak at the several kinds of carbon anode materials investigated. Although the peak current increase with increase of the square root of the sweep rate for all the kinds of carbon investigated, these dependencies with the exception of that at glassy carbon, do not show expected linear relationship. At glassy carbon, however, the dependence of peak current on the square root of the sweep rate is linear for all concentrations of alumina. Furthermore, on glassy carbon electrode, the dependence of the peak current values on alumina content in the range 1.45 – 4.23 wt.% was observed to follow a satisfactory linear relationship as shown in Figure 11.5.[10]

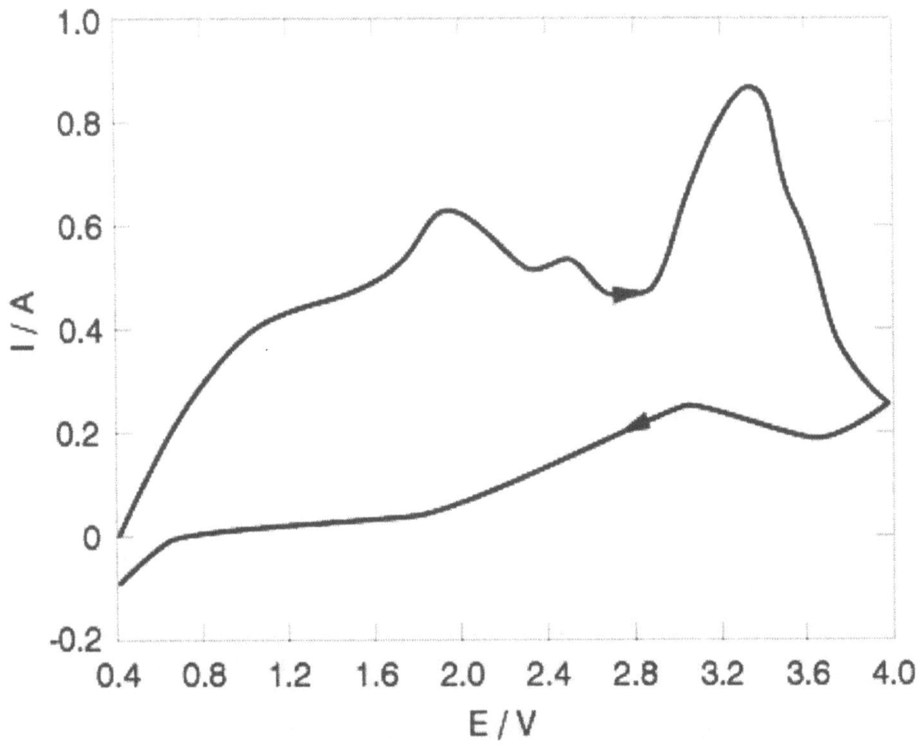

Figure 11.3. Cyclic voltammogram for anodic oxidation process at the graphite ATJ material (s=80 V/s, 30 g Al_2O_3 in alumina – cryolite melt, 1000 °C)[10]

This result indicates that the process is diffusion controlled under these conditions. At graphite materials the results were not reproducible, nor was there any observable linear relationship between current maxima values and Al_2O_3 concentration. The background current behavior also depends on the carbon material. In this way a material dependent modification of the expected ideal current response to alumina concentration is an additional but unavoidable complication.

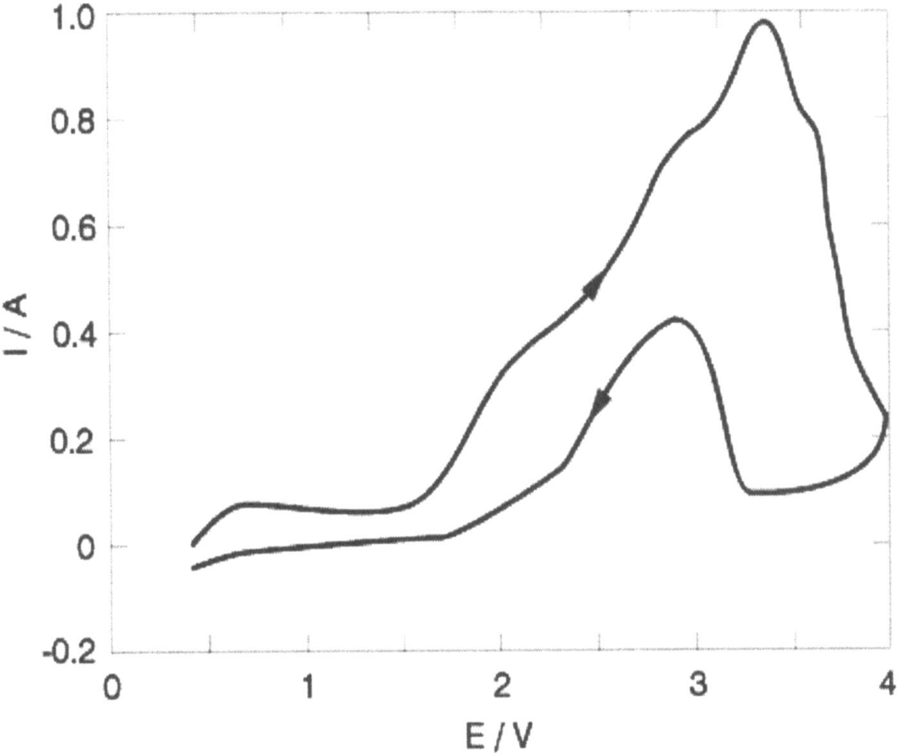

Figure 11.4. Cyclic voltammogram for anodic oxidation process at the glassy carbon (s=80 V/s, 30 g Al_2O_3 in alumina – cryolite melt, 1000 °C) [10]

The AC impedance method has also been applied to and investigated for the study of the anodic processes in the electrolysis of alumina-cryolite melts at carbon.[11] This method is used to investigate the discharge of the oxyfluoroalumnate ion at a graphite electrode. An increase of the applied anodic overvoltage leads to a variation of the shape of the complex-plane impedance diagram. For zero overvoltage with a residual alumina concentration of 0.46 wt.%, the plot of imaginary part of the impedance (*Z*'') versus the real part of the impedance (*Z*') follows a linear relationship having a 45 ° slope, which is recognized as diffusion impedance. With further increase of the overvoltage, the plots of *Z*' versus *Z*'' show inflections of the straight line and these are more pronounced as the overvoltages applied to the electrode are increased. An inductive loop appears at low frequencies. The change in the shape of the *Z*' versus *Z*'' plot with increase of the overvoltage was explained by the influence of reactions producing bubbles of gaseous compounds of oxygen with carbon which disturb and reduce the thickness of the diffusion layer.

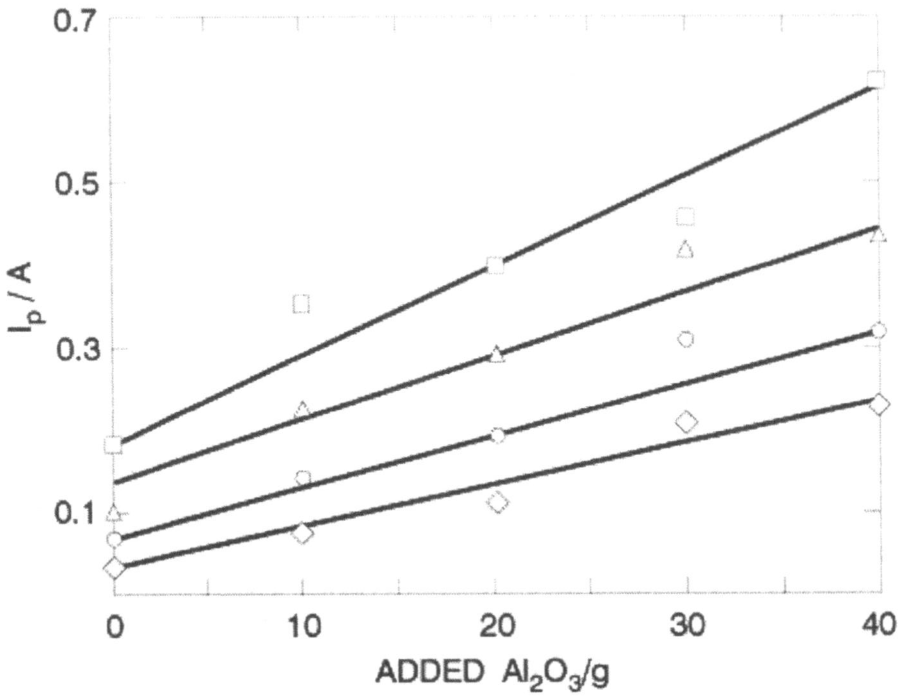

Figure 11.5. Dependence of peak current in cyclic voltammetry at glassy carbon on alumina content in the melt for different sweep s (potential range 0.4 to 2.25 V against aluminum reference electrode in the melt) (▲) 60, (●) 40, (◆) 20 and (■) 10 V/s. [10]

In spite of the intensive studies of alumina-cryolite melts by various electrochemical techniques, significant disagreements among results appeared in the published literature. These problems are consequences of nonreproducibility of results which arise due to the nature of the carbon sensor electrode materials used and, also due to a number of technical problems including the presence of dissolved aluminum metal in realistic practical systems.

11.3 THE ANODE EFFECT

The anode effect is a phenomenon that has been observed in many processes involving the electrolysis of molten salt. It is described as a blockage effect, which inhibits the current flow between the anode and the melt. Due to gas evolution, growth of bubbles and their coalescence occur covering most if not entire surface of the anode. In industrial cells during electrolysis of alumina-cryolite melt the anode effect manifests itself through an immediate increase of cell voltage from values between 4.1 and 4.3 V, during normal electrolysis to about 35 to 60 V, and sometimes even up to 130 V, depending on the current density.[7] The cell remains under the influence of the anode effect until the current is interrupted, which allows adherent gas bubbles formed at the anode surface to collapse or become detached. The effect is somewhat analogous to that observed in anodic F_2 evolution at carbon from $KF \cdot 2HF$ melts in commercial F_2 cell operation.

The reasons for appearance of the anode effect are not yet established. Chemical analysis of the anode gases shows that they contain up to 30 % fluorine compounds such as CF_4 and CF_6. The presence of fluorocarbon compound promotes dewetting of the anode surface and the growth of large bubbles. In industrial cells the anode effect arises when the alumina concentration in the melt is between 0.5 and 2 %. Thus, maintaining good control of alumina content is very important factor in avoiding the anode effect. Upon an occurrence of the anode effect, the crust on the top of the melt is broken and alumina is added.

Conditions for onset of the anode effect are associated mainly with the depletion of alumina concentration in the melt during electrolysis, increasing potential, and presence of fluorocarbon surface compounds at the carbon anode surface, causing dewetting of the anode by the electrolyte and adherence of gas bubbles.[12]

11.4 NONCONSUMABLE ANODE MATERIALS IN MOLTEN SALTS ELECTRODEPOSITION

In the electrowinning of metals from their molten salts, due to significant corrosion at higher temperatures and anodic reactions involved in the process, the graphite or other anode materials are often very easily consumed.

The search for nonconsumable, or inert anodes, has been an important research activity for a long time, especially in the electrolysis of alumina-cryolite melts, which will mostly be discussed here, although other molten salts, depending on their composition and conditions have attracted considerable attention. The requirements for materials, which could be used as anodes for in alumina-cryolte melts include resistances to attack by molten cryolite and oxygen, high electronic conductivity, mechanical strength and resistance to thermal shock. The possibility of using nonconsumable anodes in the electrowinning of alumina has become attractive for the following reasons:

(i) These anodes would not be consumed during electrolysis
(ii) The oxygen which would be formed at the anode could be utilized industrially

(iii) The problems related to contamination of the working environment, when the Hall-Héoult process is used could be reduced

(iv) The corresponding cell design would permit electrolysis with higher current efficiencies than is currently possible with carbon anodes.

(v) All above-mentioned factors could represent significant savings to the aluminum production industry.

In the case of nonconsumable anodes, the production of aluminum would be represented formally by the equation:

$$Al_2O_3 \rightarrow 3/2O_2 + 2Al \qquad (11.29)$$

(E^o=-2.19 V at 1010 °C)

In the search for an inert anode for use in the Hall-Héroult electrolysis, among many accessible materials, oxides, metals, refractory hard metals and gaseous fuel anodes have been investigated.[13]

Among the oxides, investigated as materials for anodes in electrolysis of an alumina-cryolite melt should be mentioned the cold pressed and sintered anodes of Fe_3O_4, SnO_2, Co_2O_4, NiO, CuO and Cr_2O_3, the ferrites such as $SnO_2 \cdot Fe_2O_3$ and $ZnO \cdot Fe_2O_3$, stabilized ZrO_2 as a possible inert material for production of less corrosion resistant anodes, the SnO_2 – based anodes and complex anodes Y_2O_3.

The Cu-containing cermets ($NiFe_2O_4$ spinel, NiO and metallic phase which is mostly Cu) have also been investigated as possible anode materials for the primary aluminum industry.

In the case of gaseous-fuel anodes, the production of aluminum metal is described by the following reactions:

$$Al_2O_3 + 3/4CH_4 \rightarrow 2Al = 3/2H_2O \qquad (11.29)$$

$$Al_2O_3 + 3H_2 \rightarrow 2Al + 3H_2O \qquad (11.30)$$

The hard refractory materials such borides, carbides and nitrides of the transition metals such as TiB_2, TiB_2-BN mixtures, TiC-ZrB_2, MoSi and TiCr should be mentioned.

Metal anodes such as copper, nickel, chromium, tungsten stainless steel and silver are unresistant in alumina-cryolite melts. On platinum and gold, formation of oxide films and/or corrosion occurs.

It seems that oxide materials (nickel ferrites type) are the most promising by far.[14] However, as research shows the solubility of alumina in cryolite melts is dependent on the content of alumina. In order to exhibit a slow dissolution of oxide anodes, the alumina concentration should be maintained at a relatively high level. The alumina content in the industrial cells with graphite anodes is maintained at 2 to 4 %. If the oxide type of anodes are to be used in the production, the content of alumina in the melt should probably be kept at higher levels, which should be determined by the additional studies.

11.5 FURTHER READINGS

1. Cathro K.J., Deutcher R.L., Sharma R.A. Electrowinning Magnesium from Its Oxide in a Melt Containing Neodymium Chloride. J. Appl. Electrochem. 1997; 27:404-413.
2. Polyakov E.G., Polyakova L.P., Elizarova I.R. Cathode Processes in Chloride-Fluoride Melts containing K_2NbF_7. Elektrokhimiya 1995; 31: 502-509.
3. Rosenkilde C, Vik A., Østvold T., Christensen E. Electrochemical Studies of the Molten System K_2NbF_7-Na_2O-Nb-$(LiF\text{-}NaF\text{-}KF)_{eut.}$ at 700 °C. J. Electrochem. Soc. 2000; 147: 3790-3800.
4. Katagari A., Suzuki M., Takehara Z. Electrodeposition of Tungsten in XnBr2-NaBr and ZnCl2-NaCl Melts. J. Electrochem. Soc. 1991; 138:767-773.
5. Janz G.J., Reeves R.D. "Molten Electrolytes – Transport Properties" In *Advances in Electrochemistry and Electrochemical Engineering*, Vol. 5, Tobias C.W., ed., pp. 13 – 171, New York: Interscience Publishers, 1967.
6. Haupin W.E., Frank W.B. "Electrometallurgy of Aluminum" In *Comprehensive Treatise of Eelctrochemistry*, Vol.2: Electrochemical Processing, Plenum Press, Bockris J.O'M., Conway B.E., Yeager E., White R.E., eds., pp. 301 – 325, New York: Plenum Press, 1981.
7. Grjotheim K., Krohn C., Malinovsky M., Matiasovsky K., Thonstad J. Aluminium Electrolysis, Fundamentals of the Hall-Héroult Process, Second Edition, Aluminium Verlag, Düsseldorf, 1982.
8. Djokić S.S., Conway B.E. Electroanalytical Methods for Determination of Al_2O_3 in Molten Cryolite, In *Modern Aspects of Electrochemistry*, Vol. 26, Conway B.E., Bockris J.O'M, White R.E., eds., pp. 229 – 275, New York: Plenum Press, 1994.
9. Dewing E. W., van der Kouwe E.T. Anodic Phenomena in Cryolite – Alumina Melts. I. Overpotentials at Graphite and Baked Carbon Electrode. J. Electrochem. Soc. 1975; 122: 358-363.
10. Djokić S.S., Conway B.E., Belliveau T.F. Specificity of Anodic Processes in Cyclic Voltammetry to the Type of Carbon Used in Electrolysis of Cryolite – Alumina Melts", J. Appl. Electrochem., 1994; 24: 827-834.
11. Picard G., Prat E.C. Evidencing the Electrochemical Mechanism at carbon Bath Interface by Means of Impedance Measurements: An Improved Approach to the Aluminum Reduction Process In *Light Metals*, Zabreznik R.D. ed. The Metallurgical Society, pp.507-517, Warrendale, Pennsylvania, 1987.
12. Vogt H. Effect of Alumina Concentration on the Incipience of the Anode Effect in Aluminium Electrolysis. J. Appl, Electrochem., 1999; 29: 779-788.
13. Ballehang K., Oye H.A. Inert Anodes for Aluminium Electrolysis in Hall Héroult Cells. Aluminium, 1981; 57: 146-150.
14. Olsen E., Thonstad J. Nickel Ferrite as Inert Anodes in Aluminium Electrolysis. J. Appl. Electrochem., 1999; 29: 293-311.

Chapter 12

ENVIRONMENTAL ISSUES

The metal processing industry produces various toxic gases and aqueous effluents containing ions of heavy metals, or in some case cyanides. Most of these metals are toxic. Regulations of the Environmental Protection Agency (EPA) require specific control of all air pollutants and hazardous waste.

Areas of the environmental management in the electrometallurgy field involve air pollutants, and other waste treatment and disposal. In order to minimize the negative impacts of industries involving electrowinning, electrorefinery and plating technologies to the ecosystem, an adequate treatment of environmental should carefully be taken into consideration. Safety management in the electrometallurgy is directed towards electrical hazards, explosion hazards and hazards arising due to handling and exposure to dangerous chemicals.

The electrical hazards arise from the fact that the electrometallurgical plants and plating shops operate with both direct and alternating currents. Most cells are designed in a way to minimize potential difference from the ground potential. In some plants (i.e., aluminum electrowinning from alumina-cryolite melts) strong magnetic fields are generated in cell rooms. The regulations prohibit magnetic and conductive materials in the cell rooms.

The explosion hazards depend on the system, but are often associated with plants in which, hydrogen and chlorine evolution reactions are involved in the process. Other explosion hazards may include chemicals, i.e. reactive metals such as alkali metals, magnesium etc. The regulations in this case require proper safety equipment (e.g., glasses, masks, aprons etc.).

Discussion in this chapter is divided into following sections:

(i) Environmental concerns in the electrowinning and electrorefining from aqueous solutions
(ii) Environmental concerns in molten salts electrolysis
(iii) Environmental concerns in electroplating technologies

12.1 ENVIRONMENTAL CONCERNS IN THE ELECTROWINNING AND ELECTROREFINING FROM AQUEOUS SOLUTIONS

Metals produced by the electrolysis from aqueous solutions include Cd, Cr, Cu, Mn, Co, Ni and Zn. Most of these metals are electrowon from solutions containing sulfuric acid, although metals such as cobalt and nickel can also be produced from chloride-type acidic electrolytes. Metals are deposited on cathodes, while anodic reactions, depending on the anions present in the electrolyte, usually include oxygen (sulfate solutions) or chlorine (chloride solutions).

The evolution of oxygen in the case of sulfate electrolytes is accompanied by lowering of the pH due to sulfuric acid formation. Consequently, an electrowinning process, which is carried out from a sulfate solution, results in the generation of sulfuric acid. In the case of electrowinning from chloride electrolytes, chlorine is generated at the anode and removed through suitable hoods in order to avoid environmentally objectionable fuming.

The environmental issues in the electrowinning technologies include the proper treatment of waste solutions. These solutions contain significant amounts of sulfuric acid in addition to metal that is electrowon, and traces of other heavy metals, which may have a negative impact on the environment.

After the electrolysis the sulfuric acid (e.g. processes involving electrowinning or electrorefining of Cu, Zn or Ni) is usually recovered and concentrated from residual solutions. This sulfuric acid is reused in the process. The residual solution, after the electrowinning or electrorefining is treated in combustion evaporators, and about 70 % of acid is recovered for recycle. The evaporators perform quite satisfactorily; however, disadvantages include high-energy and high-maintenance costs. Ion-exchange resins, in which ions are adsorbed, are also considered, however, these systems involve large amounts of water and the recovered acid is therefore weaker than acid in the feed material.

Another process for recovery and concentration of sulfuric acid includes electrodialysis technology.[1,2] In electrodialysis technology, the ion-exchange membranes are used for the separation process. The ions selectively permeate exchange membranes by migrating from one site to the next under the influence of electrical current. The ion-exchange membranes are arranged into continuous operation and do not require periodic stripping as with ion-exchange resins. Membranes must have a high degree of permselectivity (a selective permeability to a specific type of ion). Electrodialysis systems with monovalent anionic and cationic permselective membranes are used for the removal of monovalent ions such as chlorides, fluorides, sodium, potassium etc. The monovalent permselective membranes retain divalent ions (e.g. Zn^{2+}, Cu^{2+}, SO_4^{2-} etc.) in the dilute or electrolyte stream. The results showed a very good recovery of sulfuric acid from an acidic nickel sulfate stream by electrodialysis, where more than 80 % of the acid were recovered.

For the removal of heavy metals from residual solutions, chemical, biological, or electrochemical methods are applied.[3-6] The products or removal are either returned back in the electrowinning process or sold as various salts.

12.2 ENVIRONMENTAL CONCERNS IN THE MOLTEN SALTS ELECTROLYSIS

Environmental concerns in molten salt electrolysis (e.g. electrowinning of aluminum, magnesium etc.) depend on the metal and production conditions. In aluminum production pollution problems arise due to cell design and operation. The following facts are of great concern:

i) Fluoride emission
ii) Hydrocarbon fumes, and
iii) Dusting problems

Sources of fluoride emission include the gaseous compounds generated at the anodes and melt vapor pressure (i.e., $NaAlF_4$). Due to presence of moisture, fluorides hydrolyze, forming the HF gas. The modern aluminum plants utilize cells where the gases produced during electrolysis are collected and adequately scrubbed. Dry scrubbers are used in the treatment to catch particulates and adsorb HF on alumina, and fed back into the cell.

The hydrocarbon fumes originate from anode baking and they are generally disposed by burning. The emission of highly polluting gases during the manufacture of the carbon anodes and cathodes is carefully controlled. Dusting problems due to alumina handling are usually solved with hoods and exhaust systems, which collect the dust. The dust is separated by cyclones or filter bags etc. and recycled to the process.

The spent linings are the largest volume of waste in the production of aluminum. In order to prevent contamination of the environment, these waste materials should be properly stored to avoid leaching of toxic constituents such as cyanides and fluorides. The research is directed toward recovering valuable components from the waste and destroying the cyanide. The recovered materials are recommended for use in the manufacture of cement or mineral wool.

Other toxic constituents in the aluminum electrowinning industry include polyaromatic hydrocarbons, sulfur dioxide and hydrogen fluoride. Polyaromatic hydrocarbons are formed in industrial cells during the baking process.[7] These compounds are known as carcinogenic agents.

In addition to alumina - cryolite melts other molten salts electrolysis systems represent important health hazard requiring significant safety precautions. These systems include alkali metals e.g., lithium, sodium and potassium, magnesium etc. Inappropriate handling of alkali metals can result in serious injuries such as burns, blindness and even fatalities. The hazards arise due to the tendency of alkali metals to vigorously react with water, or to oxidize in air. In the reaction of sodium with water, hydrogen, sodium hydroxide and significant amount of heat are produced. This combination results in the explosion when air is present. The common fire extinguishers, such as water, CO_2 and CCl_4 should never be used, since all these compounds react with alkali metals. Pure metallic sodium is usually stored under organic solvents. In working with sodium or other alkali metals, face shields, hard hat, hoods and multiple layers of flame-retardant protective clothing are recommended.

In the plants where production of metals is carried out from molten chlorides, the main anodic reaction is the chlorine evolution reaction. Production and utilization of chlorine is considered safer than its transportation. The chlorine is usually collected and used in different industrial processes. Major processes utilize chlorine and create hydrogen chloride as a by-product, which is used in industrial fields. Chlorine is produced as the main anodic product in the electrolysis of fused $MgCl_2$. It is either recycled in the process or sold commercially. Chlorine is stored and transported as a liquefied gas in cylinders under pressure. Exposure to chlorine causes irritation of the eyes and the mucous membrane of the respiratory tract. LD_{50} for humans in 30 minutes is estimated at about 840 ppm, while amounts causing severe symptoms in 30 to 60 minutes are estimated at 40 to 60 ppm.

12.3 ENVIRONMENTAL CONCERNS IN THE ELECTROPLATING TECHNOLOGIES

In the electroplating technologies safety concerns are directed towards the handling and exposure to chemicals and solutions of various toxicity and waste treatment. A typical electroplating technology involves steps, which are outlined in the block diagram presented in Figure 12.1. As this block diagram shows, a typical electrodeposition technology includes steps such as degreasing, pickling, cleaning and electroplating. It is worth noting that similar types of operations and environmental concerns are involved in the electroless plating. Some of these steps are repeated several times. and order of these operations (steps) can be changed, depending on the substrate and on the metal.

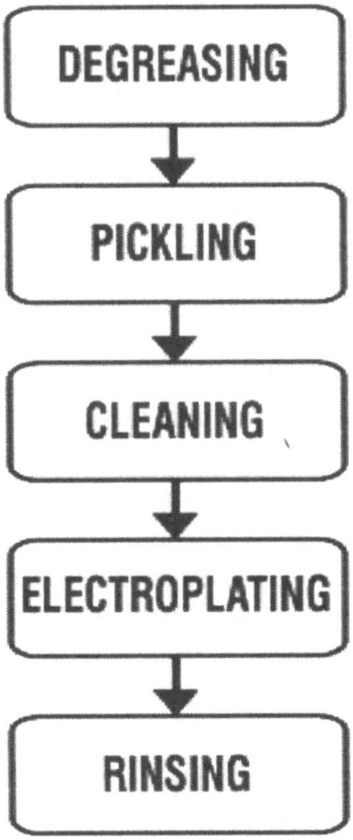

Figure 12.1. Steps involved in a typical electroplating technology

In order to remove residual chemicals on the surface of the substrate between these steps, rinsing with water is involved. Rinsing minimizes contamination, however, it increases the volume of waste solutions. The last rinse, prior to drying is very important since any traces of residues can lead to staining and even corrosion. This rinsing is carried out with hot deionized water, and sometimes with methyl/ ethyl alcohol, for faster drying. For efficient rinsing, spraying with water is very useful, and environmentally very important since it reduces the volume of waste solutions. Depending on the metal or alloy being plated, and on the nature of the substrate, operations in the electroplating technology include different solutions and substances, which are more, or less toxic. It is generally accepted that most of the substances used in this technology are dangerous up to various degrees.

The waste solutions in the electroplating technologies are divided into the following types:

(i) Solutions containing impurities immiscible with water (e.g. greases, oils) and organic solvents used,
(ii) Acidic or alkaline solutions containing heavy metals,
(iii) Solutions containing cyanides and
(iv) Solutions containing chromates.

Whenever possible preventive measures should be taken in order to minimize the volume of waste solutions. This is usually achieved with the introduction of additional equipment (e.g., tanks) as well as prolonged time of some operations (e.g., rinsing). Obviously, the preventive measures require extra investment, although this may lead to a very successful treatment. Treatment of waste solutions is usually carried out by chemical or physical methods.

The degreasing solvents (chloroform, methylene chloride, tetrachlorethylene, carbon tetrachloride etc.) are often inflammable, however, they are sources of harmful gases. In this operation, oils, separated from substrates can produce suspended and deposited solids, which may require the introduction of an additional step (e.g., filtration) in order to separate them from degreasing solvents. Emulsified oils are separated with aluminum sulfate, at a pH of about 5 to 6, while solutions are agitated with bubbled air. The scum formed at the surface is removed, and residual solution treated with other waste solutions.

Cleaning processes with biological separation are also used in the electroplating and surface finishing industry.[8] These processes operate at a low temperature and low pH and are applied on metals such as Al, Zn, Cu and Fe. The advantage of the cleaning processes with biological separation is that metals are not etched. Solutions used in the degreasing process are slightly alkaline and contain bacteria, which destroys the oil that exists on the metallic parts. Decomposed oil particles and bacteria are continuously separated from the cleaning solution as sludge in significantly smaller volumes than the associated with traditional cleaning methods.

Pickling solutions are acidic and they contain metal ions. The most commonly used reagents for the pickling process are hydrochloric and sulfuric acids. Acids containing fluorides are used for a pre-treatment of silicon containing alloys. These acids produce corrosive fumes and adequate ventilation is required. In order to protect the environment a proper disposal of waste pickling solutions is applied in the industry. Acidic waste solutions used in the pickling or non-cyanide waste acidic solutions used for zinc, nickel and copper plating are removed into separate tank or in a tank with alkaline waste solutions. In this step, the neutralization of solutions may take place, although, with an addition of lime, sodium carbonate or sodium hydroxide, the heavy metals are precipitated at pH 8.0 to 8.5. The amount of required chemicals is determined on the basis of chemical analysis of metal content (e.g. Cu, Ni, Fe, Cd, Zn, Cr etc.) in the waste solutions. After the precipitation, suspensions are filtered to separate the sludge and dispose the water.

Plating solutions may contain cyanides, chromates, heavy metals (copper, cadmium, lead, zinc and nickel). Most of these substances are toxic and also carcinogenic. Among substances used in plating technologies the following substances are considered as toxic and/or carcinogenic: cadmium, chromium, nickel, lead, mercury, cyanide and their compounds, organic solvents such as benzene, carbon tetrachloride, chloroform, toluene, xylene etc. Significant amount of research is funded to investigate proper technologies for the treatment of waste solutions, or to replace some of processes involving cyanides, chromates etc., with alternative, environmentally friendlier technologies. Among these alternatives, cyanide free solutions for gold, silver or copper electroplating should be mentioned. Also, research towards replacement hard chromium[9] and cadmium plating is carried out. Substitutes for hard chromium plating, up to certain levels include Ni-Mo, Ni-W and Ni-P plating. However these technologies involve substances, which are on the list of cancerogenic or toxic chemicals. On the other hand, physico-chemical properties of alternative coatings are frequently not comparable to those of hard chromium. For the replacement of cadmium substitutes are being sought in alternatives such as zinc, or Ni-Zn, or Sn-Zn alloys. However more research is required until cadmium plating will fully be replaced with other solutions.

Cyanides are well known as true non-cumulative protoplasmic toxic compounds. They react with enzymes of the blood that regulate oxygen transfer to cellular tissues. Exposure to cyanide solution causes severe complications and even death. The toxicity of cyanide solutions is a result of the free cyanide ion, CN^- or HCN. Hydrogen cyanide can enter the body by inhalation, oral ingestion or skin absorption.

Hydrogen cyanide undergoes an exothermic polymerization (at pH 5 to 11, especially in the presence of water or heat). This reaction can become explosively violent.

Work with cyanide solutions is carried out in a extremely well ventilated fume hood, and with special safety equipment, which includes air-masks, face-masks, rubber gloves, plastic aprons etc..

Cyanide solutions are used in plating of Cu, Zn, Au, Ag etc. Small amounts of cyanide solutions can be decontaminated in reactions with sodium hydroxide (pH > 12) and then with ferrous sulfate. The resulting ferrocyanide is relatively non-toxic.

Disposal of waste cyanide solutions is carried out with chlorine or sodium hypochlorite. With larger quantities, such as industrial waste cyanide solutions, treatment is carried out in alkaline media, according to the following reactions:

$$NaCN + 2NaOH + Cl_2 \rightarrow NaCNO + 2NaCl + H_2O \qquad (12.1)$$

The less toxic sodium cyanate, NaCNO, is further destroyed with sufficient amount of chlorine:

$$2NaCNO + 4NaOH + 3Cl_2 \rightarrow 6NaCl + N_2 + 2H_2O + 2CO_2 \qquad (12.2)$$

In this process, the pH is maintained above 10 in order to avoid formation of nitrogen trichloride or cyanogen chloride.

With smaller volumes, sodium hypochlorite is used instead of chlorine. Other methods for destruction of cyanides include hydrogen peroxide, ozone, permanganate, biological decomposition[10], removal by ion exchange and recovery, lime-sulfur reaction to give sulfate, electrochemical methods[11] etc.

Chromium plating is widely used for different engineering and decorative applications. However, compounds used in chromium plating (especially those of Cr(VI)) are recognized as very toxic and cancerogenic. Consequently, at a working place the regulations require safety glasses, rubber gloves and plastic aprons.

The treatment of waste solutions from chromium plating is based on a two-stage process. In the first, the Cr(VI) is reduced to Cr(III) and in second stage, the Cr(III) is precipitated as hydroxide and disposed as sludge. The reduction of Cr(VI) is carried out in the acidic medium (pH<3), by adding sulfuric acid. Several compounds are used as reducing agents of Cr(VI), species. These reducing agents include sodium bisulfate, sodium thiosulfate, sulfur dioxide, aluminum powder etc. After completed reduction, the waste solution containing Cr(III) species is collected for the precipitation. The precipitation is carried out with hydrated lime or sodium carbonate. The settlement of the sludge is improved by adding aluminum sulfate, ferric chloride etc. After the settlement, the sludge is separated (e.g. by decanting) and disposed. The sludge should not contain oxidizing agents (e.g. manganese dioxide, hypochlorite, chlorine or some organic oxidizing substances). These compounds can convert Cr(III) to Cr(VI). Recoveries of Cr(VI) compounds or as chromium metal are also recommended.

Physico-chemical methods of treatment of residual solution from electroplating process include evaporation, ion exchange, reverse osmosis, electrodialysis and electrolytic metal recovery.

The current status shows that in developed nations significant attention is paid towards researching and finding solutions related to the environmental concerns in the electrometallurgy field. It is expected that this research and investment in improvement of safety standards will continue to develop in the future for the safe and healthy environment in rapidly growing modern technology. It is certain that electrometallurgy related technologies will change as new information is uncovered. Adapting to these changes in terms of pollution prevention can be accomplished in an analytically sound manner.

12.4 FURTHER READINGS

1. Baltazar V., Harris G.B., White C.W. The Selective Recovery and Concentration of Sulfuric Acid by Electrodialysis. Hydrometallurgy 1992; 30:463-481.
2. Nicol M. J. Progress in Electrometallurgy research and Applications. JOM 1993; 45 (4):55-58.
3. Jütner K., Galla U., Schimeder H., Electrochemical Approaches to Environmental Problems in the Process Industry. Electrochim. Acta 2000; 45:2575-2594.
4. Njau K.N., Woude M. vd., Visser G.J., Janssen L.J.J. Electrochemical Removal of Nickel Ions from Industrial Water, Chem. Eng. Journal, 2000; 79:187-195.
5. Yu B., Zhang Y., Shukla A., Shukla S.S., Dorris K.L. The Removal of Heavy Metals by Sawdust Adsorption – Removal of Copper. J. Hazard. Materials 2000; B80:33-42.
6. Townsley C.C., Ross I. S., Atkins A.S. "Biorecovery of Metallic Residues from Various Industrial Effluents Using Filamentous Fungi." In *Fundamental and Applied Biometallurgy*, Lawrence R.W., Branion R.M.R., Ebner H.G., eds., pp. 279 – 289, Amsterdam: Elsevier, 1986.
7. Grjotheim K., Krohn C., Malinovsky M., Matiašovsky K., Thonstad J., *Aluminium Electrolysis, Fundamentals of the Hall-Héroult Process*, Second Edition, Düsseldorf: Aluminium Verlag, 1982.
8. Westerlud M., Clarin L. A Non-Dumping Water-Based Cleaning Process that Does not Require Rinsing. Plat. Surf. Finish. 1996; 83 (6):32-33.
9. Wynn P.C., Bishop C.V. Replacing Hexavalent Chromium. Plat. And Surf. Finish. 2001; 88 (2):12-14.
10. Hofseth C.S., Chapman T.W. Indirect Electrochemical Process at Rotating Disk Electrode: Catalytic Alkaline Cyanide Oxidation. J. Electrochem. Soc. 1992; 139:2525.
11. Whitlock J.L., Mudder T.I. "The Homestake water Treatment Process: Biological Removal of Toxic Parameters from Cyanidation Waste Waters and Bioassay Effluent Evaluation.", In *Fundamental and Applied Biometallurgy*, Lawrence R.W., Branion R.M.R., Ebner H.G., eds., pp. 327-339, Amsterdam: Elsevier, 1986.

Index

MIX
Papier aus verantwortungsvollen Quellen
Paper from responsible sources
FSC® C105338

If you have any concerns about our products,
you can contact us on
ProductSafety@springernature.com

In case Publisher is established outside the EU,
the EU authorized representative is:
Springer Nature Customer Service Center GmbH
Europaplatz 3, 69115 Heidelberg, Germany

Printed by Libri Plureos GmbH
in Hamburg, Germany